Lecture Notes
in Business Information Processing   288

Series Editors

Wil M.P. van der Aalst
Eindhoven Technical University, Eindhoven, The Netherlands
John Mylopoulos
University of Trento, Trento, Italy
Michael Rosemann
Queensland University of Technology, Brisbane, QLD, Australia
Michael J. Shaw
University of Illinois, Urbana-Champaign, IL, USA
Clemens Szyperski
Microsoft Research, Redmond, WA, USA

More information about this series at http://www.springer.com/series/7911

Witold Abramowicz (Ed.)

Business
Information Systems

20th International Conference, BIS 2017
Poznan, Poland, June 28–30, 2017
Proceedings

 Springer

Editor
Witold Abramowicz
Poznan University of Economics
Poznan
Poland

ISSN 1865-1348   ISSN 1865-1356 (electronic)
Lecture Notes in Business Information Processing
ISBN 978-3-319-59335-7   ISBN 978-3-319-59336-4 (eBook)
DOI 10.1007/978-3-319-59336-4

Library of Congress Control Number: 2017942985

Printed on acid-free paper

This Springer imprint is published by Springer Nature
The registered company is Springer International Publishing AG
The registered company address is: Gewerbestrasse 11, 6330 Cham, Switzerland

Preface

During the 20 years of the International Conference on Business Information Systems, it has grown to be a well-renowned event for the scientific community. Every year the conference gathers international researchers for scientific discussions on the development, implementation, and application of business information systems based on innovative ideas and computational intelligence methods. The 20th edition of the BIS conference was held in Poznań, Poland.

The BIS conference follows popular research trends, both in academic and business domains. Therefore, the theme of BIS 2017 was "Big Data Analytics for Business and Public Administration". The increasing interest in Big Data has resulted in the transformation of science, medicine, health care, engineering, business, finance, administration, and even society. Big Data Analytics helps organizations process their data in order to identify new opportunities. The results might make the business smarter by enabling faster and better decision-making, more efficient operations, and higher profits, and also lead to more satisfied customers by gauging their needs.

Big Data Analytics helps us to understand and enhance enterprises by linking many fields of information technology and business. Moreover, governments are using Big Data Analytics to gain new insights to change the accepted notions of public service, improve quality of life, and set new patient-centric standards for health care. Thus, the BIS 2017 conference continued the discussion on Big and Smart Data, that started in the previous editions.

The first part of the BIS 2017 proceedings is dedicated to Big and Smart Data research. This is followed by other research directions that were discussed during the conference, including Business and Enterprise Modelling, ICT Project Management and Process Management. Finally, the proceedings end with Smart Infrastructures as well as Applications of the newest research trends in various domains.

The Program Committee consisted of 85 members who carefully evaluated all the submitted papers. Based on their extensive reviews, a set of 24 papers were selected.

We would like to thank everyone who helped build an active community around the BIS conference. First of all, we want to express our appreciation to the reviewers for taking the time and effort necessary to provide insightful comments. We wish to thank all the keynote speakers, who delivered enlightening and interesting speeches. Last but not least, we would like to thank all authors who submitted their papers, as well as all the participants of BIS 2017.

June 2017   Witold Abramowicz

Organization

BIS 2017 was organized by Poznań University of Economics, Department of Information Systems.

Program Committee

Witold Abramowicz
(Chair)

Poznań, University of Economics and Business, Poland

PC Members

Frederik Ahlemann	University of Duisburg-Essen, Germany
Antonia Albani	University of St. Gallen, Switzerland
Rainer Alt	Leipzig University, Germany
Dimitris Apostolou	University of Piraeus, Greece
Timothy Arndt	Cleveland State University, USA
Eduard Babkin	LITIS Laboratory, INSA Rouen, France; TAPRADESS Laboratory, State University, Higher School of Economics (Nizhny Novgorod), Russia
Morad Benyoucef	University of Ottawa, Canada
Tiziana Catarci	Università di Roma La Sapienza, Italy
François Charoy	Université de Lorraine, LORIA, Inria, France
Tony Clark	Sheffield Hallam University, UK
Rafael Corchuelo	University of Seville, Spain
Beata Czarnacka-Chrobot	Warsaw School of Economics, Poland
Andrea De Lucia	University of Salerno, Italy
Josep Domingo-Ferrer	Universitat Rovira i Virgili, Spain
Suzanne Embury	University of Manchester, UK
Vadim Ermolayev	Zaporizhzhya National University, Ukraine
Werner Esswein	Technische Universität Dresden, Germany
Agata Filipowska	Poznań University of Economics, Poland
Adrian Florea	Lucian Blaga University of Sibiu, Romania
Vladimir A. Fomichov	National Research University Higher School of Economics, Russia
Johann-Christoph Freytag	Humboldt Universität zu Berlin, Germany
Naoki Fukuta	Shizuoka University, Japan
Ruediger Grimm	University of Koblenz, Germany
Volker Gruhn	Universität Duisburg-Essen, Germany
Francesco Guerra	UniMore, Italy
Hele-Mai Haav	Institute of Cybernetics at Tallinn University of Technology, Estonia

Darijus Strasunskas	POSC Caesar Association
York Sure-Vetter	Karlsruhe Institute of Technology (KIT), Germany
Jerzy Surma	Warsaw School of Economics, Poland
Genny Tortora	University of Salerno, Italy
Nils Urbach	University of Bayreuth, Germany
Herve Verjus	Université de Savoie, LISTIC, Polytech'Savoie, France
Herna Viktor	University of Ottawa, Canada
Krzysztof Węcel	Poznań University of Economics, Poland
Anna Wingkvist	Linnaeus University, Sweden
Axel Winkelmann	University of Würzburg, Germany
Guido Wirtz	University of Bamberg, Germany
Qi Yu	Rochester Institute of Technology, USA
Janusz Zawiła-Niedźwiecki	Warsaw University of Technology, Poland

Organizing Committee

Bartosz Perkowski (Chair)	Poznań University of Economics, Poland
Agnieszka Figiel	Poznań University of Economics, Poland
Barbara Gołębiewska	Poznań University of Economics, Poland
Elżbieta Lewańska	Poznań University of Economics, Poland
Włodzimierz Lewoniewski	Poznań University of Economics, Poland
Milena Stróżyna	Poznań University of Economics, Poland
Agata Szyszko	Poznań University of Economics, Poland

Additional Reviewers

Angles-Tafalla, Carles
Blanco-Justicia, Alberto
Braun, Richard
Burwitz, Martin
Dellermann, Dominik
Deufemia, Vincenzo
Di Nucci, Dario
Dittes, Sven
Ebner, Katharina
Finke, Anita
Frank, Matthias T.
Geiger, Matthias
Hernandez-Mendez, Adrian
Hoffmann, David

Hornung, Olivia
Johannsen, Florian
Kolb, Stefan
Laifa, Meriem
Leotta, Francesco
Letner, Albert
Liutvinavicius, Marius
Madlberger, Maria
Maigre, Riina
Malyzhenkov, Pavel
Marrella, Andrea
Merkle, Nicole
Olejniczak, Jarosław
Polak, Przemyslaw
Polese, Giuseppe

Ribes-González, Jordi
Ricci, Sara
Rosinosky, Guillaume
Sanko, Jelena
Savo, Domenico Fabio
Sejdovic, Suad
Sergis, Stylianos
Sonntag, Andreas
Surma, Jerzy
Thakurta, Rahul
von Entress, Matthias
Weller, Tobias
Wieczorkowski, Jędrzej

Contents

ICT Project Management

Process Management

Smart Infrastructures

Applications

Big and Smart Data

On Enriching User-Centered Data Integration Schemas in Service Lakes

Hiba Alili[1,2(✉)], Khalid Belhajjame[1], Daniela Grigori[1], Rim Drira[2],
and Henda Hajjami Ben Ghezala[2]

[1] Paris-Dauphine University, PSL Research University, CNRS, [UMR 7243],
LAMSADE, 75016 Paris, France
{hiba.alili,khalid.belhajjame,daniela.grigori}@dauphine.fr
[2] National School of Computer Sciences, University of Manouba, RIADI,
2010 Manouba, Tunisia
{hiba.alili,rim.drira,henda.benghezala}@ensi-uma.tn

Abstract. In the Big Data era, companies are moving away from traditional data-warehouse solutions whereby expensive and time-consuming ETL (Extract-Transform-Load) processes are used, towards data lakes, which can be viewed as storage repositories holding a vast amount of raw data. In this paper, we position ourselves in the recurrent context where a user has a local dataset that is not sufficient for processing the queries that are of interest to him. In this context, we show how the data lake, or more specifically the service lake since we are focusing on data providing services, can be leveraged to enrich the local dataset with concepts that cater for the processing of user queries. Furthermore, we present the algorithms we have developed for this purpose and showcase the working of our solution using a study case.

Keywords: User-centric data integration · Data provisioning service lakes · Schema enriching

1 Introduction

Big data platforms and analytic architectures have recently witnessed a wide adoption by companies and businesses world-wide to transform their increasingly growing data into actionable knowledge. In doing so, we observe a shift in the way companies are managing their data. In particular, they are moving away from traditional data-warehouse solutions [1] whereby expensive and time-consuming ETL processes are used, towards data lakes [10]. A data lake is a storage repository that holds a vast amount of raw data in its native format until it is needed. Such data is usually accessed directly through the API of big data platforms such as Hadoop and Spark, or through wrappers. We introduce in this paper a new paradigm that we call data provisioning service lake, coined by analogy to data lake, in which data can be accessed through data providing (DP) services, also known as Data as a Service (DaaS) [11].

Specifically, we consider the scenario where a user (e.g., a company employee) wishes to query a local dataset, which can be in any format, (e.g., a CSV file,

© Springer International Publishing AG 2017
W. Abramowicz (Ed.): BIS 2017, LNBIP 288, pp. 3–15, 2017.
DOI: 10.1007/978-3-319-59336-4_1

an XML document, a relational database or an RDF graph). These datasets may contain information about prospective clients of companies. However, such dataset may not be sufficient by itself to provide answers for all the user queries. Often, local datasets need to be augmented and enriched with information coming from external data sources. In such context, we propose a new data integration approach where missing information in user datasets is leveraged by invoking DP services from the service lake. The retrieved information is integrated seamlessly and transparently in the local dataset. To do so, we identify in the following three main challenges that need to be addressed in our work:

1. How to enrich the local data source schema with new concepts that are required to issue user's data queries.
2. The new concepts that are used to enrich the local schema which we term 'missing concepts' need to be populated with data instances. These data instances are retrieved by invoking data services. However, it raises the question as to how the schema mappings specifying the correspondences between the new concepts in the local schema and the data services can be defined.
3. The last challenge tackles the problem of processing user queries. Indeed, users may have different needs as to the quality and the cost (both financial and in terms of time) that they are willing to pay for their queries. Here we recall that the data services that are provided by the service lake are hosted by cloud providers. Such providers do not supply their services for free, and provide data services with varying data qualities. This raises the question as to how user queries can be processed taking into account their requirements of the user in terms of quality and cost.

In this paper, we focus on the first of these challenges. Specifically, we show how the schema of a local dataset can be enriched with new concepts given a workload specifying the user queries.

The remainder of this paper is organized as follows. We introduce the overall architecture of our solution in Sect. 2. Section 3 introduces a motivating scenario and describes the data model that we adopt. We present in detail the two algorithms that we have developed for the enrichment of local datasets in Sects. 4 and 5, and showcase how those algorithms operate using the motivating example in Sect. 6. Finally, we analyze and compare related works, and conclude the paper in Sect. 7.

2   User-Centric Data Integration in Service Lakes: Overall Architecture

Motivated by the above challenges, our goal is to propose a complete solution to leverage the missing information in user datasets in order to be able to answer the queries s/he is interested in. To do so, we explore the possibility of acquiring the missing information by invoking DP web services on the fly. In this section, we present more details about our data integration approach, giving the main steps required for answering a user' data query.

2.1   Data Provisioning Service Lakes

A data provisioning service lake is a storage repository of heterogeneous DP web services providing access to timely and high-quality information. The data returned by such services is retrieved from disparate web sources in its native format and stored in the raw data, as-is. The main idea behind service lakes is to take advantage of DP service capabilities in the lake while these services make data from web sources available through encapsulated APIs and to give minimal attention to creating schemas that define integration points between disparate provided datasets. Accordingly, instead of placing the retrieved data from different and heterogeneous web sources in a purpose-built data store, we move it into the lake, so that it may be later analyzed and mapped to the user data source schema. This facilitates and makes it possible to dynamically enrich user data sources for full query-answering purposes while eliminating the upfront costs and data ingestion.

2.2   Query-Answering Process

Figure 1 illustrates the overall process of our data integration approach.

   Given a set of queries that are of interest to the user, and given a local dataset that is provided by the user, our solution proceeds as follows:

- Step1 determines the missing information that is required to process user queries but is not provided by the local dataset. This consists mainly on browsing the schema of the local dataset and deduce the missing concepts and/or relations.

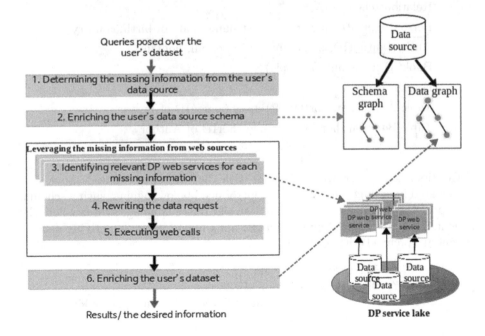

Fig. 1. An overview of the integration process

- Step 2 enriches the schema of the local dataset by defining missing elements (concepts and/or attributes) determined in the previous step.
- Step 3 identifies the set of candidate data services in the lake that can be used to populate the missing information.
- Step 4 reformulates the user's query over the relevant services' views and selects the executable query plans satisfying user's requirements in terms of data quality and cost.
- Step 5 evaluates executable query plans, which involve the call of DP services in the lake.
- Step 6 integrates the data obtained from web services into the local data source.

In the following, we focus mainly on describing the two first steps of our integration approach.

3   Data Model

In this section, we provide an illustrating example to motivate the need for an integration system where supplementary external data sources other than the ones introduced by users may be dynamically and automatically leveraged and the obtained data is seamlessly integrated with the introduced data sets. Let us consider a data set containing the following relational tables, where underlined attributes represent primary keys.

Relation schema
Person(<u>personID</u>, first name, last name, date of birth, country)
Author(<u>authorID</u>, name, university, email, domain)
Book(<u>iD</u>, title, author, topic)
Foreign Keys
Table Author: authorID references personID of Person
Table Book: author references authorID of Author

Consider now a user, who is familiar with the dataset introduced above, and is interested in issuing a set of queries, introduced below, against such a schema. Users pose their data queries using an SQL-like query language syntax where the elements that are required by the queries but missing in the underlying dataset schema are prefixed with a question mark '?'.

Q_1 : SELECT title, topic FROM Book.
Q_2 : SELECT ?iSBN, title FROM Book WHERE topic $=$ 'Webservices'.
Q_3 : SELECT title, author, ?publisher FROM ?Publisher, Book WHERE
        ?Publisher .?name $=$ Book.?publisher.

While the execution of Q_1 does not pose any problem, queries Q_2 and Q_3 can not be entirely evaluated using the introduced database. The reason being that this dataset does not provide all the necessary elements (i.e.,iSBN and publisher do not exist in any of the user tables. Also the table Publisher is not represented in the schema graph) for the evaluation of these queries, the missing information can be retrieved from external data sources. This task involves the determination of missing elements (i.e., concepts, attributes or relations) that needs to be used to enrich the local schema of the user data source to enable the evaluation of his/her queries.

Before proceeding to show how we do so, we start by introducing the data model we adopt. We represent local data sources using a graph where concepts are represented by nodes and relationships between different nodes are represented by directed edges. We adopt a graph-based model that can be used for representing data sources that are stored using different data models, including relational models, RDF, or even CSV. In this work, we distinguish between the schema level and the data level.

Schema graph is a labeled directed graph $G_S = (V, E)$, depicting the schema of the dataset, where V represents the different concepts, each one is characterized by a name and a set of attributes, and E is a set of labeled edges representing relationships between the nodes in V. We use v.`name` and v.`attributes` to denote the name and the attributes characterizing a node $v \in V$, respectively. Similarly, we use e.`label` to denote the label of an edge $e \in E$. If we consider a relational database, a node $v \in V$ would represent a relational table and v.`attributes` refers to the attributes of the relational table, while E represent referential integrity constraints between different tables in the database.

Once enriched, some of the nodes in the schema would refer to missing concepts that are populated using data services. Similarly, attributes in v.`attributes` represent the attributes that are associated with the concept.

Data graph is a directed graph $G_D = (V', E', f_{ins})$ where V' is a set of vertices representing the content (e.g., tuples in local data sources or the records retrieved by data services) of the dataset having the schema described in G_S and E' represents the relationships between the vertices in V'.

- Each node $v' \in V'$ represents an instance of a node $v \in V$ from G_S. v' is characterized by a set of attribute-value pairs of the form <name, value>. The names that appear in those attribute-value pairs refers to v.attributes.
- The edges E' in the data graph G_D are used to enforce the constraints defined within the schema graph G_S.
- $f_{ins}(v')$ is a function that given a node v' from the data graph, returns the node in the schema graph that represents the type of v'.

Continuing with the example dataset introduced above, Fig. 2 illustrates how the relational schema can be represented using our data model, whereas, Fig. 3 depicts a fragment of the data graph obtained by instantiating the schema graph in Fig. 2. Nodes B1 and B2 represent two different books: the first written by both authors A1 and A2, and the second is written only by A3.

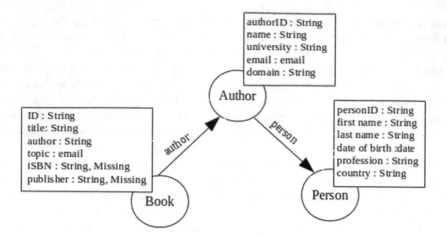

Fig. 2. Example of a schema graph

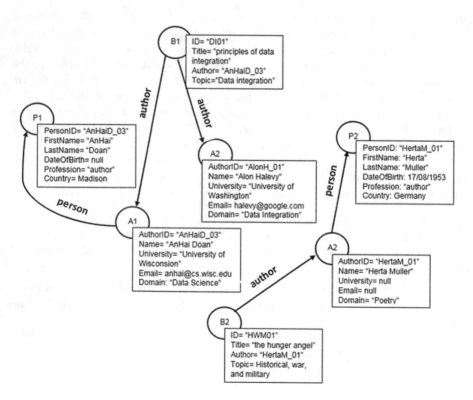

Fig. 3. A fragment of the data graph associated to schema graph introduced in Fig. 2

4   Identifying Missing Information

In our model, the user can query for data that is not yet in his local data source and that will be later leveraged from web sources on the fly. The objective of this step is to identify the missing data, more specifically the missing concepts and associated attributes that are required by the user's queries but that are not provided by his data source.

Given a set of user queries $Q_D = \{Q_1, ..., Q_n\}$ and the graph G_S representing the schema of the user data source, the Algorithm 1 processes queries one by one with respect to the order specified in the workload. We consider in this work SPJ (select-project-join) queries, which we represent using the triple (`attributes`, `concepts`, `conditions`) in order to improve readability. Such triples specify respectively the set of attributes in the select clause of the query, the set of concepts involved, and the set of conditions that appear in the where clause of the query. The algorithm parses the queries and outputs the missing concepts, MissConcepts, as well as the missing attributes, MissAttributes, characterizing existing or new concepts. A missing concept is identified with a name: $MissConcepts = \{c_1, c_2, ..., c_n\}$. As we will discuss later on, it is not always possible to identify with certainty which concept(s) a given attribute characterize. Because of this, we define a missing attribute by the triple ($name, concepts, certitude$), where $name$ is the name of the attribute, $concepts$ represents the set of concepts to which the attribute belongs, and $certitude$ is a variable that takes the value '$Certain$' or '$Uncertain$'.

If a concept in the query is not represented by a node in the schema graph, the algorithm checks if this latter was already defined as a missing concept previously (line 7, Algorithm 1). If not, it defines it as a missing concept. If the concept is represented in the schema graph, the algorithm proceeds by verifying the existence of attributes related to it (lines 4–6, Algorithm 1). A missing attribute is an attribute that does not figure in the list of attributes defined within that concept in the schema graph G_S.

As mentioned earlier, it is not always possible to identify which concepts a given attribute belongs to. This is particularly the case for join queries. For this kind of queries, an attribute may characterize a subset of concepts that are involved in the query. In order to determine the appropriate subset, we evaluate the semantic relatedness score between a given attribute and each concept from the entire set of concepts stated in the query, based on information retrieved from external sources of knowledge. In our case, we make use of the lexical database WordNet [4] and the commonsense knowledge base ConceptNet [8].

The relatedness score estimates the degree by which two words are semantically related, which is a number between 0 and 1. We select those concepts having a relatedness score with the attribute in question higher than 0.5. Note that this reduces but does not eliminate the uncertainty about the selected concepts. Because of this, our algorithm tags the missing attribute with an 'uncertain' certitude label. Thus, our method has the merit of reducing the number of concepts that needs to be examined by the user.

Algorithm 1. Searching for missing information

Require: $Q_D = Q_1, Q_2, ..., Q_n$ is a data request, $G_S = (V,E)$ is a schema graph

Ensure : MissElts= MissConcepts, MissAttributes, MissRelations

1 *MissConcepts* ← ∅, *MissAttributes* ← ∅, *MissRelations* ← ∅

2 **foreach** Q_i *in* Q_D **do**

3   **if** Q_i.*concepts involves only one concept c* **then**

4   **if** *there is a node* $v \in V$ *that corresponds to c* **then**

5   **foreach** *attribute att* $\in Q_i$.*attributes that does not belong to* v.*attributes* **do**

6   add (att, c, 'certain') to MissAttributes

7   **else if** *c was already defined in MissConcepts* **then**

8   **foreach** *attribute att* $\in Q_i$.*attributes* **do**

9   **if** *(att, concepts, certitude) was defined in MissAttributes such as* $c \in$ *concepts and certitude='uncertain'* **then**

10   replace (att, concepts, uncertain) by (att, c, 'certain')

11   **else if** *att was not defined in MissAttributes* **then**

12   add (att, c, 'certain') to MissAttributes

13   **else** add c to MissConcepts and all attributes in Q_i.*attributes* to MissAttributes;

14   **else foreach** *concept* $c \in Q_i$.*concepts that does not have any representative node* $v \in V$ **do**

15   add c to MissConcepts

16   **foreach** *condition cond* $\in Q_i$.*conditions* **do**

17   get related attribute-concept pairs $< att, c >$ in cond

18   **foreach** $< att, c >$ *such as (c is not missing from* G_S *and att is missing) or (c was defined as a missing concept however att was not defined as a missing attribute related to c)* **do**

19   add (att,c,'certain') to MissAttributes

20   **foreach** *linked concepts c1 and c2 in cond that are not related in* G_S *by an edge* $e \in E$ **do**

21   define a new edge e= (c1, c2, att1) in MissRelations where c1 represents the outgoing node, c2 is the incoming node and att1 is the label of this edge

22   **foreach** *a in* Q_i.*attributes that does not belong to any c. attributes and was not defined as a missing attribute related to c such as* $c \in Q_i$.*concepts* **do**

23   concepts ← ∅

24   **foreach** *c in* Q_i.*concepts* **do**

25   compute the relatedness score("a", "c")

26   **if** *c has a relatedness score higher than 0.5* **then**

27   add c to concepts

28   add(a, concepts, 'uncertain') to MissAttributes

29   ;

Furthermore, the fact that we process a workload of queries may help reduce the uncertainty. Though after a given query there may be uncertainty about the membership of a given attribute to a given concept, another query may confirm it. In such a case, we update the certainty tag of the attribute in question from 'uncertain' to 'certain' (lines 8–9, Algorithm 1). Nevertheless, our system always requires final confirmation from the user before proceeding to apply these modifications on the data source schema. At this step, the user can make changes if he is not entirely satisfied by the system's proposition.

5   Enriching User-Specific Data Source

If the execution of Algorithm 1 leads to a non-empty set of missing elements, the system must update the user data source at the schema level as well as at the data instances level, as described in the following paragraphs.

Schema Level: Enriching the schema graph consists in adding new concepts and attributes to those already defined in the source schema. Algorithm 2 creates for each missing concept in MissConcepts a new node in G_S and adds missing attributes to the corresponding concepts listed in *attribute.concepts* (lines 1–5, Algorithm 2). In the following, we differentiate missing elements in the graph by the label 'Missing'. The algorithm also defines new integrity constraints/semantic relations between different nodes of the graph (lines 6–7, Algorithm 2).

Algorithm 2. Enriching Schema Graph

Require: $G_S = (V, E)$, MissElts= MissConcepts, MissAttributes, MissRelations

Ensure : G_S (the enriched Schema Graph)

1 **foreach** $c \in MissConcepts$ **do**
2 $\quad$ add a new node named c with the label 'M' to V

3 **foreach** $att \in MissAttributes$ **do**
4 $\quad$ **foreach** $c \in attribute.concepts$ **do**
5 $\quad\quad$ define a new attribute att additionally to initial attributes defined within the concept c, having 'String' as a type and 'Missing' as a state

6 **foreach** $rel \in MissRelations$ **do**
7 $\quad$ add a new edge e to E outgoing from rel.OutNode, incoming to rel.InNode and labeled with rel.label

---

Data Instances Level: Unlike schema enrichment which comes immediately after the identification of missing concepts and attributes, data graph enrichment can only be performed once data services are invoked and the missing data is retrieved by them. That is why mapping generation between web service call results and the concepts in the schema graph is addressed in later steps outside the scope of this paper. In fact, the results returned by a data service call are used to populate the concepts and associated attributes in the data graph of the local data source.

Once the system has finished the evaluation of all the queries in the workload and enriched the local data source, it removes all 'missing' tags from the schema graph, thereby preparing the environment for future user interrogations.

6 Case Study

Continuing with the relational database (cf. Figure 2) introduced in the motivating scenario, we illustrate in this section how Algorithms 1 and 2 respectively operate on a sequence of three queries: Q_2, then Q_3, and finally Q_4. Q_2 and Q_3 are defined above in Sect. 3 and we will define Q_4 below.

Q_2 involves only the relational table 'Book', that we consider as a concept in our model, in order to get titles and iSBNs of all books stored in the table. Algorithm 1 first searches for iSBN in the list of attributes of 'Book'. It does not find it, therefore it defines it as a missing attribute (iSBN, Book, certain) in MissAttributes. Then, it proceeds by processing the next query in the workload, Q_3. Algorithm 1 first verifies the existence of all the concepts involved in Q_3 in the schema graph G_S, concluding that Publisher is not represented in G_S. It also examines $Q_3.conditions$ to identify that an integrity constraint between the relational tables 'Book' and 'Publisher' was not be represented in the initial schema, and that the missing attribute 'name' must be added within the concept 'Publisher', whereas 'publisher' should be defined additionally to Book.attributes.

Now we apply Algorithm 1 to Q_4:

$$Q_4 : \texttt{SELECT ?director, ?writers, ?starsFROM ?Movies}$$

All requested information in Q_4 is not represented in the database, be they relational tables ('Movies') or attributes. As a consequence, Algorithm 1 defines Movies as a missing concept, director, writers and stars as missing attributes.

In the second step, Algorithm 2 updates the schema graph by representing missing elements defined earlier by Algorithm 1 in MissElts. As explained in Sect. 5, the representation of missing concepts is done before the definition of missing attributes. Therefore, two new nodes representing respectively the relational tables 'Movies' and 'Publisher' are added to the schema graph, labeled with the character 'M' to denote that they basically represent missing concepts. Then, iSBN and publisher are added to the set of attributes Table.attributes and finally all of the attributes director, writers and stars are defined as missing attributes characterizing the concept 'Movies'. Furthermore, the algorithm creates a new integrity constraint 'publisher' between the relations 'Book' and 'Publisher'.

All of these modifications are depicted in Fig. 4: nodes and relations in gray represent respectively missing concepts and missing relations, and the attributes with a gray background represent the missing attributes.

Consider a schema graph of n nodes, and a data query requesting m attributes related to at most l concepts and under q conditions. Algorithm 1 runs in $\mathcal{O}(m.l.n)$ time, while Algorithm 2 clearly runs in $\mathcal{O}(m.l)$ time.

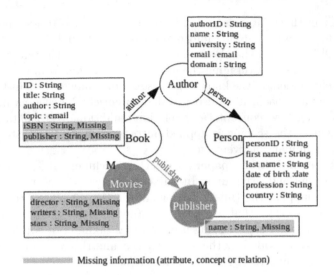

Missing information (attribute, concept or relation)

Fig. 4. An enrichment example of the schema graph introduced in Fig. 2

7   Related Works and Concluding Remarks

Our work has taken shape in the context of a rich and interesting literature focused on data integration. Over the past decades, considerable academic and commercial efforts have been made to deal with data integration, most of which are being surveyed in [6,13] such as KARMA [12], TSIMMIS [5], MOMIS [3] and SIMS [2]. Several integration systems are built on the notion of mediators, instructing the system exactly how to retrieve elements from the source data sources. This requires constructing a global schema on which global queries are posed by users, however, if any new sources join the system, considerable effort may be necessary to update the mediator. Other efforts as [2,7] construct a general domain model (under an information manifold) that encompasses the relevant parts of the data sources scheme where the description of different data sources is done independently from the queries that are subsequently asked on them. Then, the integration problem is shifted from how to build a single integrated schema to map between the domain and the data source descriptions.

While these approaches reduce the user's effort to perform data integration tasks, users queries must be formulated over the mediated schema (either the integrated schema or the domain model), therefore, users are required to pick up complementary data sources to interrogate in order to get sufficient answers to their queries. However, the interaction is not guaranteed to yield a non-empty result set.

Our work differs from past integration systems in that we propose an active data integration approach where queries are posed over the user data source schema s/he is interested in. Furthermore, users can query information that does not exist in their datasets and it is apt to our system to enrich the initial schema and leverage the missing information from External data sources.

ANGIE [9] is perhaps the closest work to ours in that it attempts to enrich knowledge databases by leveraging the missing information from web sources. In ANGIE, the enrichment of the knowledge base is done only at data instance level, whereas, in our approach, it is also possible to enrich the schema of the data source by defining new concepts and relations additionally to data instances. The enrichment of the schema is applied automatically by the system without user demand or human intervention.

The work presented in this paper tackles the enrichment of local data sources with new concepts and attributes. In our ongoing work, we are devising new techniques for mapping the (missing) elements in the schema of the local data source with data services that are able to populate such concepts. We are also examining new means for processing users queries taking into consideration the quality of the data provided by the data services, and the cost (in terms of time and financial cost) incurred by data service calls.

References

1. Anisimov, A.A.: Review of the data warehouse toolkit: the complete guide to dimensional modeling. SIGMOD Rec. **32**, 101–102 (2003)
2. Arens, Y., Chee, C.Y., Hsu, C., Knoblock, C.A.: Retrieving and integrating data from multiple information sources. Int. J. Coop. Inf. Syst. **2**, 127–158 (1993)
3. Beneventano, D., Bergamaschi, S., Castano, S., Corni, A., Guidetti, R., Malvezzi, G., Melchiori, M., Vincini, M.: Information integration: the MOMIS project demonstration. In: Proceedings of the 26th International Conference on Very Large Data Bases, pp. 611–614 (2000)
4. Budanitsky, A., Hirst, G.: Evaluating wordnet-based measures of lexical semantic relatedness. Comput. Linguist. **32**, 13–47 (2006)
5. Chawathe, S.S., Garcia-Molina, H., Hammer, J., Ireland, K., Papakonstantinou, Y., Ullman, J.D., Widom, J.: The TSIMMIS project: integration of heterogeneous information sources. In: IPSJ, pp. 7–18 (1994)
6. Halevy, A.Y., Rajaraman, A., Ordille, J.J.: Data integration: the teenage years. In: Proceedings of the 32nd International Conference on Very Large Data Bases, pp. 9–16 (2006)
7. Levy, A.Y., Rajaraman, A., Ordille, J.J.: Querying heterogeneous information sources using source descriptions. In: Proceedings of the 22th International Conference on Very Large Data Bases, pp. 251–262 (1996)
8. Liu, H., Singh, P.: Conceptnet: a practical commonsense reasoning toolkit. BT Tech. J. **22**, 211–226 (2004)
9. Preda, N., Kasneci, G., Suchanek, F.M., Neumann, T., Yuan, W., Weikum, G.: Active knowledge: dynamically enriching RDF knowledge bases by web services. In: Proceedings of the ACM SIGMOD International Conference on Management of Data, pp. 399–410 (2010)
10. Quix, C.: Managing data lakes in big data era. In: Proceedings 5th International Conference on Cyber Technology in Automation, Control and Intelligent Systems, pp. 820–824 (2015)

11. Truong, H.L., Dustdar, S.: On analyzing and specifying concerns for data as a service. In: 4th IEEE Asia-Pacific Services Computing Conference, pp. 87–94 (2009)
12. Tuchinda, R., Knoblock, C.A., Szekely, P.A.: Building mashups by demonstration. Trans. Web **5**, 16: 1–16: 45 (2011)
13. Ziegler, P., Dittrich, K.R.: Three decades of data integration - all problems solved? In: Jacquart, R. (ed.) Building the Information Society. IFIP International Federation for Information Processing, vol. 156, pp. 3–12. Springer, Toulouse (2004)

Design Principles for Digital Occupational Health Systems

Maedeh Yassaee[(✉)]

Institute for Information Management, University of St. Gallen (HSG),
Unterer Graben 21, 9000 St. Gallen, Switzerland
maedeh.yassaee@unisg.ch

Abstract. Advancements in low-cost and unobtrusive wearable computing devices have prompted employers to begin providing their employees with wearable technology as a part of corporate wellness programs. While the adoption of wearable health-tracking systems might improve employees' wellbeing, the introduction of such systems in organizational settings might also instigate certain tensions, in particular those between privacy and wellbeing, and work and private life. This study was based on an analysis of these tensions; following the design science research paradigm, design principles were derived to minimize such strain.

Keywords: Digital occupational health · Personal health monitoring systems · Design principles · Design science research

1 Introduction

The miniaturization of sensors and electronic circuits has played a key role in advancing low-cost and unobtrusive personal health monitoring systems (PHMS). These systems now feature a wide range of health-related services outside of a clinical setting [1]. Such services emphasize the provision of self-care features to individuals at any stage of the care cycle, enabling the prevention of sickness, early diagnosis of a variety of ailments, and better management of chronic disease [2]. Most of the world's population spends at least one-third of their adult life at work [3]; thus, PHMS have the potential to offer applications that are particularly useful in work environments. To help improve employees' overall health and control the cost of medical care, a growing number of companies have committed to providing wearable devices that offer employees various forms of psychosocial support [4–6]. In this study, the term digital occupational health system (DOHS) is used to refer to digitized health monitoring systems designed for use in work environments and distributed as a means of promoting the health and wellbeing of the greater workforce. The technologies required to enable DOHS goals can be grouped into three main categories: wearable and ambient sensors for collecting physiological, movement, and environmental data; communications hardware and software for relaying data to a remote center; and data analysis techniques for extracting relevant information [7, 8].

DOHS shows considerable promise for extracting meaningful information, providing managers with group performance metrics and employees with self-performance

© Springer International Publishing AG 2017
W. Abramowicz (Ed.): BIS 2017, LNBIP 288, pp. 16–27, 2017.
DOI: 10.1007/978-3-319-59336-4_2

evaluations, offering health and wellbeing enhancement recommendations, and implementing a greater level of security at work. Yet these systems also raise new challenges. Trust could be the main obstacle for adopting such systems in the workplace. Being monitored by wearable and ambient sensors may result in employees fearing for their privacy. Employers' inability to gain their employees' trust regarding their use intentions could hamper the overall level of acceptance [9, 10]. Thus, systems such as these should provide technical and social means of ensuring that employees' data are safe, and that there is no means of abusing the data produced by the system. Another challenge in introducing and adopting such systems in daily work environments is a possible blurring of the boundary between work and private life, thereby causing social tension. Even though the goal of these systems is to manage and reduce psychosocial risk factors, related social strain could actually provoke stress in the work environment. Work stress can cause employee burnout [11] and diminished organizational commitment and performance [12], so this is a major risk that must be carefully considered.

In response to this novel context, this study argues that researchers should reconsider the social aspects of the design and implementation of these types of systems. This research contributes to the literature on this topic by deriving design principles that will help DOHS gain wider acceptance and lead to a greater level of added value for employees. This work is organized as follows: Sect. 2 presents the methodology, which includes the identification and evaluation of design principles from an information systems design science research perspective. In Sect. 3, the construction of DOHS design principles is presented. In Sect. 4, this study is concluded by outlining the research and practice implications.

2 Methodology

In this study, following the design science research (DSR) approach [13–16] a set of design principles are constructed. The DSR approach is based on a problem-solving paradigm and aims to design purposeful artefacts (i.e., "design principles," "technological rules," and "patterns" "constructs," "methods," "models," "instantiation," and "design theory") [14, 17, 18]. Design principles have been defined as "design decisions and design knowledge that are intended to be manifested or encapsulated in an artefact, method, process or system" [19]. The validation and justification of principles, should be grounded theoretically, internally, and empirically [20]. Theoretical grounding involves the use of external theories and knowledge. Internal grounding is control of internal cohesion and consistency of the design principles. Empirical grounding consists of observations of its utilization and effects.

This study will follow the reference process proposed by Carlsson, Henningsson, Hrastinski and Keller [21] for constructing DOHS design principles:

Identifying scope, problem situations and the desired outcomes. The first step, within the process of deriving the design principles, is structuring the problem at hand to identify a class of goals, which directs the process [22, 23].

Reviewing extant theories and knowledge. Corresponding with the goal of the principles, this activity is concerned with the identification of and refinement of justificatory knowledge. This knowledge can be constitute the kernel theories (theories from natural or social sciences) [24] or practitioner-in-use theories [25]. *Proposing/refining design principles.* During the process of deriving the design principles, a transition from the kernel theories to the context of information systems (IS) design results in an increase in specialization (or concretization) of the theories' constructs [26]. Design principles provide a rationale by relating the specialized independent variables (cause) to IS design requirements or goals (effect).

Testing design principles. To test the effects of the proposed principles, an IS artefact can be instantiated following the design principles, and then tested if the instantiated IS artefact satisfies the requirements. To instantiate the design principles, design items need to be defined as the IS features, that are, a particular instantiation of the specialized independent variable [27]. Design items are chosen from a set of alternatives and are thus subject to reasoned preferences [27].

DSR evaluation can be performed either ex ante (before) or ex post (after) the design of the IS artefact, as well as artificially or naturalistically [28]. Artificial evaluation is not limited to a specific technology solution in experimental settings, but instead can include simulated settings where the technology solution (or its representation) can be studied under substantially artificial conditions. Naturalistic evaluation explores the performance of a constructed solution technology in a real environment (i.e., within the organization) [28]. In this study, an *ex ante* artificial evaluation is conducted to test the effectiveness of design principles by potential end users. This evaluation will potentially reduce cost by repairing technical issues before any actual implementation of the design principles in DOHS.

3 DOHS Principles

3.1 Perception of Privacy Risk

Much of the value of the services offered by DOHS rests in the confidential and personal data about the health, identity, and practices of employees. Therefore, the possibility that this personal data might be used by the employer or a third party for discriminatory purposes is a threat to employees' privacy. Employees' perceptions regarding this risk could lessen their willingness to accept the technology [9, 10]. In addition, organizations need to consider employee privacy when incorporating these systems into the workplace because such integration could lead to legal issues. Therefore, the concept of information privacy must be cautiously addressed when specifying DOHS's technical and organizational requirements. On the technical side, designers should base their considerations on privacy-aware monitoring architecture and the adaptation of established authentication techniques. On the organizational side, decision makers must understand that a radical shift in the way employees think about these systems is needed.

The adoption of these systems is an incremental process of influencing individuals' perceptions of risks to their privacy. Throughout this process, employees need to be properly educated on what is and is not being monitored, what data are collected, and how those data are secured. Correspondingly, beyond the technical requirements, this research seeks to understand the effects of different functionalities and features that may influence employees' perceptions of privacy risk. Individuals' decisions regarding their privacy involve complex psychological processes wherein they engage with multiple considerations [29]. Consequently, a variety of theories have been employed in the effort to gain a deeper understanding of the factors that influence their perceptions [29]. Procedural fairness [30], social presence [31], and social response [32] theories are all models that have been adopted to illustrate the impact of institutional factors on privacy concerns. This study is grounded in these theories, and the design principles are formulated to positively influence employees' risk perception.

Procedural fairness, also known as procedural justice, refers to an individual's perception that a particular activity in which they are participating is conducted fairly [30]. It has been argued that the following constructs facilitate fairness: informing the individual about different activities of the interaction; seeking his or her consent to get involved in the activity; and providing s/he the power [33]. In the context of DOHS design the specialization (or concretization) of the procedural fairness theory results in the following design principles and corresponding design items to apply the principle:

Design Principle: DOHS should feature social fairness (notice, consent, and controllability of the employees' personal information) to reduce employees' privacy-based risk perception.

Design Item: Noticing the employee regarding their personal data collection, use, dissemination, and maintenance.

Design Item: Seeking employees consent for the collection, use, dissemination, and maintenance of employees' data.

Design Item: Providing mechanisms which employees can control the access, correction, and redress regarding DOHS's use of data.

Social presence theory proposes that the elevated level of social presence through richer media increases trust and approval of the content communicated [34]. For the case of privacy risk perception, people generally feel a stronger level of trust when they engage in face-to-face or video-supported communication because it allows them to use signs such as eye contact, body gestures, and facial expressions. Adapting this theory to the context of DOHS, the relevant design principle and the applicable design items would be the following:

Design Principle: Richer media should be used instead of text-based privacy statements to reduce employees' privacy-based risk perception.

Design Item: Using human embodiment (e.g., the supervisor) to announce the privacy statement.

Design Item: Using a rich media (e.g., videos) to announce privacy policies in addition to the text version of privacy statements.

Finally, social response as another institutional factors adopted in information privacy literature is about the tendency to disclose in response to a prior disclosure which is known as the principle of reciprocity [35]. In order to achieve this reciprocity for the case of DOHS, it is important for employers to openly communicate and share how they are going to use the data for the benefit of employees – and not against them – and regularly communicate the outcome of their DOHS use. The design principle and the design items based on this theory would be the following:

Design Principle: DOHS should feature a medium that facilitates an open sharing and communication of an organization's approach to their use of DOHS, to reduce employees' privacy-based risk perception.

Design Item: Giving access to employees a demo of employers interface (dashboard) to follow which aspect of employee's health and his environment have been monitored and how it has been used.

Design Item: Providing a list of actions that have been considered to be taken to improve the employees' wellbeing in the organization based on the data gathered by DOHS.

3.2 Work/Life Integration

The integration of work and personal life through the use of DOHS could result in conflict [37]. Electronic integration of the professional and personal is in contrast with many individuals' preference of keeping their public and private lives separate [38]. Findings of previous studies on employees' concerns related to this issue indicate that the use of these types of devices may also cause role conflict and work interruptions [39].

Role Conflict: By altering the scope of the activities undertaken in the work environment, DOHS could make it difficult for employees to balance their public and private roles; the result would be role stress, triggered by role overload and conflict [36]. Role conflict has been defined as incompatibilities among the demands of the employee's work environment, such as contradictory expectations and inadequate resources for performing tasks [37]. Adoption of DOHS could result in role conflict, in which an employee must find a balance between conflicting work and leisure demands. Using DOHS at work means that an employee would have to use worktime to take care of their personal wellbeing, which is not usually defined as a work task. To prevent or at least manage this conflict, DOHS would need to create a border between the employee's private (their responsibility for their own health and wellbeing) and professional roles while at work [38]. Following the model of coping with role conflict [39], this border could be created in two steps: structural and personal role redefinition.

Structural role redefinition can be accomplished through "communication with [the] role sender and [by] negotiating a new set of expectations, which will be mutually agreed upon" [39]. Within this step, organizations must define an acceptable time limit for interactions with the system, which will serve as a temporal border [38] for DOHS use in the work environment. Personal role redefinition can be achieved by changing one's attitude towards role expectations, avoiding overlapping roles, or setting priorities among and within those roles. It can also be achieved by blocking DOHS influences that fall outside of an accepted temporal border and, at the same time, allowing a controlled amount of flow for necessary interactions [38].

Design Principle: Organizations should define the temporal border for the use of DOHS and limit the interaction to necessary interactions during work time.

Design Item: Noticing the employee about the limited time of interaction with system (e.g. checking the dashboard and other dedicated wellbeing features on the system).

Design Item: Limiting the DOHS interaction with employees to necessary alerts out of the accepted time span.

Work Interruption: While employees' interactions with DOHS resulting from intentional acquisition (for instance, by checking their performance on their personal dashboard) will be limited, they can still receive information without actively looking for it. Such passive interactions (alerts, recommendations, reminders, etc.) might demand non-work activities (e.g., taking a break, drinking water, competing with col-leagues, etc.), and thus could interrupt work-related tasks. Repeated interruptions can be distracting, adding to the required level of related cognitive effort; this, in turn, could lead to an almost automatic dismissal of most alerts, including those that are safety-critical [40].

This study proposes certain DOHS design principle geared towards managing work interruptions, following the "Interruption Evaluation Paradigm" applied in human/computer interaction (HCI) [41]. The Interruption Evaluation Paradigm is an attempt at managing interruptions based on the social or cognitive context of the person being interrupted, as well as factors related to the content of the interruption. Only the most severe warnings are allowed to be sent and, thus, interrupt work [42]. The cognitive context includes all aspects of the receiver's mental level of involvement in a task [42]. The social context includes all aspects of the receiver's immediate environment, as understood in a social sense; this would include the place the individual is in, the people present within that place, and the social nature of the activity occurring [42]. The following design principle and design items are based on adapting this paradigm to DOHS applications.

Design Principle: DOHS should support the prioritization and filtering of interactions based on different levels of severity of the content (the relational context) and the employee's social and cognitive context, in order to reduce unnecessary interruptions.

Design Item: Filtering the low-severity alerts when employee is cognitively or socially overloaded.

Design Item: Putting the user in control of managing interruptions (e.g. the format, block the interaction in specific time, etc.).

3.3   Validation: Testing the Design Principles' Effectiveness

Data Collection

An *ex ante* artificial evaluation of principles can be performed by means of one particular instantiation. There are several prototyping techniques for instantiating a design architecture. Prototypes are defined as the means of examining design problems and evaluating solutions [43]. The right prototyping technique depends on what that technique is meant to emphasize; they vary from high fidelity, "a finished looking (or behaving) prototype," to low fidelity, "such as storyboarding and paper-based prototyping." Low fidelity prototyping techniques are considered to be most effective when the goal is to describe what an artefact could do for a user, rather that how it would look [43]. Therefore, a low fidelity prototype was most effective for this study, since the goal was to assess how potential DOHS end users would examine the proposed design items, rather than testing the technical features of the system.

In this study, storyboarding was adopted as a low fidelity prototyping technique, in order to instantiate the design architecture proposed by the design items. Storyboarding helped to direct the focus of the audience to the scenarios communicated, and kept them from being distracted by technical and logistical details. In addition, the stories stimulated their imagination and helped them to fill in missing details the designers did not include. The focus of each story was the user, what they did and perceived, and what the experience meant to them [44]. The storyboards provided a design space for the narrative visualization of users' interactions with this type of system, as well as the critical contextual aspects over time [45]. Key elements of any storyboard are the inclusion of people, their actions, and emotions, the depiction of time, inclusion of text, and a level of detail [46].

In order to verify the effectiveness of the proposed principles, a survey study was conducted using these storyboards. Each relevant design item from the principles was presented as a scenario (see Fig. 1). The privacy risk perception storyboards were administered first. Next, respondents were asked about the effectiveness of the storyboards in presenting role conflict coping strategies. Finally, the work interruption management storyboards were administered. For each, respondents were asked to rate the effectiveness of the scenarios on a 1 to 5 scale (1 being the least effective, and 5 being the most).

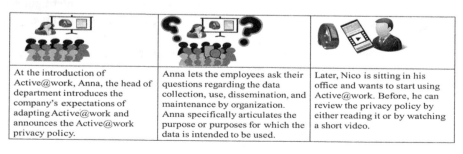

| At the introduction of Active@work, Anna, the head of department introduces the company's expectations of adapting Active@work and announces the Active@work privacy policy. | Anna lets the employees ask their questions regarding the data collection, use, dissemination, and maintenance by organization. Anna specifically articulates the purpose or purposes for which the data is intended to be used. | Later, Nico is sitting in his office and wants to start using Active@work. Before, he can review the privacy policy by either reading it or by watching a short video. |

Fig. 1. Social presence storyboard.

Respondents were recruited through the researchers' website and by e-mail, resulting in a sample of 78 responses. The sample characteristics are summarized in Table 1. Out of the total sample, 44.78% were female and 50% were male; 5.20% did not indicate a gender. Half of the respondents ranged between 35 and 55 years of age. The respondents were mainly employed in engineering (24.35%) or IT-related (33.33%) positions. Most were regular office workers (44.78%) or low-level managers (20.51%).

Table 1. Sample characteristics (n = 78)

Characteristics	N	%	Characteristics	N	%
Gender			*Age*		
			Under 25	4	5.12
Female	35	44.78	26 to 35	32	41
Male	39	50	36 to 45	29	37.17
NA	4	5.20	46 to 55	9	11.53
Job level			Over 55	1	1.20
			NA	3	3.84
Executive	3	2.84	*Job function*		
Vice president	2	2.56	IT	26	33.33
Manager	11	14.10	Support services	6	7.69
Associate	2	2.56	Marketing/Sales	7	8.97
Team leader	16	20.51	Engineering	19	24.35
Team member	35	44.78	Finance	2	2.56
Intern	2	2.56	Administration	5	6.41
Other	5	6.41	Other	12	15.38
NA	2	2.56	NA	1	1.20

Results

In general, all of the storyboards were perceived as effective for intervening in privacy risk, role conflict, and task interruption issues. On average, the respondents assessed the effectiveness of all of the storyboards as moderately high (all were ranked above 3.40). Among the three scenarios proposed to reduce employees' apprehension of privacy risk, procedural fairness (PF) received the highest ranking (with a mean of 3.98). Social presence (SP) and social response (SR) were also perceived to be effective, and their means were 3.42 and 3.55, respectively. The two proposed coping strategies for managing role conflict received similar rankings to one another. On average, structural redefinition (SR) was rated 3.37, and personal redefinition (PD) was ranked 3.40. With regards to interruption management principles, automated interruption management (AI) was rated lower (with a mean of 3.51) than manual interruption management (with a mean of 3.89).

One way to assess the comparability of a multi-item survey is to assess whether items that are supposed to measure the same construct correlate with one another. It should be noted that correlations among principles designed to reduce employees' apprehension associated with privacy risk were relatively high (all above 0.43). The correlation between SD and PD was also high (0.61). However, the correlation between AI and MI was relatively low (0.31), due to somewhat different ways of approaching interruption prevention. In an automated interruption management scenario, employees are passively involved, while in manual interruption management, empowering employees to control interruptions forces them to be actively involved. The mean scores, standard deviations, extracted variances, and inter-item correlation estimates are all summarized in Table 2.

Table 2. Inter-item correlations, means, standard deviations and variances

	SP	SR	PF	SD	PD	AI	MI	Mean	SD	Variance
SP	1.00							3.42	1.01	1.03
SR	0.43	1.00						3.55	1.08	1.17
PF	0.53	0.47	1.00					3.98	1.01	1.03
SD	0.28	0.35	0.17	1.00				3.37	1.15	1.34
PD	0.29	0.42	0.30	0.61	1.00			3.40	1.12	1.26
AI	0.22	0.40	0.25	0.31	0.46	1.00		3.51	1.14	1.30
MI	0.22	0.30	0.34	0.33	0.49	0.31	1.00	3.89	0.98	0.96

4   Discussion and Conclusions

This study proposed a set of principles for the design of DOHS, following a DSR paradigm. These design principles are expected to reduce privacy concerns and the additional mental pressure caused by such systems that – if left unchecked – would significantly diminish an employee's willingness to use such devices while at work. The effectiveness of these principles was tested by querying potential end users.

Grounded in kernel theory from social science, this study attempted to reduce apprehension related to perceived risks to employees' privacy, by using technical and organizational features to embed more social responses, presence and fairness interventions in DOHS. To avoid role conflict, employers should commit not only to providing the devices, but also to allocating an acceptable amount of time for employees to interact with those devices. Systems designers should provide features for limiting access to the devices outside the acceptable time window. However, even though active interaction (e.g., checking the dashboard, playing games, and other dedicated wellbeing features) can be limited, passive interaction (e.g., receiving alerts and recommendations) should not. Such passive interactions, however, should not interrupt employees' work activities. Therefore, as evidenced by the Interruption Evaluation Paradigm of HCI, there is a need for two different levels of interruption management. One should feature an automated reduction of excessive alerts, based on an analysis of the user's context and the importance of the interruption. The other should give the employee full power to manage and control interruptions, when needed.

Being limited to a primarily conceptual level, the focus of this study is on the theoretical underpinings of the design principles. Therefore, further research instantiating these principles to actual DOHS implementations will need to be undertaken.

Acknowledgments. This research was supported by the European Commission and Swiss State Secretariat for Education, Research, and Innovation (SERI), under the Ambient Assisted Living (AAL) Joint Program.

References

1. Mittelstadt, B., Fairweather, B., Shaw, M., McBride, N.: The ethical implications of personal health monitoring. Int. J. Technoethics **5**, 37–60 (2014)
2. Tartarisco, G., Baldus, G., Corda, D., Raso, R., Arnao, A., Ferro, M., Gaggioli, A., Pioggia, G.: Personal health system architecture for stress monitoring and support to clinical decisions. Comput. Commun. **35**, 1296–1305 (2012)
3. World Health Organization: Global Strategy on Occupational Health for All: the Way to Health at Work, Recommendation of the Second Meeting of the WHO Collaborating Centres in Occupational Health, 11–14 October 1994, Beijing, China (1995)
4. Giddens, L., Gonzalez, E., Leidner, D.: I track, therefore I am: exploring the impact of wearable fitness devices on employee identity and well-being. In: Twenty-Second Americas Conference on Information Systems, San Diego (2016)
5. http://www.forbes.com/sites/parmyolson/2014/06/19/wearable-tech-health-insurance/
6. Vyas, D., Fitz-Walter, Z., Mealy, E., Soro, A., Zhang, J., Brereton, M.: Exploring physical activities in an employer-sponsored health program. In: Proceedings of the 33rd Annual ACM Conference Extended Abstracts on Human Factors in Computing Systems, pp. 1421–1426. ACM (2015)
7. Atallah, L., Lo, B., Ali, R., King, R., Yang, G.-Z.: Real-time activity classification using ambient and wearable sensors. IEEE Trans. Inf. Technol. Biomed. **13**, 1031–1039 (2009)
8. Patel, S., Park, H., Bonato, P., Chan, L., Rodgers, M.: A review of wearable sensors and systems with application in rehabilitation. J. Neuroeng. Rehabil. **9**, 1 (2012)
9. Pavlou, P.A.: Consumer acceptance of electronic commerce: integrating trust and risk with the technology acceptance model. Int. J. Electron. Commer. **7**, 101–134 (2003)
10. Guo, X., Zhang, X., Sun, Y.: The privacy–personalization paradox in mhealth services acceptance of different age groups. Electron. Commer. Res. Appl. **16**, 55–65 (2016)
11. Fisher, C.D., Gitelson, R.: A meta-analysis of the correlates of role conflict and ambiguity. J. Appl. Psychol. **68**, 320 (1983)
12. Jackson, S.E., Schuler, R.S.: A meta-analysis and conceptual critique of research on role ambiguity and role conflict in work settings. Organ. Behav. Hum. Decis. Process. **36**, 16–78 (1985)
13. Winter, R.: Design science research in Europe. Eur. J. Inf. Syst. **17**, 470–475 (2008)
14. Gregor, S., Hevner, A.R.: Positioning and presenting design science research for maximum impact. MIS Q. **37**, 337–355 (2013)
15. Wieringa, R.J.: Design science as nested problem solving. In: 4th International Conference on Design Science Research in Information Systems and Technology, p. 12. Association for Computing Machinery, Malvern (2009)
16. Peffers, K., Tuunanen, T., Rothenberger, M., Chatterjee, S.: A design science research methodology for information systems research. J. Manag. Inf. Syst. **24**, 45–77 (2007)

17. Hevner, A.R., March, S.T., Park, J., Ram, S.: Design science in information systems research. MIS Q. **28**, 75–105 (2004)
18. Winter, R., Albani, A.: Restructuring the design science research knowledge base - a one-cycle view of design science research and its consequences for understanding organizational design problems. In: Baskerville, R., de Marco, M., Spagnoletti, P. (eds.) Designing Organizational Systems: An Interdisciplinary Discourse, pp. 63–81. Springer, Heidelberg (2013)
19. Gregor, S.: Design theory in information systems. Australas. J. Inf. Syst. **10**, 14–22 (2002)
20. Goldkuhl, G.: Design theories in information systems - a need for multi-grounding. J. Inf. Tech. Theor Appl. **6**, 59–72 (2004)
21. Carlsson, S.A., Henningsson, S., Hrastinski, S., Keller, C.: Socio-technical IS design science research: developing design theory for is integration management. IseB **9**, 109–131 (2011)
22. Mandviwalla, M.: Generating and justifying design theory. J. Assoc. Inf. Syst. **16**, 314 (2015)
23. Gregory, R.W., Muntermann, J.: Heuristic theorizing: proactively generating design theories. Inf. Syst. Res. **25**, 639–653 (2014)
24. Simon, H.A.: The Sciences of the Artificial. MIT Press, Cambridge (1996)
25. Sarker, S., Lee, A.S.: Using a positivist case research methodology to test three competing theories-in-use of business process redesign. J. Assoc. Inf. Syst. **2**, 7 (2002)
26. Kuechler, W., Vaishnavi, V.: A framework for theory development in design science research: multiple perspectives. J. Assoc. Inf. syst. **13**, 395 (2012)
27. Niehaves, B., Ortbach, K.: The inner and the outer model in explanatory design theory: the case of designing electronic feedback systems. Eur. J. Inf. Syst. **25**, 303–316 (2016)
28. Venable, J.R., Pries-Heje, J., Baskerville, R.L.: FEDS: a framework for evaluation in design science research. Eur. J. Inf. Syst. **25**, 77–89 (2016)
29. Li, Y.: Theories in online information privacy research: a critical review and an integrated framework. Decis. Support Syst. **54**, 471–481 (2012)
30. Lind, E.A., Tyler, T.R.: The Social Psychology of Procedural Justice. Springer, New York (1988)
31. Reis, H.T., Shaver, P.: Intimacy as an interpersonal process. Handb. Pers. Relat. **24**, 367–389 (1988)
32. Short, J., Williams, E., Christie, B.: The Social Psychology of Telecommunications. Wiley, London (1976)
33. Culnan, M.J., Armstrong, P.K.: Information privacy concerns, procedural fairness, and impersonal trust: an empirical investigation. Organ. Sci. **10**, 104–115 (1999)
34. Guerin, B.: Mere presence effects in humans: a review. J. Exp. Soc. Psychol. **22**, 38–77 (1986)
35. Gouldner, A.W.: The norm of reciprocity: a preliminary statement. Am. Sociol. Rev. **25**, 161–178 (1960)
36. Ragu-Nathan, T., Tarafdar, M., Ragu-Nathan, B.S., Tu, Q.: The consequences of technostress for end users in organizations: conceptual development and empirical validation. Inf. Syst. Res. **19**, 417–433 (2008)
37. Rizzo, J.R., House, R.J., Lirtzman, S.I.: Role conflict and ambiguity in complex organizations. Adm. Sci. Q. **15**, 150–163 (1970)
38. Clark, S.C.: Work/family border theory: a new theory of work/family balance. Hum. Relat. **53**, 747–770 (2000)
39. Hall, D.T.: A model of coping with role conflict: the role behavior of college educated women. Adm. Sci. Q. **17**, 471–486 (1972)
40. Wipfli, R., Lovis, C.: Alerts in clinical information systems: building frameworks and prototypes. Stud. Health Technol. Inform. **155**, 163–169 (2010)

41. Milewski, A.E.: Interruption management and telephone call screening. Int. J. Hum. Comput Interact. **20**, 19–33 (2006)
42. Grandhi, S., Jones, Q.: Technology-mediated interruption management. Int. J. Hum. Comput Stud. **68**, 288–306 (2010)
43. Houde, S., Hill, C.: What do prototypes prototype. Handb. Hum. Comput Interact. **2**, 367–381 (1997)
44. Carroll, J.M.: Making Use: Scenario-Based Design of Human-Computer Interactions. MIT Press, Cambdrige (2000)
45. Hackos, J.T., Redish, J.: User and Task Analysis for Interface Design. Wiley, New York (1998)
46. Truong, K.N., Hayes, G.R., Abowd, G.D.: Storyboarding: an empirical determination of best practices and effective guidelines. In: Proceedings of the 6th Conference on Designing Interactive Systems, pp. 12–21. ACM (2006)

Estimating the Quality of Articles in Russian Wikipedia Using the Logical-Linguistic Model of Fact Extraction

Nina Khairova[1]([✉]), Włodzimierz Lewoniewski[2], and Krzysztof Węcel[2]

[1] National Technical University "Kharkiv Polytechnic Institute",
NTU "KhPI" 2, Kyrpychova str., Kharkiv 61002, Ukraine
`khairova@kpi.kharkov.ua`
[2] Poznań University of Economics and Business,
Al. Niepodległości 10, 61-875 Poznań, Poland
`{wlodzimierz.lewoniewski,krzysztof.wecel}@ue.poznan.pl`

Abstract. We present the method of estimating the quality of articles in Russian Wikipedia that is based on counting the number of facts in the article. For calculating the number of facts we use our logical-linguistic model of fact extraction. Basic mathematical means of the model are logical-algebraic equations of the finite predicates algebra. The model allows extracting of simple and complex types of facts in Russian sentences. We experimentally compare the effect of the density of these types of facts on the quality of articles in Russian Wikipedia. Better articles tend to have a higher density of facts.

Keywords: Russian Wikipedia · Article quality · Fact extraction · Logical equations

1 Introduction

Nowadays, in order to make correct financially significant economic decisions, a large amount of information and knowledge should be analyzed. Useful information can be found both in specialized economic sources and in Web-resources of general nature. In recent years Wikipedia has become one of the most important sources of knowledge throughout the world. In the ranking of the most popular websites this online encyclopedia with more than 44 million articles in almost 300 languages[1] occupies the 5th place in the world. Many articles of this multilingual encyclopedia contain information about the various types of products, e.g. cars, movies, video games, cell phones. Information in Wikipedia is also used to automatically enrich various public databases (such as DBpedia).

Russian-language edition of Wikipedia is one of the major language versions of the online encyclopedia. For instance, the largest language version, which

---

[1] https://meta.wikimedia.org/wiki/List_of_Wikipedias.

© Springer International Publishing AG 2017
W. Abramowicz (Ed.): BIS 2017, LNBIP 288, pp. 28–40, 2017.
DOI: 10.1007/978-3-319-59336-4_3

is English Wikipedia, contains five million articles, while Russian Wikipedia contains one million articles.

The number of articles is continually rising, and authors of the articles may not have an official confirmation of their expertise in a given domain. Sometimes the authors are anonymous. Additionally, there is no process of obligatory expert reviewing of the Wikipedia articles does not exist. All changes to the article immediately are visible on the site. Therefore, in order to provide the computer encyclopedia with qualitative information, which is reliable for making business decisions, its articles' quality must be evaluated.

Generally, the quality of the Wikipedia article is estimated manually in accordance with the Wikipedia policies, guidelines and community rules in a particular language version. Today, there exist techniques of automatic evaluation of the quality of articles that are mostly based on using different quantitative characteristics (article length, number of images, number of links and others). However, qualitative characteristics are rarely used to evaluate the quality of the Wikipedia articles. There are at least two reasons for that. First, text has relatively rich semantics [1]. Second, qualitative grammatical and stylistic characteristics of the text of the article depends on an article's language [2,3].

We suggest to use qualitative characteristics of the density of simple and complex facts to automatically estimate the quality of articles in Russian Wikipedia. In order to identify a fact in a text, we developed the logical-linguistic model of fact extraction from Russian sentences.

2   Related Work

Nowadays, there exist quite a lot of the approaches to measuring the quality of textual information [4–6]. Among other things in scientific works, various methods for automatic distinguishing of high-quality Wikipedia articles are written. Most of them use various quantitative features of the article as independent variables and the article quality class as a dependent one. Usually, the quality of Web content is assessed with such metrics as objectivity [7], content maturity and readability [8]. At the same time, current approaches to the automatic assessment of documentation quality are mostly based on statistical models or on some formalisation of grammar. For instance, Blumenstock [9] proposes to use word count as a simple metric for capturing quality indicator of Wikipedia articles, Lipka and Stein [2] exploit an article's character trigram distribution for the automatic assessment of information quality. Online service WikiRank[2] used different quantitative parameters of articles (text length, the number of images, references etc.) to calculate the so-called relative quality of the same article in various language versions of Wikipedia.

However, it is obvious that quality of texts may depend not only on grammatical features of a document but also on its semantic characteristics [10]. The reason is that text informativeness mostly depends directly on semantics. Most

---

[2] http://wikirank.net.

applications that use semantic characteristics to assess the quality of textual documents are based on knowledge from ontologies such as WordNet [11]. In this approach, it is necessary to have explicitly expressed relationships such as meronymy and hypernymy between entities in the text. In [12,13], it is proposed to use the number of facts and the factual density as features to identify high quality articles in English Wikipedia. Lex et al. consider the fact in the form of a triplet with two entities and a relationship between them. Authors used the ReVerb Open Information Extraction framework to extract facts from the articles in English Wikipedia [14].

Today there exist a lot of different techniques for information extraction and, in particular, for facts extraction. The most of them are domain-specific or focus on a small number of relations in specific preselected domains [14]. More advanced Information Extraction systems use a domain-independent architecture and sentence analyzer. Nevertheless these systems demand either a large hand-tagged corpus to create a training set or knowledge from ontologies such as WordNet [15]. Anyway, every modern facts extraction system depends on the language of texts which are analyzed and the vast majority of these systems is focused on English, Spanish and German [3,16].

In our study we consider densities of simple and complex facts as features to measure the quality of articles in Russian Wikipedia. In order to extract facts from Russian texts we propose to use the built logical-linguistic model.

3   Formal Model of Fact Extraction

In order to build a model we use logical-algebraic equations of the finite predicates algebra (FPA) [17], which can describe any finite and determined relations. These mathematical tools of the FPA have been successfully used for building different Artificial Intelligence and natural language models [18]. Basic predicates of the FPA are the predicates of recognition of the element a by the variable x_i:

$$x_a^i = \begin{cases} 1, & \text{if } x_i = a \\ 0, & \text{if } x_i \neq a \end{cases} \quad (1 \leq i \leq n), \tag{1}$$

where a is any of the elements of universe U. In our model, the universe U contains various elements of the language system: lexemes, morphemes, sentences, grammatical and semantic features of Russian words etc. We then introduce to the universe the subsets of grammatical and semantic features of words in Russian sentences $M = \{X, Y, Z\}$, where X is the finite subset of the characteristic of animacy, Y is the finite subset of semantic features of nouns and Z is the finite subset of morphological features that describe the grammatical cases of Russian nouns.

Let us write the grammatical cases of Russian nouns per the predicates of recognition of the element (1)

$$P(z) = z^{nom} \vee z^{gen} \vee z^{dat} \vee z^{acc} \vee z^{inc} \vee z^{loc}, \tag{2}$$

where *nom*, *gen*, *dat*, *acc*, *ins*, *loc* are nominative, genitive, dative, accusative, instrumental and prepositional cases of Russian nouns respectively. Similarly, we can write semantic features of the nouns that represent the participants of the sentences:

$$P(x) = z^{anim} \vee z^{inan}, \tag{3}$$

$$P(y) = y^{device} \vee y^{hum} \vee y^{tool} \vee y^{pc:hue} \vee y^{spacee} \vee y^{time:moment} \vee y^{time:period} \vee y^{s:loc} \vee y^{others}, \tag{4}$$

where $P(x)$ is predicate that describes the feature of animacy of the noun (index *anim* means animate, index *inan* means inanimate); $P(y)$ is predicate that describes others particular semantic feature of the noun (*device*, *tool*, *space*, *time* : *moment*, *time* : *period*, index *hum* means belonging to the semantic class "person", index *pc* : *hum* means belonging to the semantic class "part of the body" and index *s* : *loc* means belonging to the semantic class "destination"). Such choice of semantic categories is motivated by the necessity of the correct and complete description of the seven considered semantic roles. The labelling corresponds to semantic labelling in Russian National corpus.

Let us introduce the system of the predicates $P_k(x, y, z)$ over Cartesian products $P(x) \times P(y) \times P(z)$:

$$P_k(x, y, z) = \gamma_k(x, y, z) \wedge P(x) \wedge P(y) \wedge P(z), \tag{5}$$

where the predicates $\gamma_k(x, y, z)$, $k \in [1, h]$ represent a complete set of semantic roles of the sentence participants of the facts that we consider, where h is the number of semantic roles of the facts in our model. We base on the assumption of Fillmore [19] that there is an action and participants of the action at the semantic level of a sentence. These are represented by a verb and nouns at the grammar level respectively. Every participant plays certain semantic role (a.k.a. deep case) in the action.

The predicate $\gamma_k(x, y, z)$ holds if the specific grammatical and semantic features of the noun in a Russian sentence define the specific semantic role of the sentence participant and the predicate is false otherwise. Therefore, the predicate excludes morphological and semantic features of the noun that are not inherent in the specific semantic role.

In our study, we consider the simple and the complex types of facts. The simple fact consists of the Subject and the Predicate[3]. In grammar, the simple fact is represented by the smallest grammatical clause. A typical clause is a group of words that includes a verb (or a verb phrase) and a noun (or a noun phrase) [20]. The complex fact apart from the Subject and the Predicate consists of the Object or others participants of an action. In grammar, the complex fact is represented by a sentence with a verb (or a verb phrase) and a few nouns or noun phrases.

---

[3] We use 'Subject', 'Object' and 'Predicate' with the first upper-case letters to denote the element of a fact triplet Subject -> Predicate -> Object.

We define the semantic role of the Subject of a fact via the predicate γ_1:

$$\gamma_1(x, y, z) = x^{anim} z^{nom} \vee x^{inan} z^{nom} (y^{device} \vee y^{tool} \vee y^{pc:hue}). \tag{6}$$

The predicate $\gamma_1(x, y, z)$ shows grammatical and semantic features of a noun in Russian sentence that denotes the Subject of a fact. We also explicitly distinguish the semantic role of Object of a fact via the predicate γ_2:

$$\gamma_2(x, y, z) = z^{acc}(x^{inan} \vee x^{anim}) \tag{7}$$

We also explicitly distinguish the semantic roles of other parts related to the fact via a set of the predicates $\{\gamma_3, ..., \gamma_7\}$. Grammatical and semantic characteristics of the beneficiary of an action is defined by the following predicate:

$$\gamma_3(x, y, z) = z^{dat} y^{hum} x^{anim} \tag{8}$$

The predicate γ_4 denotes semantic and grammatical features of the action tool or the action reason:

$$\gamma_4(x, y, z) = z^{ins} x^{inan} (y^{tool} \vee y^{pc:hum} \vee y^{device}) \tag{9}$$

We distinguish the attributes of location, time and destination of the action via the predicates γ_5, γ_6, γ_7 respectively:

$$\gamma_5(x, y, z) = z^{loc} x^{inan} (y^{space} \vee y^{s:loc}) \tag{10}$$

$$\gamma_6(x, y, z) = x^{inan} (z^{acc} y^{time:moment} \vee z^{loc} y^{time:period}) \tag{11}$$

$$\gamma_7(x, y, z) = z^{acc} x^{inan} y^{space} \tag{12}$$

Based on the above predicates, we can define the simple fact and the complex fact as follows.

Definition 1. The *simple fact* in a Russian sentence is the smallest grammatical clause that includes a verb and a noun, where the semantic and grammatical features of the noun have to denote the Subject of the fact according to the Eq. (6).

Definition 2. The *complex fact* in Russian texts is a grammatical sentence that includes a verb and a few nouns. Among these nouns, one has to play the semantic role of the Subject (6), semantic and grammatical characteristics of the other nouns have to satisfy one or more Eqs. (7–12).

Using some definitions from the recent works on measuring the quality of Web content [12,13] we can also denote density of simple and complex facts in the Russian Wikipedia article.

Definition 3. The *density of simple facts* in the Russian Wikipedia article is defined as the number of simple facts divided by the number of words in the article.

Definition 4. The *density of complex facts* in the Russian Wikipedia article is defined as the number of complex facts divided by the number of words in the article.

4   Experiments and Results

Our dataset includes about 31,000 present articles (December 2016) from the most popular domains from Russian Wikipedia. Table 1 shows the distributions of the analyzed articles according to domains of Russian Wikipedia. There is no generally accepted standard classification of articles quality in Wikipedia community. The classification schemes vary in language versions. For instance, Belarus version uses three quality classes, whereas German one uses only two classes. In Russian Wikipedia, there are seven quality classes that can show the "maturity" of an article. They are (in decreasing order): Featured, Good, Solid, Full, Developed, Developing and Stub.

Table 1. The distributions of the analyzed articles according to domains of Russian Wikipedia.

Domain	All articles	NeedsWork articles				GoodEnough articles		
		Stub	Developing	Developed	Full	Solid	Good	Featured
Adm. division	28691	811	289	40	9	10	6	1
Album	14039	5153	760	212	46	75	109	35
Company	9343	318	385	100	18	15	14	3
Film	25148	92	157	53	8	73	46	27
Filmmaker	23155	468	251	66	11	42	34	4
Football player	28905	330	486	36	9	37	58	12
Human settlement	183411	1407	2153	135	24	22	22	7
Military person	32728	564	1237	298	48	412	55	6
Musician	19313	381	416	82	16	32	34	14
Officeholder	35969	1650	844	283	101	416	159	44
Person	40829	874	898	310	60	230	81	27
River	31008	166	103	22	6	19	7	2
Scientist	32327	1363	3337	421	46	176	74	40
Writer	16158	281	494	196	32	31	31	23

According to previous studies [12,21,22], we distinguish two groups to evaluate the quality of the Russian Wikipedia articles. The first group includes Featured, Good, Solid classes and it is called **GoodEnough** group articles. The second group includes Full, Developed, Developing and Stub classes and is referred to as **NeedsWork** group. We consider that the articles in the first group are of higher quality than the articles in the other group. The main reason for such conclusion is the following. To receive any estimates from the first group of estimates, the article must be subjected to a complex procedure involving discussion and voting of the users of Wikipedia.

Using the capabilities of API Wikipedia we have created two corpora of plain articles texts of selected domains. The first corpus contains articles from Russian Wikipedia that are assigned to Featured, Good, Solid categories. The second corpus contains articles from Russian Wikipedia that are assigned to Full, Developed, Developing and Stub categories.

Before the application of our model of fact extraction, we apply the pymorphy2[4], the library for morphological analysis of the Russian language. Our algorithm uses the OpenCorpora dictionary[5].

In order to estimate the quality of articles in Russian Wikipedia based on our logical-linguistic model of fact extraction, we focus on two approaches. In the first approach, we determine the average densities of simple and complex facts in each category of each domain of our corpora.

The second approach is based on the hypothesis that subjectivity in an article has a large impact on the quality of Wikipedia text. For instance, according to a widely accepted standard that all editors English Wikipedia should normally follow, all content on Wikipedia must be written from a neutral point of view.

In the second approach, before we determine the average densities we excluded facts that may comprise some subjective assessment of the authors. In order to solve this problem, we have created the set of Russian verbs V that have certain semantic component of subjectivity. The set includes 120 speech verbs (such as *tell, recall, dictate* and others), 154 feelings verbs and 103 emotions verbs (such as *wish, rejoice, worry* and others). We designate these verbs as the mental verbs. If a simple or a complex fact includes Predicate that is represented by a verb from the set V, we exclude the fact from the number of facts in a calculation of density of facts. As a result of this procedure the number of simple facts was decreased by 7.37% and the number of complex facts was decreased by 6.86% in GoodEnough articles group. The number of simple facts was decreased by 10.5%, the number of complex facts was decreased by 9.46% in NeedsWork articles group. This supports our hypothesis that the higher quality Wikipedia articles the less subjective they are. The results of these studies are shown in the Table 2.

Table 2 shows the dependence of the simple facts density and complex facts density from quality categories and domains of the articles. The table compares the results of the first and the second approaches. The table compares the results of two approaches. In the first approach, we calculate the average of densities of simple and complex facts in each category of each domain of our corpora. In the second approach, we carry out similar calculations, excluding facts with the so-called mental verbs.

Table 3 shows mean, standard deviation and median of simple and complex facts density in the articles of two corpora. Figure 1 shows four curves for the densities of simple and complex facts in different domains of two our corpora. The

---

[4] https://pymorphy2.readthedocs.io.
[5] http://opencorpora.org.

Table 2. Simple and complex facts density in Russian articles Wikipedia corpora (DSF - density of simple facts, DCF - density of complex facts)

Domain	With mental verbs				Without mental verbs			
	GoodEnough		NeedsWork		GoodEnough		NeedsWork	
	DSF	DCF	DSF	DCF	DSF	DCF	DSF	DCF
Administrative division	0.043	0.041	0.040	0.034	0.041	0.039	0.040	0.033
Album	0.046	0.041	0.021	0.019	0.040	0.036	0.019	0.018
Company	0.040	0.038	0.033	0.031	0.038	0.036	0.032	0.030
Film	0.051	0.045	0.039	0.036	0.044	0.040	0.035	0.032
Filmmaker	0.046	0.043	0.022	0.021	0.042	0.040	0.021	0.020
Football player	0.051	0.048	0.038	0.036	0.048	0.045	0.037	0.035
Human settlement	0.043	0.040	0.033	0.031	0.041	0.038	0.031	0.029
Military person	0.033	0.031	0.028	0.027	0.032	0.030	0.027	0.026
Musician	0.043	0.039	0.028	0.026	0.038	0.035	0.026	0.025
Officeholder	0.043	0.040	0.031	0.029	0.039	0.036	0.029	0.028
Person	0.043	0.040	0.031	0.029	0.039	0.036	0.029	0.027
River	0.044	0.041	0.038	0.035	0.041	0.039	0.036	0.033
Scientist	0.030	0.028	0.024	0.023	0.027	0.026	0.022	0.021
Writer	0.039	0.036	0.029	0.027	0.035	0.032	0.027	0.025

Table 3. Mean, standard deviation and median (DSF - density of simple facts, DCF - density of complex facts)

Parameter	With mental verbs				Without mental verbs			
	GoodEnough		NeedsWork		GoodEnough		NeedsWork	
	DSF	DCF	DSF	DCF	DSF	DCF	DSF	DCF
Mean	0.041	0.037	0.028	0.026	0.037	0.034	0.026	0.025
Median	0.042	0.038	0.027	0.026	0.038	0.035	0.025	0.024
Std. deviation	0.012	0.010	0.016	0.015	0.010	0.009	0.016	0.014

plain lines represent the densities of simple and complex facts in the GoodEnough group for higher quality articles. The dotted lines represent the densities of simple and complex facts in the NeedsWork group of articles.

We found that the densities of simple and complex facts in higher quality articles corpus are higher than the similar densities in the lower quality articles corpus for all domains. From this observation, we conclude that the densities of simple and complex facts, along with the article length, can be a good feature to separate higher quality articles of Russian Wikipedia from lower quality ones. Besides, data in Table 3 shows that standard deviations of complex facts

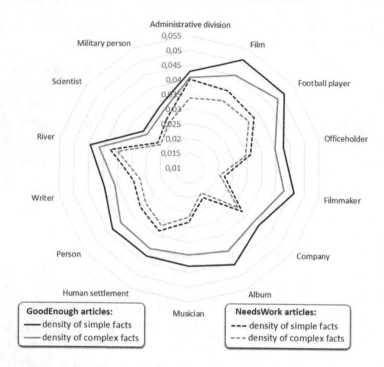

Fig. 1. Densities of simple and complex facts in different domains of the GoodEnough and NeedsWork groups of articles.

distributions are less than standard deviations of simple facts distributions for all groups of articles. It means that values of densities for complex facts are closer to means than for simple facts. It should be noted also that the densities of simple and complex facts depend on a particular domain, though the ratio of the densities of simple and complex facts in the two corpora is retained.

Additionally, we can see that the density of complex facts is a more discriminative feature than the density of simple facts for distinction of higher quality articles of Russian Wikipedia.

Figures 2, 3, 4 and 5 show the distributions of the articles of both corpora according to the densities of simple and complex facts. Figure 2 represents the distribution of the articles of two groups according to the density of simple facts including mental verbs. Figure 3 represents the similar distribution according to the density of complex facts including mental verbs. Analogously, Figs. 4 and 5 show the distributions of the articles of both corpora according to the densities of simple and and complex facts respectively, excluding the mental verbs. The separation of distributions calculated with the so-called mental verbs and the one without mental verbs helps in understanding the impact of neutral point of view on quality of the article.

Since the numbers of articles in GoodEnough and NeedsWork groups are different, we normalise them by representing the article rate in the respective corpus. We can see that the articles from GoodEnough corpus have relatively

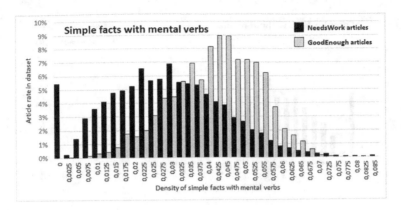

Fig. 2. Distributions articles of GoodEnough and NeedsWork groups according to the density of simple facts including mental verbs.

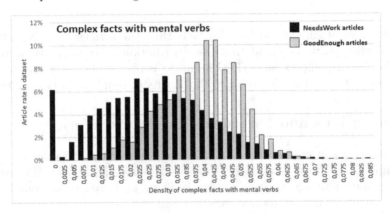

Fig. 3. Distributions articles of GoodEnough and NeedsWork groups according to the density of complex facts including mental verbs.

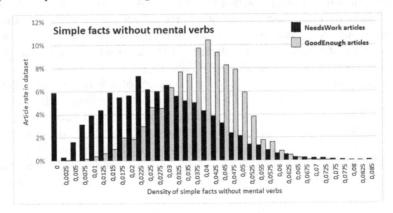

Fig. 4. Distributions of articles of GoodEnough and NeedsWork groups according to the density of simple facts excluding the mental verbs.

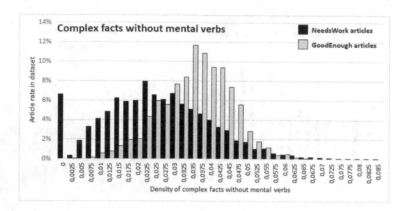

Fig. 5. Distributions of articles of GoodEnough and NeedsWork groups according to the density of complex facts excluding the mental verbs.

higher densities of simple and complex facts than the articles from NeeedWork corpus. Additionally, we found that the distribution of the articles according to the density of complex facts (Figs. 3 and 5) is more demonstrative than the distribution of the articles according to the density of simple facts (Figs. 2 and 4).

5   Conclusions and Directions for Future Work

In this paper we leveraged the semantic categories of the densities of simple and complex facts to determine the quality of Wikipedia articles. In order to calculate number of simple and complex facts we proposed to use our logical-linguistic model of fact extraction from Russian texts.

The performed experiment showed that density of simple facts and density of complex facts, which were selected as a result of the model application, indeed characterise the quality level of articles in Russian Wikipedia. Additionally, elimination of facts with the so-called mental verbs allowed us to better distinguish the quality of articles, as the reduction rate of density of facts was higher in articles of lower quality. However, the influence of the facts whose Predicates have a flavour of subjectivity on the quality of Wikipedia articles requires further study.

The results of the paper can increase precision of the quality classification of articles in Russian Wikipedia. The obtained features, along with others, can be used in supervised learning algorithms that have shown their effectiveness in other studies related to the automatic evaluation of the Wikipedia articles quality. One should note that regarding distinction of higher quality articles of Russian Wikipedia, the density of complex facts is a more discriminative feature than the density of simple facts. Furthermore, the density of facts (complex or simple) excluding the mental verbs is a more discriminative feature than the density of facts (complex or simple) including the mental verbs.

We suggest that trends and dependencies between qualitative characteristics of the density of simple and complex facts and the quality of Wikipedia articles also cover other languages. However, we should develop the specific logical-linguistic model of fact extraction for every language [3].

Additionally, in our logical-linguistic model, we consider grammatical and semantic features of words in Russian sentences. However, consideration of the semantic characteristics in the conducted experimental research is limited. This is due to using the pymorphy2 library, which uses a limited number of tags. In the future, we aim to consider influence of all semantic features of Russian words on the result of the implementation of the logical-linguistic model of fact extraction in more details.

References

1. Anderka, M.: Analyzing and predicting quality flaws in user-generated content: the case of Wikipedia. PhD, Bauhaus-Universitaet Weimar Germany (2013)
2. Lipka, N., Stein, B.: Identifying featured articles in wikipedia: writing style matters. In: Proceedings of the 19th International Conference on World Wide Web, pp. 1147–1148 (2010)
3. Khairova, N.F., Petrasova, S., Gautam, A.P.S.: The logical-linguistic model of fact extraction from English texts. In: Dregvaite, G., Damasevicius, R. (eds.) ICIST 2016. CCIS, vol. 639, pp. 625–635. Springer, Cham (2016). doi:10.1007/978-3-319-46254-7_51
4. Arthur, J.D., Stevens, K.T.: Document quality indicators: a framework for assessing documentation adequacy. J. Softw. Maint. Res. Pract. 4(3), 129–142 (1992)
5. Knight, S.A., Burn, J.: Developing a framework for assessing information quality on the world wide web. Informing Sci. J. 8, 159–172 (2005)
6. Shpak, O., Löwe, W., Wingkvist, A., Ericsson, M.: A method to test the information quality of technical documentation on websites. In: 2014 14th International Conference on Quality Software, pp. 296–304, October 2014
7. Lex, E., Juffinger, A., Granitzer, M.: Objectivity classification in online media. In: Proceedings of the 21st ACM Conference on Hypertext and Hypermedia, HT 2010, pp. 293–294. ACM, New York (2010)
8. Weber, N., Schoefegger, K., Bimrose, J., Ley, T., Lindstaedt, S., Brown, A., Barnes, S.-A.: Knowledge maturing in the semantic mediawiki: a design study in career guidance. In: Cress, U., Dimitrova, V., Specht, M. (eds.) EC-TEL 2009. LNCS, vol. 5794, pp. 700–705. Springer, Heidelberg (2009). doi:10.1007/978-3-642-04636-0_71
9. Blumenstock, J.E.: Size matters: word count as a measure of quality on wikipedia. In: WWW, pp. 1095–1096 (2008)
10. Wingkvist, A., Ericsson, M., Löwe, W.: Making sense of technical information quality - a software-based approach measuring the quality of technical data depends on developing models from which metrics can be extracted and analyzed. Using an open source tool the authors describe one approach to this (2012)
11. Fellbaum, C.: Wordnet: An Electronic Lexical Database. MIT Press, Cambridge (1998)
12. Lex, E., Voelske, M., Errecalde, M., Ferretti, E., Cagnina, L., Horn, C., Stein, B., Granitzer, M.: Measuring the quality of web content using factual information. In: Proceedings of the 2nd Joint WICOW/AIRWeb Workshop on Web Quality - WebQuality 2012, p. 7 (2012)

13. Horn, C., Zhila, A., Gelbukh, A., Kern, R., Lex, E.: Using factual density to measure informativeness of web documents. In: Proceedings of the 19th Nordic Conference of Computational Linguistics (NODALIDA 2013). NEALT Proceedings Series 16, Oslo University, Norway, 22–24 May 2013, Number 085, pp. 227–238. Linköping University Electronic Press (2013)

14. Etzioni, O., Banko, M., Soderland, S., Weld, D.S.: Open information extraction from the web. Commun. ACM **51**(12), 68–74 (2008)

15. Eugene, A., Luis, G.: Extracting relations from large plain-text collections. In: Proceedings of ACM 2000 (2000)

16. Fader, A., Soderland, S., Etzioni, O.: Identifying relations for open information extraction. In: Proceedings of the Conference on Empirical Methods in Natural Language Processing, pp. 1535–1545. Association for Computational Linguistics (2011)

17. Bondarenko, M., Shabanov-Kushnarenko, J.: The intelligence theory. In: SMIT, Kharkiv, p. 576 (2007)

18. Petrasova, S., Khairova, N.: Automatic identification of collocation similarity. In: 2015 Xth International Scientific and Technical Conference, Computer Sciences and Information Technologies (CSIT), pp. 136–138, September 2015

19. Fillmore, C.J.: The case for case. In: Bach, E., Harms, R. (eds.) Universals in Linguistic Theory. Holt, Rinehart, and Winston, London (1968)

20. Osborne, T., Gross, T.: Constructions are catenae: construction grammar meets dependency grammar. Cogn. Linguist. **23**(1), 165–216 (2012)

21. Węcel, K., Lewoniewski, W.: Modelling the quality of attributes in wikipedia infoboxes. In: Abramowicz, W. (ed.) BIS 2015. LNBIP, vol. 228, pp. 308–320. Springer, Cham (2015). doi:10.1007/978-3-319-26762-3_27

22. Lewoniewski, W., Węcel, K., Abramowicz, W.: Quality and importance of wikipedia articles in different languages. In: Dregvaite, G., Damasevicius, R. (eds.) ICIST 2016. CCIS, vol. 639, pp. 613–624. Springer, Cham (2016). doi:10.1007/978-3-319-46254-7_50

Business and Enterprise Modelling

Automatic Discovery of Object-Centric Behavioral Constraint Models

Guangming Li[✉], Renata Medeiros de Carvalho, and Wil M.P. van der Aalst

Eindhoven University of Technology,
P.O. Box 513, 5600 MB Eindhoven, The Netherlands
{g.li.3,r.carvalho,w.m.p.v.d.aalst}@tue.nl

Abstract. Process discovery techniques have successfully been applied in a range of domains to automatically discover process models from event data. Unfortunately existing discovery techniques only discover a behavioral perspective of processes, where the data perspective is often as a second-class citizen. Besides, these discovery techniques fail to deal with object-centric data with many-to-many relationships. Therefore, in this paper, we aim to discover a novel modeling language which combines data models with declarative models, and the resulting *object-centric behavioral constraint model* is able to describe processes involving *interacting instances* and *complex data dependencies*. Moreover we propose an algorithm to discover such models.

Keywords: Process mining · Object-centric modeling · Process discovery · Cardinality constraints

1 Introduction

Process discovery is one of the most challenging process mining tasks. However, state of the art techniques can already deal with situations where each process instance is recorded as a case with ordered events and each event is related to exactly one case by a case identifier [1]. Examples of algorithms that consider process instances to derive models include the Inductive Miner, ILP Miner, Heuristic Miner and Declare Miner, distributed as ProM plugins.[1] All examples extract models from behavior-centric logs (e.g., XES logs). Moreover, there are already over 20 commercial software products supporting process mining (e.g., Disco, Celonis, ProcessGold, QPR, etc.).

However, when it comes to data-centric/object-centric processes supported by CRM and ERP systems, most of the existing discovery techniques fail. Such systems have one-to-many and many-to-many relationships between data objects that makes it impossible to identify a unique process instance notion to group traces. If we enforce such a grouping anyway, it leads to convergence and divergence problems. Besides, the discovered models using existing approaches are

---

[1] http://www.processmining.org/prom/start.

© Springer International Publishing AG 2017
W. Abramowicz (Ed.): BIS 2017, LNBIP 288, pp. 43–58, 2017.
DOI: 10.1007/978-3-319-59336-4_4

often based on business process modeling languages such as Petri nets, BPMN diagrams, Workflow nets, EPCs, and UML activity diagrams. They typically consider process instances in isolation, ignoring interactions in between. Moreover, they cannot model the data perspective in a precise manner. Data objects can be modeled, but the more powerful constructs (e.g., cardinality constraints) used in Entity-Relationship (ER) models [5], UML class models [9] and Object-Role Models (ORM) [10] cannot be reflected at all in today's process models. As a result, data and control-flow need to be described in separate diagrams.

Numerous approaches in literature tried to solve the problems mentioned above. Various techniques of *colored Petri nets*, i.e., Petri nets where tokens have a value, are employed to add data to process models [7,8,13,14,23]. These approaches do not support explicit data modeling, i.e., there is no data model to relate entities and activities. The earliest approaches that explicitly related process models and data models were proposed in the 1990s [11,22]. One example is the approach by Kees van Hee [11] who combined Petri nets, a specification, and a binary data model. Other approaches such as data-aware process mining discovery techniques [17,21] extend the control-flow perspective with the data perspective. They discover the control-flow perspective of processes, using one of the process discovery techniques available today (e.g., inductive mining techniques), and then the data perspective (e.g., read and write operations, decision points and transition guards) using standard data mining techniques. These techniques mainly focus on control-flow perspective, considering the data perspective as a second-class citizen. Artifact-centric approaches [6,12,15,18] (including the work on proclets [2]) attempt to describe business processes in terms of so-called business artifacts. Artifacts have data and lifecycles attached to them, thus relating both perspectives. There are a few approaches to discover artifact-centric models from data-centric processes [16,19,20]. However, these force users to specify artifacts as well as a single instance notion within each artifact, and tend to result in complex specifications that are not fully graphical and distribute the different instance types over multiple diagrams.

This paper uses a novel modeling language, named *Object-Centric Behavioral Constraint (OCBC)*, that combines declarative language (*Declare* [4]), and data/object modeling techniques (ER, UML, or ORM) [3]. Cardinality constrains are used as a unifying mechanism to tackle data and behavioral dependencies, as well as their interplay. Besides motivating that the novel language is useful for modeling data-centric processes, we also propose an algorithm for discovering OCBC models from event data lacking a clear process instance notion. By doing this, we demonstrate that this novel modeling language has potential to be used as an alternative to mainstream languages for all kinds of process mining applications.

The remainder is organized as follows. Section 2 presents a process to introduce OCBC models. Section 3 illustrates the ingredients of OCBC models. Our discovery algorithm is proposed in Sect. 4. Section 5 shows some experimental results showing the validity of our approach and implementation and Sect. 6 concludes the paper.

2   Motivation Example

In this section, the Order To Cash (OTC) process, which is the most typical business process supported by an ERP system, is employed to illustrate OCBC models. The OTC process has many variants and our example is based on the scenario in *Dolibarr*.[2]

Figure 1 shows an OCBC model which describes the OTC process in Dolibarr. The top part shows behavioral constraints. These describe the ordering of activities (*create order, create invoice, create payment*, and *create shipment*). The bottom part describes the structuring of objects relevant for the process, which can be read as if it was a UML class diagram (with six object classes *order, order line, invoice, payment, shipment*, and *customer*). Note that an order has at least one order line, each order line corresponds to precisely one shipment, each order refers to one or more invoices, each invoice refers to one or more payments, each order, shipment or invoice refers to one customer, etc. The middle part relates activities, constraints, and classes.

The notation will be explained in more detail later. However, to introduce the main concepts, we first informally describe the 9 constructs highlighted in Fig. 1. Construct ③ indicates a one-to-one correspondence between *order* objects and *create order* events. If an object is added to the class *order*, the corresponding activity needs to be executed and vice versa. ①, ② and ⑤ also represent the one-to-one correspondence. ④ shows a one-to-many relation between *create order* events and *order line* objects. ⑥ expresses that each *create invoice* event is followed by one or more corresponding *create payment* events and each *create payment* activity is preceded by one or more corresponding *create invoice* events. A similar constraint is expressed by ⑦. ⑧ demands that each *create order* event is followed by at least one corresponding *create shipment* event.

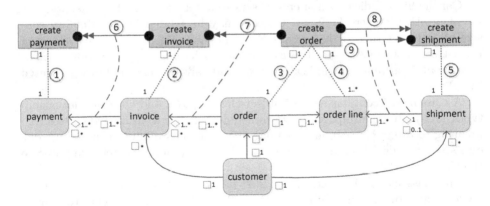

Fig. 1. A small *Object-Centric Behavioral Constraint* (OCBC) model.

---

[2] Dolibarr ERP/CRM is an open source (webpage-based) software package for small and medium companies (www.dolibarr.org). It supports sales, orders, procurement, shipping, payments, contracts, project management, etc.

⑨ denotes that each *create shipment* event is preceded by precisely one corresponding *create order* event. Note that *one payment* can cover *multiple invoices* and *multiple payments* can be executed for *a particular invoice* (i.e., one payment only covers a part of the invoice). Obviously, this process has one-to-many and many-to-many relations, and it is impossible to identify a single case notion.

The process described in Fig. 1 cannot be modeled using conventional notations (e.g., BPMN) because (a) four different types of instances are intertwined and (b) constraints in the class model influence the allowed behavior. Moreover, the OCBC model provides a *full* specification of the allowed behavior in a *single diagram*, so that no further coding or annotation is needed.

3   Object-Centric Behavioral Constraint (OCBC) Modeling Language

After introducing OCBC models based on a typical real-life process, we describe the data perspective and the behavioral perspective, and show how OCBC models relate both perspectives. See [3] for the formal definition of the OCBC language.

3.1   Modeling Data Cardinality Constraints

In this paper, the term "object" is different from it used in other fields, such as software engineering. In general, objects are data elements generated and used by information systems. These are grouped in classes and have some attributes. For example, a record in the "order" table can be considered as an object of class "order". Each value (e.g., a customer name "Mary") in the record can be considered as an attribute of the object.

Cardinalities indicates non-empty sets of integers, i.e., "1..*" denotes the set of positive integers $\{1, 2, ...\}$. Objects may be related and cardinality constraints help to structure dependencies. As shown in Fig. 2(a), we use a subset of mainstream notations to specify a *class model* with temporal annotations such as "eventually" cardinalities (indicated by $\Diamond$) and "always" cardinalities (indicated by $\Box$).[3]

A class model contains a set of object classes (OC) and a set of relationship types (RT). Relationship types are directed (starting from source classes and pointing to target classes) and each one defines two cardinality constraints: one on its source side (close to the source class) and one on its target side (close to the target class).[4]

The class model depicted in Fig. 2(a) has three object classes, i.e., $OC = \{a, b, c\}$ and two relationship types, i.e., $RT = \{r_1, r_2\}$. r_1 points to b from

---

[3] $\Box$ indicates the constraint should hold at any point in time and $\Diamond$ indicates the constraint should hold from some point onwards.

[4] For the sake of brevity, we omit redundant cardinalities in the graph. For instance, "$\Box 1$" implies "$\Diamond 1$" and therefore "$\Diamond 1$" can be removed in this case.

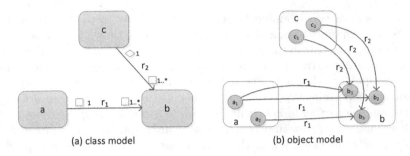

(a) class model (b) object model

Fig. 2. Example of a class model and corresponding object model.

a, which indicates a is the source class, b is the target class, and a and b are related through r_1. The annotation "$\square 1..*$" on the target side of r_1 indicates that for each object in a, there is always at least one corresponding object in b. "$\diamond 1$" on the source side of r_2 indicates that for each object in b, there is eventually precisely one corresponding object in c. A class model defines a "space" of possible *object models*, i.e., concrete collections of objects and relations instantiating the class model.

An object model includes a set of objects (*Obj*) and a set of object relations (*Rel*). More precisely, an object relation can be viewed as a tuple consisting of a class relationship type, a source object and a target object. For instance, (r_1, a_1, b_1) is an object relation, with r_1 as its name, a_1 as the source object, b_1 as the target object, and a_1 and b_1 are related through r_1. Note that each object has a corresponding object class, e.g., a_1 corresponds to the object class a.

Figure 2(b) shows an object model. The objects are depicted as grey dots: $Obj = \{a_1, a_2, b_1, b_2, b_3, c_1, c_2\}$. Among them, a_1 and a_2 belong to object class a; b_1, b_2 and b_3 belong to object class b; c_1 and c_2 belong to object class c. There are three relations corresponding to relationship r_1 (e.g., (r_1, a_1, b_1)), and three relations corresponding to relationship r_2 (e.g., (r_2, c_1, b_1)).

3.2 Modeling Behavioral Cardinality Constraints

A process model can be viewed as a set of *constraints*. For example, in a procedural language like Petri nets, places correspond to constraints: removing a place may allow for more behavior and adding a place can only restrict behavior. In this paper, we will employ a graphical notation inspired by *Declare*, a declarative workflow language [4].

Fig. 3. An example behavioral model with two behavioral cardinality constraints.

Figure 3 shows two example behavioral constraints: con_1 and con_2. Each constraint corresponds to one constraint type. Table 1 shows eight examples of constraint types. Constraint con_1 is a *response* constraint and constraint con_2 is a *unary-response* constraint. The graphical representations of the eight example constraint types are shown in Fig. 4. Besides the example constraint types, we allow for any constraint type that can be specified in terms of the *cardinality of preceding and succeeding target events relative to a collection of reference events*. As a shorthand, one arrow may combine two constraints as shown in Fig. 5. For example, constraint con_{56} states that after creating an order there is precisely one validation and before a validation there is precisely one order creation.

Table 1. Examples of constraint types, inspired by *Declare*. Note that a constraint is defined with respect of a reference event.

Constraint	Formalization
response	$\{(before, after) \in \mathbb{N} \times \mathbb{N} \mid after \geq 1\}$
unary-response	$\{(before, after) \in \mathbb{N} \times \mathbb{N} \mid after = 1\}$
non-response	$\{(before, after) \in \mathbb{N} \times \mathbb{N} \mid after = 0\}$
precedence	$\{(before, after) \in \mathbb{N} \times \mathbb{N} \mid before \geq 1\}$
unary-precedence	$\{(before, after) \in \mathbb{N} \times \mathbb{N} \mid before = 1\}$
non-precedence	$\{(before, after) \in \mathbb{N} \times \mathbb{N} \mid before = 0\}$
co-existence	$\{(before, after) \in \mathbb{N} \times \mathbb{N} \mid before + after \geq 1\}$
non-co-existence	$\{(before, after) \in \mathbb{N} \times \mathbb{N} \mid before + after = 0\}$

Given some *reference event e* we can reason about the events *before e* and the events *after e*. One constraint type may require that the number of corresponding events of one particular reference event before or after the event lies within a particular range (e.g., before $\geqslant 0$ and after $\geqslant 1$ for *response*). For instance,

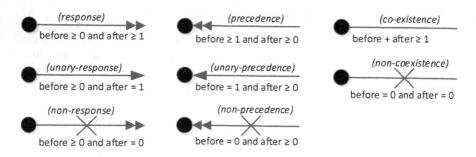

Fig. 4. Graphical notation for the example constraint types defined in Table 1. The dot on the left-hand side of each constraint refers to the *reference events. Target events* are on the other side that has no dot. The notation is inspired by *Declare*, but formalized in terms of cardinality constraints rather than LTL.

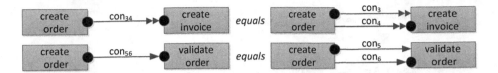

Fig. 5. An arrow with two reference events (•) can be used as a shorthand. Constraint con_{34} (con_{56}) corresponds to the conjunction of constraints con_3 and con_4 (resp. con_5 and con_6).

constraint con_1 specifies that each A event should be succeeded by at least one corresponding B event and constraint con_2 specifies that each B event should be succeeded by precisely one C event.

A *behavioral constraint model* is a collection of activities and constraints. More precisely, a constraint corresponds to a constraint type, a reference activity and a target activity. Figure 3 displays a behavioral model consisting of two constraints (con_1 and con_2) and three activities (A, B and C). Each constraint has a dot referring to the *reference activity*. The corresponding *target activity* can be found on the other side. For example, the reference activity of con_2 is B (see dot) and the target activity of con_2 is C. The shape (e.g., a double-headed arrow) of each constraint indicates the constraint type. For instance, con_1 has a dot on the left side and a double-headed arrow on the right side, which means the corresponding constraint type is *response*, the reference activity is A and the target activity is B.

3.3 Object-Centric Behavioral Constraints

Section 3.1 focused on structuring objects and formalizing cardinality constraints on object models (i.e., classical data modeling) while Sect. 3.2 focused on control-flow modeling and formalizing behavioral constraints *without* considering the structure of objects. This subsection relates *both perspectives* by combining control-flow modeling and data modeling to fully address the challenges described in the introduction.

We use so-called *AOC* relationships (denoted by a dotted line between activities and classes) and constraint relations (denoted by a dashed line between behavioral constraints and classes or class relationships) to combine the behavioral constraint model in Fig. 3 with the class model in Fig. 2, resulting in the complete example OCBC model in Fig. 6. For better understanding, we attach a scenario on the model. For example, activity A corresponds to *create order* activity while class a corresponds to class *order*.

The example model has four AOC relationships, i.e., $AOC = \{(A, a),\ (A, b), (B, b), (C, c)\}$.[5] Note that A refers to object classes a and b while b refers

---

[5] In this paper, we use the upper-case (lower-case) letters to express activities (classes), and use the upper-case (lower-case) letters with a footnote to express events (objects).

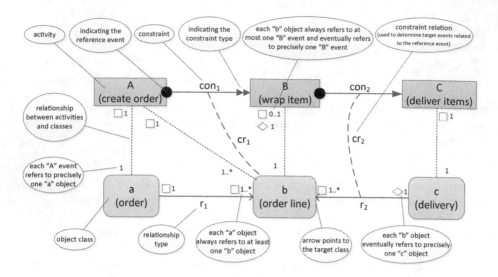

Fig. 6. An example model illustrating the main ingredients of OCBC models.

to activities A and B. This shows that OCBC models are capable of modeling one-to-many and many-to-many relationships between events and objects. AOC relationships also have cardinalities. The $\square$ ($\diamondsuit$) cardinalities on the activity side define how many events there always (eventually) need to be for each object. The cardinalities on the class side (without $\square$ or $\diamondsuit$ symbols) define how many objects there need to be for each event when the event occurs.

Constraint relations define the *scope* of each constraint thereby relating reference events to selected target events. If a constraint relation connects a constraint to a class, events are correlated through objects of this class. Consider the constraint relation cr_1 between con_1 and b. Let A_1 be one reference event for con_1 (i.e., one *create order* event) and A_1 refers to a set of b objects (i.e., *order line* objects). Each B event (i.e., *wrap item* event) that refers to at least one object in the set is the target event of A_1 for cr_1. If a constraint relation connects a constraint to a relationship, the target events are related to the reference event through object relations (of this relationship) in the object model. Consider the constraint relation cr_2 between con_2 and r_2. Let B_1 be one reference event for con_2 (i.e., one *wrap item* event) and B_1 refers to b objects (i.e., *order line* objects) which are related to c object (i.e., *delivery* objects) through r_2 relations. Each C event (i.e., *deliver items* event) that refers to at least one one of these c objects (i.e., *delivery* objects) is the target event of B_1 for cr_2. Note that, indicated by the example model, B_1 refers to precisely one b object that is related to one c object, which means B_1 has precisely one target event.

4   Discovery of Object-Centric Behavioral Models

In this section, we specify a new format of logs that are object-centric, and propose a novel algorithm to discover OCBC models based on such logs.

4.1   Object-Centric Event Logs

A process is merely a collection of *events* without assuming some case or process instance notion, and the corresponding event log provides a *snapshot of the object model after each event*, where the object model represents the state of the process. Such a log can be extracted from real-life IT systems. For instance, the Oracle database provides *change tables* to record any modification in the database. With these tables, it is possible to reconstruct any previous state of the database. Besides, without the change tables, it is still possible to produce such a log by exploiting explicit change logs in systems like SAP.

In a log, each event corresponds to an object model (in the "Object Model" column) which represents the state of the process just after the execution of the event. Besides, each event corresponds to an *activity* and may have additional *attributes*, e.g., the time at which the event took place. Moreover, events are atomic and ordered (indicated by the "Index" column). In order to relate the behavioral perspective and the data perspective (i.e., events and objects), each event also refers to at least one object (in the "Reference" column). Logs of this format are called object-centric event logs (denoted as XOC logs in remainder).

Table 2 gives an example XOC log containing 7 events. Event A_1 corresponds to the first occurrence of activity A, has one attribute att_1 whose value is v_1 and refers to three objects: a_1, b_1, and b_2. The corresponding object model of A_1 consists of three objects and two object relations. Table 2 also illustrates the evolution of the object model. After the occurrence of some event, objects may have been added, and relations may have been added or removed.[6] Note that the example log has the same scenario as indicated by the model in Fig. 6, e.g., activity A means activity *create order* and object a_1 means an *order* object.

4.2   Discovery Algorithm

The algorithm takes an XOC log as well as a set of possible behavioral constraint types as input, which means users can specify the constraint type set based on their needs. In Fig. 3 the *response* and *unary-response* types were used, but the user can select from a range of possible types that can be discovered. Next, we explain the discovery process based on the example log.

---

[6] We assume that objects cannot change class or be removed at a later stage to avoid referencing non-existent objects. Objects can be marked as deleted but cannot be removed (e.g., by using an attribute or relation).

Table 2. An example XOC log

Index	Event	Activity	Attributes	References	Object model	
					Objects	Relations
1	A_1	A	$\{att_1 = v_1\}$	$\{a_1, b_1, b_2\}$	$\{a_1, b_1, b_2\}$	$\{(r_1, a_1, b_1), (r_1, a_1, b_2)\}$
2	B_1	B	$\{att_2 = v_2\}$	$\{b_1\}$	$\{a_1, b_1, b_2\}$	$\{(r_1, a_1, b_1), (r_1, a_1, b_2)\}$
3	B_2	B	$\{att_2 = v_3\}$	$\{b_2\}$	$\{a_1, b_1, b_2\}$	$\{(r_1, a_1, b_1), (r_1, a_1, b_2)\}$
4	A_2	A	$\{att_1 = v_4\}$	$\{a_2, b_3\}$	$\{a_1, a_2, b_1, b_2, b_3\}$	$\{(r_1, a_1, b_1), (r_1, a_1, b_2), (r_1, a_2, b_3)\}$
5	B_3	B	$\{att_2 = v_5\}$	$\{b_3\}$	$\{a_1, a_2, b_1, b_2, b_3\}$	$\{(r_1, a_1, b_1), (r_1, a_1, b_2), (r_1, a_2, b_3)\}$
6	C_1	C	$\{att_3 = v_6, att_4 = v_7\}$	$\{c_1\}$	$\{a_1, a_2, b_1, b_2, b_3, c_1\}$	$\{(r_1, a_1, b_1), (r_1, a_1, b_2), (r_1, a_2, b_3), (r_2, c_1, b_1)\}$
7	C_2	C	$\{att_3 = v_8, att_4 = v_9\}$	$\{c_2\}$	$\{a_1, a_2, b_1, b_2, b_3, c_1, c_2\}$	$\{(r_1, a_1, b_1), (r_1, a_1, b_2), (r_1, a_2, b_3), (r_2, c_1, b_1), (r_2, c_2, b_2), (r_2, c_2, b_3)\}$

4.2.1   Discovery of Class Models

In general, the class model is discovered based on the object models in the input log. Figure 2(a) shows the discovered class model from the example log, where $OC = \{a, b, c\}$ and $RT = \{r_1, r_2\}$.

OC can be learned by incorporating all classes of all objects in the object models of all events. For instance, a is a discovered class since object models contain objects of class a, e.g., a_1. RT can be learned through observing object relations in object models of each event. r_1 (having a as the source class and b as the target class) is discovered since there exist object relations involving r_1, e.g., (r_1, a_1, b_1), and each of them has a object as the source object and b object as the target object.

For each relationship, its "always" ("eventually") cardinalities can be derived through integrating the number of related objects of each reference object in the object model of each (the last) event.[7] For instance, the discovered "always" cardinality on the source side of r_1 is "1" since in the object model of each event, each b object has precisely one related a object, e.g., b_1 and b_2 have one related a object a_1. The discovered "eventually" cardinality on the source side of r_1 is also "1" since in the object model of the last event (i.e., C_2), b_1 and b_2 have one related a object a_1 while b_3 has one related a object a_2 (the "eventually" cardinality is omitted on the graph for simplicity).

Note that the directly discovered "always" and "eventually" cardinalities on the target side of r_1 should be $\{1, 2\}$, since a_1 has two related b objects (b_1 and b_2) while a_2 has one related b object (b_3). We use a strategy to extend $\{1, 2\}$ to $\{1, 2, ...\}$, which will be explained later.

---

[7] In terms of cardinalities on the source (target) side of a relationship, the objects in the target (source) class are reference objects.

4.2.2 Discovery of AOC Relationships

After the class model is discovered, we can mine AOC relationships based on the objects referred to by each event.[8] The idea is that if an event refers to an object, the activity of the event refers to the class of the object. For instance, since event A_1 refers to three objects a_1, b_1, and b_2, activity A refers to class a and b, which means two AOC relationship (A, a) and (A, b) can be discovered as shown in Fig. 6.

For each AOC relationship, its cardinalities on the class side can be achieved by incorporating numbers of referred objects by each event. Consider the cardinality on the class side of (A, b). Since A_1 has two referred b objects (b_1 and b_2) while A_2 has one referred b object (b_3), the directly discovered cardinality is $\{1, 2\}$ and it is extended to $\{1, 2, ...\}$. Similarly, the "always" ("eventually") cardinalities on the activity side can be achieved by incorporating numbers of events referring each reference object just after every (the last) event happens. Consider the cardinality on the activity side of (B, b). Since b_1 and b_2 are not referred by any B event after the first event A_1 just happens, 0 is an element of the "always" cardinality. After the second event B_1 just happens, b_1 is referred by B_1, which adds a new element "1" into the "always" cardinality. After we check all events, the discovered "always" cardinality is $\{0, 1\}$. In terms of the "eventually" cardinality on the activity side of (B, b), we just check the moment when the last event just happens. Since each b object is referred by precisely one B event (i.e., b_1 is referred by B_1, b_2 is referred by B_2 and b_3 is referred by B_3), the discovered "eventually" cardinality is $\{1\}$.

4.2.3 Discovery of Behavioral Models

Based on the discovered class model and AOC relationships, we can relate events by objects and discover the constraints between activities. More precisely, each pair of activities referring to the same class or two related classes may have potential constraints in between. The class or the relationship between the two related classes serves as the intermediary to relate events. Note that each potential constraint, e.g., *con*, between an activity pair, e.g., (A, B), takes A as the reference activity and B as the target activity, and corresponds to a constraint relation which connects the constraint and its intermediary. The constraint relation can identify the target events of each reference event for *con* (cf. Section 3.3). If the relation between each reference event and its target events satisfies the restriction indicated by a constraint type, e.g., *response* (cf. Section 3.2), the potential constraint *con* becomes a discovered constraint which takes *response* as the constraint type. Consider the activities A and B in Fig. 6 (assuming the model does not have behavioral constraints) and the example log. Since both A and B refer to b, they have potential constraints in-between. If we assume A is

---

[8] There is a reference relation between an event (e.g., A_1) and an object (e.g., a_1) if and only if the event refers to the object, denoted as (A_1, a_1). The reference relations accumulate along with the occurrence of events. For instance, after A_1 happens, the set of reference relations is $\{(A_1, a_1), (A_1, b_1), (A_1, b_2)\}$; after B_1 happens, the set of reference relations is $\{(A_1, a_1), (A_1, b_1), (A_1, b_2), (B_1, b_1)\}$.

the reference activity, then there are two reference events A_1 and A_2, where A_1 is followed by two target events B_1 and B_2, and A_2 is followed by one target event B_3. This relation satisfies the requirement indicated by constraint type *response*, resulting in a discovered constraint con_1.

4.3 Discussion of Model Metrics

Based on the above three steps, we can discover a model similar to the one shown in Fig. 6 from the example log. The fitness of the discovered model is 1. As we mentioned, we need heuristics to extend the directly discovered cardinalities, e.g., when to replace $\{1, 2, 5, 8\}$ by 1..* ? Since the directly discovered cardinalities only contain the actual numbers observed in the log, their quality depends on the size of the log, i.e., if the log is not large enough to contain complete cardinalities in the process, the discovered model is overfitting. In order to improve generalization, we can extend cardinalities to allow more possibilities. An extreme example is to extend all directly discovered cardinalities to "*", which allows all possibilities and makes the model to be underfitting.

The difference between the discovered model and the one shown in Fig. 6 is that the former one has more behavioral constraints (e.g., a constraint with B as its reference activity and A as its target activity). In this sense, discovered models tend to have too many behavioral constraints, since our algorithm discovers all allowed constraints between each activity pair. This often makes discovered models spaghetti-like. In order to get more understandable models, we can remove less important constraints based on the specific situation. For instance, implied constraints can be removed without losing fitness and precision.[9] Note that, in general, filtering a model tends to improve (at least remain) fitness (i.e., more behavior fits the model), decrease complexity (i.e., the model has fewer edges), improve generalization (i.e., more behavior is allowed) and degrades precision (i.e., unobserved behaviors in the log may become allowed). Based on the specific need, one needs to balance between such concerns. Our plugin introduced in next section allows for seamless navigation possibilities to balance fitness, precision and simplicity.

5   Experiments

The discovery algorithm was validated based on logs extracted from data generated by the Dolibarr ERP/CRM system when executing the OTC (Order to Cash) process. More precisely, the data was extracted from 6 tables in the database of Dolibarr. For instance, "llx_commande" table records customer orders while "llx_facture" table consists of invoices. Based on the tables, we derived 4 activities (*create order*, *create invoice*, *create shipment* and *create payment*) and 6 object classes (i.e., one table corresponds to one object class) to be included

---

[9] The implied constraint by one constraint has the same reference activity, the same target activity and refers to the same class or relationship as the constraint as well as allowing more behavior than the constraint.

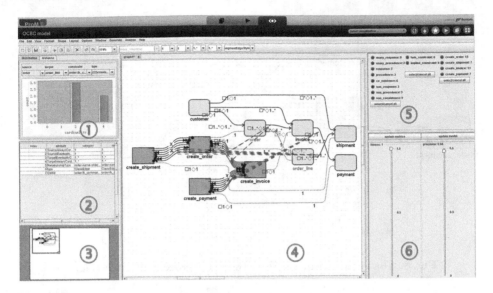

Fig. 7. The interface of the "OCBC Model Discovery" Plugin. (Color figure online)

in the XOC logs.[10] We instrumented the ERP/CRM system in such a way that we could extract data executed by real and simulated users of the system.

Our algorithm has been implemented in the "OCBC Model Discovery" Plugin in ProM.[11] Fig. 7 shows the interface of the plugin and a discovered model (in panel ④) from an XOC log. Panel ① presents the distribution of cardinalities and the instances related to one selected constraint (highlighted in red in panel ④). Panel ② shows the metrics of constraints such as confidence and support (this discussion is beyond the paper). It is possible to zoom in/out models through operating panel ③.

As discussed in last section, we can filter the discovered models to get a better understanding. Using the filter panels, it is possible to filter behavioral constraints based on constraint types (the plugin discovers all constraints of 9 common types by default) and activity names through panel ⑤, or based on the regulation of fitness and precision (the method for computing fitness and precision is not covered by this paper) through panel ⑥. For instance, if the desired action is to inspect the *unary-response, response, unary-precedence* and *precedence* constraints between *create order* and *create shipment* activities, one can uncheck the other boxes (all boxes are checked by default). The filtered model for the example log is shown in Fig. 8.

In the filtered model, there exist a *response* and a *unary-precedence* constraints between *create order* and *create shipment* activities. The constraints

---

[10] These tables and logs can be found at https://svn.win.tue.nl/repos/prom/Packages/ OCBC/Trunk/tests/testfiles/logs&models/OCBCModelDiscovery.

[11] Download *ProM 6 Nightly builds* from http://www.promtools.org/prom6/nightly/ and update the *OCBC package*.

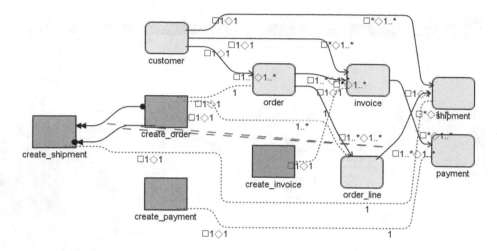

Fig. 8. The model discovered from the OTC process after filtering.

indicate one *create order* event is followed by one or more corresponding *create shipment* events while one *create shipment* event is always preceded by precisely one corresponding *create order* event. Investigating the Dolibarr system and its tables, it is possible to affirm that the process behavior of the system is according to these statements. The system allows creating multiple shipments for one order, but does not allow a shipment to contain products from multiple orders (as shown in Fig. 8). Although the discovered model in Fig. 7 is more complex than the real model (designed based on the real process) in Fig. 1, we can easily get the same insights after filtering appropriately.

6   Conclusion

In this paper we introduced *Object-Centric Behavioral Constraint* (OCBC) modeling language to graphically model control-flow and data/objects in a truly integrated manner. This novel language uses cardinality constraints to *describe data and behavioral perspectives in a single diagram* which overcomes the problems of existing data-aware approaches that separate the data (e.g., a class model) and behavioral (e.g., BPMN, EPCs, or Petri nets) perspectives. In OCBC models, different types of instances can interact in a fine-grained manner and the constraints in the class model guide behavior.

In this paper, we proposed an algorithm to discover OCBC models from object-centric event logs. Currently, the discovered models perfectly fit the source logs (i.e., there is no noise in logs or we do not distinguish noise). In future, we will extend the algorithm to better deal with infrequent and incomplete behavior. Besides, some metrics such as fitness, precision and generalization will be proposed to evaluate discovered models. Also, we will improve our approach to deal with larger scale logs in more complex scenarios, i.e., enabling the approach to discover compact models in a scalable manner (e.g., remove redundancies).

Moreover, this paper serves as a starting point for a new line of research. Next to model discovery and its support tools (OCBC Model Editor and OCBC Model Discovery Plugin) in *ProM*, we also support conformance checking. Based on OCBC models, many deviations which cannot be detected by existing approaches can be revealed.

References

1. van der Aalst, W.M.P.: Process Mining: Data Science in Action. Springer, Heidelberg (2016)
2. van der Aalst, W.M.P., Barthelmess, P., Ellis, C.A., Wainer, J.: Proclets: a framework for lightweight interacting workflow processes. Int. J. Coop. Inf. Syst. **10**(4), 443–481 (2001)
3. van der Aalst, W.M.P., Li, G., Marco, M.: Object-centric behavioral constraints. Corr technical report, arXiv.org e-Print archive (2017). https://arxiv.org/abs/1703.05740
4. van der Aalst, W.M.P., Pesic, M., Schonenberg, H.: Declarative workflows: balancing between flexibility and support. Comput. Sci. Res. Devel. **23**(2), 99–113 (2009)
5. Chen, P.: The entity-relationship model - toward a unified view of data. ACM Trans. Database Syst. **1**(1), 9–36 (1976)
6. Cohn, D., Hull, R.: Business artifacts: a data-centric approach to modeling business operations and processes. IEEE Data Eng. Bull. **32**(3), 3–9 (2009)
7. Genrich, H.J.: Predicate/transition-nets. In: Brauer, W., Reisig, W., Rozenberg, G. (eds.) Advances in Petri Nets 1986 Part I: Petri Nets: Central Models and Their Properties. LNCS, vol. 254, pp. 207–247. Springer, Heidelberg (1987). doi:10.1007/BFb0046841
8. Genrich, H.J., Lautenbach, K.: The analysis of distributed systems by means of predicate/transition-nets. In: Kahn, G. (ed.) Semantics of Concurrent Compilation. LNCS, vol. 70, pp. 123–146. Springer, Heidelberg (1979). doi:10.1007/BFb0022467
9. Object Management Group: OMG Unified Modeling Language 2.5. OMG (2013)
10. Halpin, T., Morgan, T.: Information Modeling and Relational Databases. Morgan Kaufmann Publishers Inc., San Francisco (2008)
11. van Hee, K.M.: Information System Engineering: A Formal Approach. Cambridge University Press, Cambridge (1994)
12. Hull, R., et al.: Business artifacts with guard-stage-milestone lifecycles: managing artifact interactions with conditions and events. In: International Conference on Distributed Event-Based Systems (DEBS 2011). ACM (2011)
13. Jensen, K.: Coloured petri nets. In: Brauer, W., Reisig, W., Rozenberg, G. (eds.) Advances in Petri Nets 1986 Part I: Petri Nets Central Models and Their Properties. LNCS, vol. 254, pp. 248–299. Springer, Heidelberg (1987). doi:10.1007/BFb0046842
14. Jensen, K.: Coloured Petri Nets: Basic Concepts, Analysis Methods and Practical Use. Springer, Heidelberg (1996)
15. Lohmann, N.: Compliance by design for artifact-centric business processes. In: Rinderle-Ma, S., Toumani, F., Wolf, K. (eds.) BPM 2011. LNCS, vol. 6896, pp. 99–115. Springer, Heidelberg (2011). doi:10.1007/978-3-642-23059-2_11
16. Lu, X., Nagelkerke, M., van de Wiel, D., Fahland, D.: Discovering interacting artifacts from ERP systems. IEEE Trans. Serv. Comput. **8**(6), 861–873 (2015)

17. de Leoni, M., van der Aalst, W.M.P.: Mining, data-aware process: discovering decisions in processes using alignments. In: Proceedings of the 28th Annual ACM Symposium on Applied Computing, pp. 1454–1461. ACM (2013)
18. Nigam, A., Caswell, N.S.: Business artifacts: an approach to operational specification. IBM Syst. J. **42**(3), 428–445 (2003)
19. Nooijen, E.H.J., Dongen, B.F., Fahland, D.: Automatic discovery of data-centric and artifact-centric processes. In: Rosa, M., Soffer, P. (eds.) BPM 2012. LNBIP, vol. 132, pp. 316–327. Springer, Heidelberg (2013). doi:10.1007/978-3-642-36285-9_36
20. Popova, V., Fahland, D., Dumas, M.: Artifact lifecycle discovery. Int. J. Coop. Inf. Syst. **24**(01), 1–44 (2015)
21. Rozinat, A., Aalst, W.M.P.: Decision mining in ProM. In: Dustdar, S., Fiadeiro, J.L., Sheth, A.P. (eds.) BPM 2006. LNCS, vol. 4102, pp. 420–425. Springer, Heidelberg (2006). doi:10.1007/11841760_33
22. Verkoulen, P.A.C.: Design, integrated information systems: an approach based on object-oriented concepts and petri nets. Ph.D. thesis, Eindhoven University of Technology, Eindhoven (1993)
23. Zervos, C.R.: Coloured petri nets: their properties and applications. Ph.D. thesis, University of Michigan, Michigan (1977)

Semi-automated Model-Based Generation of Enterprise Architecture Deliverables

Juan Pablo Sáenz, Steve Cárdenas, Mario Sánchez$^{(\boxtimes)}$, and Jorge Villalobos

Systems and Computing Engineering Department,
Universidad de Los Andes, Bogotá, Colombia
{jp.saenz79,sx.cardenas10,mar-san1,jvillalo}@uniandes.edu.co

Abstract. As part of Enterprise Architecture projects, models are built using different languages and tools to document and analyze the state of business and IT. However, models are just intermediate assets: deliverables are the actual outputs, but they are typically hand built using information from the models, and following the structures specified in EA methods. This requires manual effort, is error-prone, and results in artifacts that might be out-of-date very quickly. This paper addresses this by making a proposal to support the semi-automated generation of EA deliverables using a scripting Domain Specific Language for the creation of deliverable templates.

Keywords: Enterprise Architecture · Enterprise Modeling · Domain specific modeling languages · Architecture deliverable

1 Introduction

A fundamental element to Enterprise Architecture (EA) are Enterprise Models: they provide a method to structure, abstract, and analyze the complexity inherent to each organization. Besides, they largely satisfy visualization and communication needs and contribute to the effective understanding of the organization in terms of its domains (typically they are categorized as business, application, technology and information [1]). By means of Enterprise Models, the enterprise as a whole is represented through various models [2], each one structured according to some meta-model. Additionally, each model intends to describe, as accurately and completely as possible, some particular aspect of the organization that is of concern to some stakeholder.

Enterprise Modeling refers to the use of a *modeling language* to coherently specify and describe components of an organization along with their relationships [3]. Each Enterprise Model has its own audience, purpose, scope and level of detail. Also, there may be various modeling tools available to build them, ranging from mere drawings to sophisticated web-based Enterprise Modeling tools [4].

J.P. Sáenz—The author is currently a Ph.D. student at the Politecnico di Torino.

© Springer International Publishing AG 2017
W. Abramowicz (Ed.): BIS 2017, LNBIP 288, pp. 59–73, 2017.
DOI: 10.1007/978-3-319-59336-4_5

However, these models are not completely disjoint. There are always *common concepts* shared between them, and having integrated or global Enterprise Models would represent a definitive advantage by means of offering a unifying perspective. This would lead to improved cross-cutting analysis offering more valuable findings. Unfortunately, in today's state of the art, this is not typically possible because the different notations, meta-models, and tools used to create the models keep them from being integrated.

In fact, models are just an intermediate asset in Enterprise Architecture projects: *deliverables* are the actual outputs containing architects' findings, observations, and analyses, and typically serve to present the current, intermediate, or desired state of the enterprise as a whole or in part. Deliverables are thus supposed to be concrete but partial views of the models, containing the same information but presented in more manageable ways that may vary according to their purpose, scope, and their level of granularity. However, the construction of deliverables normally requires *extensive human intervention* to extract the necessary information from the models and to build the corresponding artifacts. On top of that, the fact that *models are fragmented* make this work even harder and limits the possibility of doing cross-cutting analyses.

This situation motivated the development of the proposal described in this paper, in which the information of several Enterprise Models (built across different modeling tools) is gathered, integrated and analyzed, in a cross-cutting manner, to *automatically generate EA deliverables*. To face the problem of tool heterogeneity, in this proposal the holistic view of the Enterprise Models is achieved through *meta-model mapping* instead of using a deep weaving. To face the problem of the lack of automation, we propose a set of functions to describe *deliverable templates*, and a set of functions for using analysis methods whose results are embedded in the deliverables. All those functions are invoked by means of a DSL.

The rest of the paper is structured as follows. Section 2 discusses Enterprise Architecture deliverables. Section 3 describes the strategy employed in our proposal, including the template language and the analysis language. This section also presents the mapping between the three meta-models that we used for our experiments. Then Sect. 4 presents a case study to demonstrate the applicability of the proposal, Sect. 5 describes the related works, and Sect. 6 concludes the paper.

2   Enterprise Architecture Deliverables

The creation of Enterprise Architecture deliverables relies on Enterprise Models and typically spans infrastructure components, business applications, business processes, information models, and the many relationships among them [5]. Enterprise Models' purpose is to accurately capture and represent the current, intermediate, or desired state of the Enterprise, in order to support analysis and decision-making processes. However, building these models is not a trivial task. On the contrary, it faces major challenges due to the high complexity of the today's organizations.

Figure 1 illustrates the typical process for creating Enterprise Architecture deliverables. It stems from the selection of specific perspectives of the enterprise that are of interest and should be included in the deliverables. For each one, some meta-models are selected (e.g., BPMN, ArchiMate, SysML) and it is possible for these to share concepts and relationships. Then, some tools that support the selected meta-models are used to build the corresponding models. Once these are built, they are expressed in the form of artifacts (catalogs, diagrams, text or matrices), that are finally integrated into Enterprise Architecture deliverables. The format and structure of said deliverables depend on the needs of stakeholders, the specific concerns of the enterprise, and the specifics of the Enterprise Architecture Framework in use.

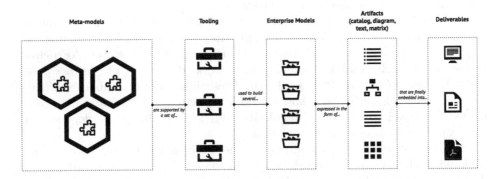

Fig. 1. Enterprise Architecture documentation process

Existing Enterprise Architecture documentation approaches struggle with the information volume and rapidly changing requirements within organizations [6]. Moreover, they rely on a high degree of manual work with very little automation during the documentation and maintenance of Enterprise Models [7]. As a result, Enterprise Architecture documentation endeavors are regarded as time-consuming, cost intensive, and error-prone [8]. For example, analysts may make mistakes when copying information from the models to the deliverables, may omit information, or may not update the deliverables as fast as models are updated.

The context outlined above has motivated various research efforts oriented to automation mechanisms that would improve Enterprise Architecture documentation process [9]. In [10], based on the application of a practitioner survey and a literature review, challenges regarding EA documentation automation were identified and grouped into four high-level categories: data challenges, transformation challenges, business and organizational challenges, and tooling challenges.

The transformation challenges that they identified are the following: the need to consolidate ambiguous concepts imported from the productive systems in the organization; the need to ensure actuality and consistency of collected data from the productive systems; and the need to avoid duplication of EA elements imported from different productive systems of the organization. Meanwhile, tooling

challenges are mainly related to the fact that available tools do not support importing, editing, and validating model data for automated EA documentation.

Moreover, attention should be drawn to the fact that Enterprise Architecture
documentation occurs within an EA framework. On these frameworks, a set of
deliverables (the contractual or formal work products of an architecture project)
is specified, in accordance with the methodology proposed.

The TOGAF Architecture Development Method (ADM) [11] for instance, is
a generic method that can be used by enterprises in a wide variety of industry
types for developing and managing the life-cycle of an EA. When following the
ADM, the first step is to modify or extend the proposed generic methodology
in order to suit the specific needs of each organization, considering its architecture discipline maturity, its architecture principles, and the previously adopted
enterprise-frameworks, among other factors.

The ADM establishes a Preliminary Phase whose purposes include doing
any necessary work to initiate and adapt the generic method by defining an
organization-specific framework that could be using either the TOGAF deliverables or the deliverables of another framework, depending on the needs of
a specific Enterprise. In other words, the Preliminary Phase is about defining
"where, what, why, who, and how we do architecture" of the enterprise concerned. Moreover, the level of granularity addressed in this phase depends on
the scope and goals of each organization.

Among the deliverables produced at the Preliminary Phase, there is the
Tailored Architecture Framework, whose purpose, as its name suggests, is to
derive a tailored architecture method, together with a set of expected deliverables and artifacts, as well as its deployed tools, and interfaces with governance
models and other frameworks. This is why each EA project follows their customized methodology with its own set of deliverables and artifacts; each one
of which is built through the meta-models, the models, the viewpoints and the
tools that better satisfy the stakeholder's visualization and analysis needs.

Adherence to an EA framework adds complexity to the whole EA documentation process. It implies that the set of deliverables and their content varies
among enterprises, architecture projects, and methodologies. Which also means
that the set of Enterprise Models is variable too, and the artifacts to be included
in the deliverables are not predefined.

3   Automated Generation of Enterprise Architecture Deliverables

Taking into account the challenges mentioned in the previous section, the proposal presented in this paper aims to automate the generation of EA deliverables from diverse and complementing Enterprise Models. The core of the proposal are the methods to gather information from said models, map common
entities and relationships, perform cross-cutting queries and analyses, construct
artifacts, and finally output all of these results in deliverables that follow precise structures. For experimentation purposes, we selected three modeling tools

with their corresponding meta-models: Bizagi Business Process modeler, which is based on BPMN and XML Process Definition Language (XPDL); the Iteraplan Enterprise Architecture Management tool, which has its own meta-model; and Archi, an open source ArchiMate modeling tool. These tools were chosen because they target aspects of the organization which are present in typical Enterprise Architecture projects, and because they share some common concepts that make cross-cutting analysis useful and interesting.

Our approach for generating EA deliverables is illustrated in Fig. 2 and comprises the following steps: (a) The meta-modeler composes a set of functions intended to perform queries and analysis over the Enterprise Models, and to create artifacts based on these analyses. These functions are invoked through a DSL that we developed and named Pollux. (b) Then, the enterprise architect defines a deliverable template in terms of chapters, sections, and artifacts. This is done through a different DSL that we named Castor. The architect also embeds query and analysis functions invocations into the deliverable template. (c) The resulting deliverable template, along with its embedded functions, is then inputted into the Enterprise Architecture Deliverables Generator (EADG) engine which executes the corresponding queries and analysis over the models. (d) Subsequently, the EADG builds the artifacts (catalogs, matrices, and diagrams) that are represented as texts, tables and images. Finally, (e) based on the deliverable template and the artifacts produced by the engine, the document generator composes the deliverable and exports it in several formats, depending on the stakeholder's visualization requirements.

We now describe the four main components of our approach: the template language (Castor), the query language (Pollux), the mapping procedure between meta-models, and the EADG.

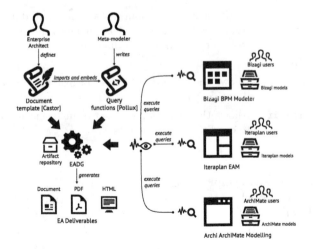

Fig. 2. EA deliverable generation process

3.1   Castor: The Template Language

Castor is a set of functions along with a Domain Specific Language intended to define the structure of deliverables based on information extracted from Enterprise Models. The actual content for the deliverable is also extracted from the models through the usage of query and analysis functions (Pollux set of functions).

The main function of Castor is to specify deliverables' structures in terms of chapters and sections. Castor also has a set of directives for: importing external files, such as text and images; invoking the query and analysis functions; building concrete EA artifacts with the data brought from the EA models; including template fragments to favor reuse; and embedding control flow directives. Table 1 presents the whole set of instructions available in the Castor DSL, along with their description.

Table 1. Castor DSL instructions

General functions	
`importFunction`	Imports the functions of a given library, which is associated to a certain Enterprise Modeling tool
`importSource`	Defines the set of Enterprise Models to be included into the Enterprise Architecture Deliverables generation
`template`	Defines the name of the resulting output file, which corresponds to the generated deliverable
Text functions	
`chapter - section - text`	Inserts a new chapter (into the deliverable), section (into the chapter) or text fragment, into the deliverable
Artifact functions	
`catalog - matrix - image`	Displays the corresponding artifact with data gathered from one or several models. Images typically correspond to Enterprise Modeling diagrams
Flow control functions	
`forEach`	Allows the execution of a set of instructions in an iterative way
`if - else - else if`	Allows the execution of a set of instructions according to conditional statements

Listing 1 presents an example of the usage of Castor. First of all, it is necessary to load a set of models that will provide the information for the deliverable (line 1), as well as the set of query and analysis functions to pull content from the previously models (line 2). In the example, a library of functions to communicate with *Archi Modeler* is imported along with a model called *archisurance* (corresponding the ArchiSurance case study from Archi). Line 1 also defines an alias for the model: `$mArchi`. Finally, in line 4 some basic features for the template are defined (title and output directory).

In the ensuing lines, the structure of the deliverable is specified by means of containers which can be primary, for chapters and sections, or secondary for the case of tables and lists. In Listing 1, line 5 defines a new Chapter in the template by providing its name, and line 6 defines a Section within that chapter also with a given title.

Within the containers, artifacts of different kinds can be included by means of specific directives. These serve to include text, images, and template fragments, or to embed artifacts (matrix, catalogs, and diagrams). In the listing, lines 7 and 8 demonstrate the inclusion of text content by means of the invocation of the function `getInformationView()` and `getDocumentation()` on the Archi model.

```
1  importModel"archisurance" as $mArchi
2  importFunction "./archi.jar" as $fArchi
3
4  template {name:"Archisurance deliverable", output_dir: "./deliverable"}
5  chapter {title:"Target Business Architecture"}
6  section {title:"Business goals and objectives"}
7  text {contents:
8  $fArchi.getInformationView($mArchi,"Goal and Principle View").
      getDocumentation()}
9
10 forEach $fArchi.GetProcess() as $process:
11   text {contents: $process.name}
12   if($fBpm.ProcessExists($process.name) == true)
13     catalog {function: $fBpm.GetActivities($process.name)}
14   else:
15     text {contents: "The process does not contains activities"}
16   end
17 /forEach
18
19 catalog {function: $fArchi.getViewElements($mArchi, "Goal and Principle
      View")}
```

Listing 1. Castor sample code

On top of the above, there is an additional kind of directives intended to provide a more dynamic control over the elements in the deliverable, based on the contents of the EA models. These directives are expressed through conditionals and cycles. Cycles allow the composition of instructions such as *"create a section in the document for each business process brought from Bizagi BPM Modeler, including its name and description"* (Listing 1, lines 10 to 17). Using conditionals, it becomes possible to express statements such as *"embed into the document the name of a business process and, in case it has activities, their names and descriptions. Otherwise, if the process does not have activities associated, display a message informing the situation"* (Listing 1, lines 12–16).

3.2   Pollux: The Query Language

Pollux is both a language and a set of functions to communicate with EA modeling tools to perform queries and analysis over the models, and to generate artifacts using their results. These functions are classified into query and analysis functions, and artifact generation functions. The first ones are responsible for extracting information from the models and applying filters based on any criteria specified by the user; the second group of functions is responsible for building artifacts. This means generating matrices, catalogs or diagrams based on the information extracted by the query and analysis functions. Lines 20 and 21 of Listing 1 illustrate this by invoking the function to create a catalog and giving this information the result of querying the Archi model using the `getViewElements()` function.

Through the use of these functions, it is possible to build artifacts by gathering and integrating information from different models that might be built across different tools. This feature enables the recognition of relationships between the elements of several models, that otherwise would go unnoticed. For instance, to determine which server supports a given business process. Therefore, this feature significantly supports and enhances the analysis tasks of the enterprise architect.

Analysis functions are generally related to quantitative analysis over the elements of a model when there is enough information into it to perform the function [12]. For instance, the response times of the business processes and applications.

Table 2. Pollux DSL instructions for Archi

Views	
`getViewInformation`	Gets the information (properties) of a certain view
`getViews`	Gets a list of all the views into the Archi Enterprise Model
`getViewImage`	Gets the graphical representation (diagram) associated to a certain view
`getViewElements`	Gets the elements present in a certain view
Layers	
`getElement`	Gets an element along with its attributes, properties, and relationships
`getElementsByType`	Gets the set of elements of a certain type
`getElementsByLayer`	Gets the elements belonging to a certain layer. For instance, Business layer, Application layer, Technology layer
`getProcess`	Gets the elements whose type is process
Canvas	
`getContainersFromView`	Delivers a list of the containers present in a certain Enterprise model
`getViewElementsByContainer`	Delivers a list of elements or notes that are placed inside a certain container

It is important to mention that all the functions were developed as implementations of a common interface. This means that the current libraries can be extended to incorporate new functions or to support communication with new tools that may be incorporated in the future. These functions are packaged into a JAR (Java ARchive) that later will be imported into the deliverable template. Table 2 exemplifies the basic set of functions available to gather and deal with Archi Enterprise Models.

3.3   Mapping Between Meta-models

In order to be able to perform queries and analysis in a cross-cutting manner and generate meaningful artifacts, it is necessary to be able to find relationships between models built with different tools. To support this, it was necessary to identify common concepts between the meta-models by means of establishing a mapping between equivalent elements. In the case of the sample tools that we selected, this included a mapping between ArchiMate and BPMN, one between BPMN and Iteraplan, and one between Iteraplan and ArchiMate. Table 3 describes, at first, the mapping that was made between entities in ArchiMate and entities in BPMN. It is important to highlight that beyond those entities that are specific to the meta-models, other entities belonging to the tool, were included. That is the case of the Bizagi element called *Entity*, which does not belong to BPMN.

Further on, the mapping between the entities from ArchiMate and Iteraplan is presented. In this regard, it is noticeable that the mapping between the entities does not always comply a one-to-one correspondence. In fact, several Iteraplan business concepts do not have any associated entity in the ArchiMate business layer meta-model.

Finally, the last section of the table presents the mapping between Iteraplan and Bizagi meta-models. There are few mappings as the Iteraplan meta-model is more focused on project management at a lower level of detail while Bizagi offers a high level of detail over the business process. In the Case Study section, an example of a cross-cutting query and analysis function that depends on this mapping proposal is presented.

3.4   Generation of the Enterprise Architecture Deliverables

Once the deliverable template is defined along with the Pollux functions and the external files, the EADG engine generates the document which is consistent with a meta-model, proposed by us, to represent the different kind of components that might be present in a text processing tool. Entities such as document, document body, container element, content element, table, list, image, text, table row, table cell and list item are included into this meta-model. Lastly, the document generator composes the deliverable, and according to the stakeholder's visualization preferences, exports it in Microsoft Word, PDF or HTML format.

Table 3. Mapping between modeling tools meta-models

ArchiMate business layer	Bizagi element
Business process	Business process diagram, pools, lanes
Function	Task, sub-process
Business event	Event
Business object	Entity
Business role	Lane, organizational role
ArchiMate application layer	
Application function	Service, task, script task
Data object	Entity
ArchiMate technology layer	
Device	
Artifact	
ArchiMate business layer	*Iteraplan (business)*
	Business domain
	Information system domains
	Architectural domains
Business process	Business process
Business role	Business unit
Product, business service	Product
	Business mapping
Business function	Business function
Business object	Business object
ArchiMate application layer	*Iteraplan (application)*
Relation, application interface	Interface
Application component, application collaboration	Information system
Application service, application function	IT service
ArchiMate technology layer	*Iteraplan (technology)*
Technical component	System software, artifact
Infrastructure element	Node, device, network
ArchiMate (implementation/migration)	
Gap	Project
Iteraplan (business)	*Bizagi element*
Business process	Business process diagram, pools, lanes
Business function	Task, sub-process
Business object	Entity
Iteraplan (application)	
Business object	Entity

4   Case Study

To illustrate the proposal presented in this paper, a case study was developed on the basis of three inputs: an EA academic project; an EA framework specifically tailored for this project; and the set of three different Enterprise Modeling tools, mentioned in the previous section (Bizagi BPM Modeler, Iteraplan EAM, and Archi ArchiMate modeling).

The experimentation consisted in generating Enterprise Architecture deliverables for Editorial de los Alpes (EDLA), an academical exercise to simulate a publishing house that is responsible for carrying out the complete textbook production process, from the selection of the author to the distribution to the points of sale. The main organizational goals of the publishing house are to break into the digital market, lower the operational costs and increase the annual sales.

The meta-model of EDLA is composed of 13 domains: applications, business motivation model, business process architecture, business partners, financial structure, human resources, information, infrastructure, business assets, organizational structure, products, services, and technology. There are 109 entities modeled into these domains and the model built upon this meta-model, has about 1000 elements and more than 1500 relationships.

A deliverable of 48 pages was produced and EADG was able to generate and insert 26 customized artifacts (catalogs, matrices, diagrams, and texts), by executing query and analysis functions, some of them in a cross-cutting manner, over several models built on Bizagi BPM Modeler, Iteraplan EAM, and Archi ArchiMate modeling, as shown in Figs. 3 and 4.

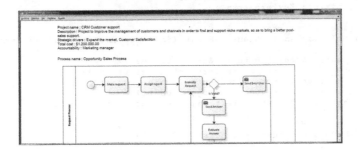

Fig. 3. Project information from Iteraplan and Business Process diagram from Bizagi

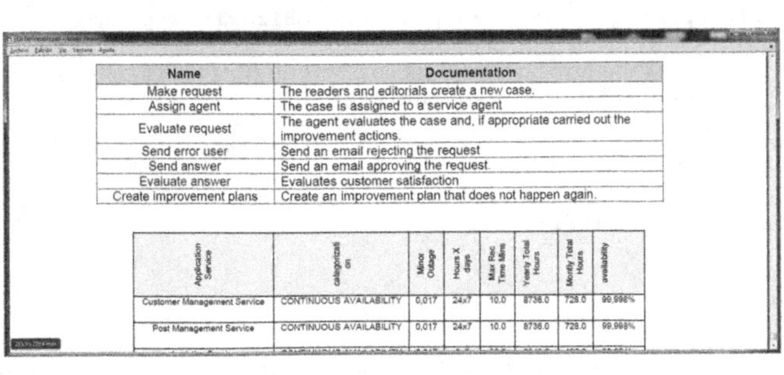

Fig. 4. Catalog with tasks from Bizagi and Application services from Archi

The Enterprise Models built on each tool were the following. In Archi, our Enterprise Architecture research group developed all the models regarding business motivators, business canvas, business capabilities, competitors, infrastructure, value chain, and stakeholders. In Bizagi all the business process were modeled. And in Iteraplan, the landscape diagram, cluster diagram, nesting cluster, information flow, and portfolio flow were generated based on the information corresponding to the 14 building blocks.

The sample code in Listing 2 illustrates the capability of our tool to perform cross-cutting queries and analysis by extracting certain information from several models built across different tools. In this example, at first, a query is performed over the projects registered in Iteraplan and their information is embedded as text into the deliverable (Listing 2, lines 15–21). Then, their business processes are taken from Bizagi, embedding the corresponding diagram and catalog of tasks (Listing 2, lines 24–32). Finally, the application services that support each task are gathered from Archi and embedded into the deliverable as a catalog (Listing 2, line 35).

```
 1  importFunction"./archi.jar" as $fArchi
 2  importModel "archisurance" as $mArchi
 3
 4  importFunction "./process.jar" as $fBizagi
 5  importModel "process" as $mBizagi
 6
 7  importFunction "./iteraplan.jar" as $ fIteraplan
 8  importModel "http://eadg.edu.net/iteraplan/" as $mIteraplan
 9
10  template { name :"EDLA deliverable", output_dir : "./ deliverable"}
11  chapter { title :"Project Management"}
12  section { title :"Key projects"}
13
14  /* Iterate over the projects in Iteraplan */
15  <forEach $mIteraplan.getElementsByType($mIteraplan,"project" as $project:$>
16
17    Project name: $<text{contents:$project.name}$>
18    Description: $<text{contents:$project.description}$>
19    Strategic drivers: $<text{contents:$project.strategicDrivers}$>
20    Total cost: $<text{contents:$project.costs}$>
21    Accountability: $<text{contents:$project.accountability}$>
22
23    /* Iterate over the business process of the current project in Bizagi */
24    <forEach $project.businessProcess as $bp:$>
25
26      Process name: $<text{contents:$bp.name}$>
27
28      /* Display the diagram of the current business process */
29      <image{id:"bpImg", uri:$bp.getDiagram($mBizagi, $bp.name)}$>
30
31      /* Embed a catalog with the tasks of the current business process */
32      <catalog{function:$bp.getTasks(mBizagi)}$>
33
34      /* Embed a catalog with the Application services (modeled in Archi)
             that supports the business process */
35      <catalog{function:$mArchi.getRelatedElements("businessProcess", bp.name
            ,"applicationService")}$>
36
37    </forEach$>
38  </forEach$>
```

Listing 2. Cross-cutting function over the three tools code sample

The cross-cutting query involves elements that belong to different models and were built in different tools. By means of the meta-model mapping presented above, it is possible to move across the elements and their relationships.

5 Related Work

Jackson et al. [13] state that a review document or any other paper artifact can be interpreted as a view of the system or engineering model. Based on this premise, they document the development of a tool and a method to produce sophisticated artifacts as views and by-products of integrated models. In their proposal, any paper artifact is assumed as a serialized model that describes and narrates a set of views of the main engineering model. The document structure is completely designed and modeled as in UML, using a custom profile, while the content is provided through query and analysis functions whose inputs are pointers to the main engineering model. Then, the software produces a DocBook XML file, which may be used to generate the final document in PDF or HTML format.

A methodology for automatic software documentation generation is proposed in [14]. In this, the automatic generation is achieved through the design of a documentation model and the definition of mapping relationships between its elements and the system model. Based on this artifacts, the mapping process is able to extract graphical and textual information from UML and SmartC models, organizes them according to the documentation model and generates the software documentation into a Microsoft Word document.

Buschle et al. [15] focused in automating the collection of the data used to build the organization-wide models. The paper illustrates how a vulnerability scanner can be utilized for data collection aimed to automatically create Enterprise Architecture models, mainly covering infrastructure aspects (Application and Technology layer of the organization). Moreover, the outcomes from the vulnerability scanner are taken by an EA analysis tool, responsible for asserting certain system quality attributes. Scanners do not deliver complete EA models (especially those regarding Business Layer).

A particular Enterprise Service Bus (ESB) implementation can be used to extract EA relevant information according to [16,17]. This statement is motivated by the idea that an ESB can be considered "the nervous system of an enterprise interconnecting business applications and processes as an information source". Moreover, based on the application of a survey, the paper concludes that the SAP PI ESB seems to be suitable and a reasonable start point for an automated EA documentation endeavor.

All the proposals outlined above share concerns regarding the time consuming, cost intensive and error prone nature of EA documentation and maintenance. Their solution approaches mainly focus on the automatic generation of Enterprise Models based on the execution of scanning tools over IT systems that are already deployed. Likewise, in [10] transformation challenges and tooling challenges are related to the need to ensure actuality and consistency of collected data from the productive systems, as well as the inability of the available tools to support importing, editing and validating model data for automated EA documentation.

On the contrary, our proposal focuses on automating EA deliverables generation based on a set of Enterprise Models that are already built using different tools and according to several meta-models. Moreover, in addition to the possibility to automatically gather information from these models in order to compose the deliverables, our proposal allows the integration between them through a meta-model mapping, which enables, as well, cross-cutting analysis over different kinds of Enterprise Models. These Enterprise Models are not restricted to Application and Technology layers, they also represent business logic that may not be gathered from the productive systems of the organization.

6   Conclusions

This paper presents an approach for automatically generating Enterprise Architecture deliverables by integrating multiple Enterprise Models built across different tools. A Domain Specific Language was developed to invoke the query and analysis functions that are embedded into a deliverable template. Likewise, a set of query and analysis functions were developed in order to communicate and gather information from the Enterprise Models. The information extracted from these models is embedded into the deliverables in the form of artifacts, such as catalogs, matrices, and diagrams. Nevertheless, the value-added of this proposal does not completely lie on the capability to automatically generate the Enterprise Architecture deliverables by embedding information extracted from many Enterprise Models. Besides that, there is the opportunity to perform cross-cutting queries and analysis by means of a mapping proposal between common entities and relationships from several meta-models. To illustrate our approach, Bizagi Business Process modeler, Iteraplan Enterprise Architecture Management tool, and Archi ArchiMate modeler meta-models were mapped, and an Enterprise Architecture deliverable was generated. We think that the Enterprise Architecture documentation may substantially benefit from our proposal, given the fact that currently this process relies on a high degree of manual work, and is regarded as time-consuming, cost intensive, and error-prone. Moreover, the cross-cutting queries and analysis enable the recognition of relationships between the elements of several models, that otherwise would go unnoticed.

References

1. Lankhorst, M.: Enterprise Architecture at Work, vol. 10. Springer, Heidelberg (2009)
2. Lankhorst, M.M.: Enterprise architecture modelling—the issue of integration. Adv. Eng. Inform. **18**(4), 205–216 (2004). Enterprise Modelling and System Support
3. Jonkers, H., Lankhorst, M.M., van Buuren, R., Hoppenbrouwers, S., Bonsangue, M.M., van der Torre, L.: Concepts for modeling enterprise architectures. Int. J. Coop. Inf. Syst. **13**, 257–287 (2004)

4. Matthes, F., Buckl, S., Leitel, J., Schweda, C.M.: Enterprise architecture management tool survey 2008. Technical report (2008)
5. Buckl, S., Matthes, F., Roth, S., Schulz, C., Schweda, C.M.: A conceptual framework for enterprise architecture design. Interface **3**, 1–4 (2008)
6. Roth, S., Hauder, M., Farwick, M., Breu, R., Matthes, F.: Enterprise architecture documentation: current practices and future directions. In: 11th International Conference on Wirtschsftsinformatik, March 2013, pp. 1–15 (2013)
7. Winter, K., Buckl, S., Matthes, F., Schweda, C.M.: Investigating the state-of-the-art in enterprise architecture management methods in literature and practice. In: MCIS (2010)
8. Kaisler, S., Armour, F., Valivullah, M.: Enterprise architecting: critical problems. In: Proceedings of the 38th Annual Hawaii International Conference on System Sciences, no. C, pp. 1–10 (2005)
9. Lucke, C., Krell, S., Lechner, U.: Critical issues in enterprise architecting - a literature review. In: 16th Americas Conference on Information Systems (AMCIS) 2010, pp. 1–11 (2010)
10. Hauder, M., Matthes, F., Roth, S.: Challenges for automated enterprise architecture documentation. In: Aier, S., Ekstedt, M., Matthes, F., Proper, E., Sanz, J.L. (eds.) PRET/TEAR -2012. LNBIP, vol. 131, pp. 21–39. Springer, Heidelberg (2012). doi:10.1007/978-3-642-34163-2_2
11. The Open Group: TOGAF Version 9 (2009)
12. Florez, H., Sánchez, M., Villalobos, J.: A catalog of automated analysis methods for enterprise models. SpringerPlus **5**(1), 406 (2016)
13. Jackson, M., Delp, C., Bindschadler, D., Sarrel, M., Wollaeger, R., Lam, D.: Dynamic gate product and artifact generation from system models. In: Aerospace Conference 2011, pp. 1–10. IEEE, March 2011
14. Wang, C., Li, H., Gao, Z., Yao, M., Yang, Y.: An automatic documentation generator based on model-driven techniques. In: 2010 2nd International Conference on Computer Engineering and Technology (ICCET), vol. 4, pp. V4-175–V4-179, April 2010
15. Buschle, M., Holm, H., Sommestad, T., Ekstedt, M., Shahzad, K.: A tool for automatic enterprise architecture modeling. In: Nurcan, S. (ed.) CAiSE Forum 2011. LNBIP, vol. 107, pp. 1–15. Springer, Heidelberg (2012). doi:10.1007/978-3-642-29749-6_1
16. Grunow, S., Matthes, F., Roth, S.: Towards automated enterprise architecture documentation: data quality aspects of SAP PI. In: Morzy, T., Härder, T., Wrembel, R. (eds.) Advances in Databases and Information Systems. AISC, vol. 186. Springer, Heidelberg (2013)
17. Buschle, M., Ekstedt, M., Grunow, S., Hauder, M., Matthes, F., Roth, S.: Automating enterprise architecture documentation using an enterprise service bus. In: Americas conference on Information Systems vol. 6, pp. 4213–4226 (2012)

Towards Automated Process Model Annotation with Activity Taxonomies: Use Cases and State of the Art

Michael Fellmann[✉]

Institute for Computer Science, University of Rostock,
Albert-Einstein-Str. 22, 18059 Rostock, Germany
Michael.Fellmann@Uni-Rostock.de

Abstract. In business process modeling, semi-formal models typically rely on natural language to express the labels of model elements. This can easily lead to ambiguities and misinterpretations. To mitigate this issue, the combination of process models with formal ontologies or predefined vocabularies has often been suggested. A cornerstone of such suggestions is to annotate elements from process models with ontologies or predefined vocabularies. Although annotation is suggested in such works, past and current approaches still lack strategies for automating the annotation task which is otherwise labor intensive and prone to errors. In this paper, first an example for use cases is given and then a comprehensive overview of the state of the art of annotation approaches is presented. The paper at hand thus may provide a starting point and basis for researchers engaged in (semi-)automatically linking semi-formal process models with more formal knowledge representations.

Keywords: Business process · Semantic annotation · Automatic matching

1 Introduction

In business process modeling, semi-formal modeling languages such as BPMN are used to specify which activities occur in which order within business processes. Whereas the order of the activities is specified using constructs of the respective modeling language, the individual semantics of a model element such as "Check order" is bound to natural language. As long as models are created and read by humans only and a commonly agreed (potentially restricted) language is used, the usage of the natural language is no serious limitation. However, if models have to be interpreted by machines, e.g. for offering modeling support, search on a semantic level, content analysis in merger and acquisition scenarios and for re-using implementation artifacts linked to process elements (e.g. web services), a machine processable semantics of modeling elements is required [1]. In the past, several approaches tried to formalize the semantics of individual model elements by annotating elements of ontologies or other predefined vocabularies that to some degree formally specify the semantics of a model element. However, such approaches up to date suffer from a major limitation: Annotation is a highly manual and tedious task. The user has to select suitable elements of an

W. Abramowicz (Ed.): BIS 2017, LNBIP 288, pp. 74–90, 2017.
DOI: 10.1007/978-3-319-59336-4_6

ontology by browsing the ontology or doing a keyword-based search in the labels of
the ontology. Even if the system is capable of presenting some annotation suggestions,
e.g. based on lexical similarity of labels, the user has to make sure that annotations
match the appropriate context in the process model by inspecting the structure of the
ontology that typically is organized in a hierarchy. For example, if the ontology contains two activities labelled with "Accept invitation", it is important whether this
activity is part of the hiring process (where the applicant accepts e.g. a job interview) or
the planning process for business trips (where the employee accepts an invitation of a
business partner). In other words, the semantic context of an element that is to be
annotated must be considered. Since no highly automated context-sensitive approach
for process model annotation is available so far, this contribution is meant to facilitate
developing, comparing and optimizing such approaches. To bootstrap systematic
research in this direction, use cases for automated annotation approaches are described
and existing annotation approaches are reviewed. With this, interest in a very promising
research topic should be raised; both in regard to scientific outcome as well as practical
usefulness.

The remainder is structured as follows. Section 2 provides use cases for automatic
process model annotation. In Sect. 3, existing annotation approaches are reviewed. In
Sect. 4, a conclusion and short outlook on research opportunities is provided.

2   Use Cases for Automated Annotation

In the following, application scenarios leveraging an automated process model annotation are presented.

Modeling Support. If process elements are automatically annotated with elements of
an ontology or taxonomy containing a set of predefined activities, this knowledge can
be exploited to help the modeler completing his or her modeling task. This is illustrated
by Fig. 1 showing a process fragment (bottom) being automatically annotated with a
task ontology (top). This knowledge can then be exploited to provide modeling suggestions (right). The advantage of using this knowledge is that the suggestions for the
following model element are not only derived on basis of one (or more) previous model
element(s). Rather, they can be based on the knowledge representation that is linked to
the model element via annotation. For example, in the knowledge representation it may
be specified that after offering the job, potential candidates should be selected. The key
difference to approaches based on e.g. suggesting activities retrieved from similar
models such as the work by Koschmider [2] is that in this way *normative knowledge* is
used, i.e., how an enterprise *should* act. Besides modeling support, automated annotation also provides the basis for leveraging information from knowledge representations that may provide additional value. For example, the PCF taxonomy [3] contains
key performance indicators for all of the activities it contains (in the industry independent version approx. 1000 activities). Also, information to enact a process in the
workflow environment may be linked to the set of specified reusable activities. *All in
all, new ways of modeling support and of providing additional assistance in the*

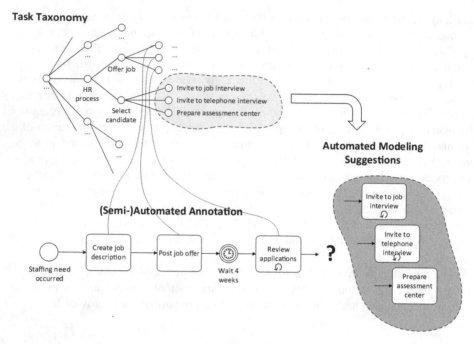

Fig. 1. Automated modeling suggestions

model-based design of process supporting information systems are possible due to an automatic process model annotation.

Process Retrieval. Current repositories are equipped mainly with keyword-based search mechanisms or rely on process query languages such as BPMN-Q [4]. These instruments allow searching the process space using natural language as well as structural and behavioral information. However, they lack to restrict search to broader content or topics of a process corresponding with the distinct functional areas in an enterprise, in short with the *business topic*. Although it may be possible to manually assign descriptors to models and in fact manual annotation approaches have been discussed recently [5], this imposes an extra effort on modelers having to focus on delivering high-quality models in a timely manner. Moreover, descriptors must be kept up to date if the model is adapted. Hence computing the business subject of a process model automatically based on activities that are annotated automatically creates an additional value. It can be re-computed from time to time to keep the information up to date.

How the automatic annotation of processes may improve the retrieval of processes from a repository is shown in Fig. 2. The user types in the keyword "review" in the search form (top). Since reviewing activities can occur in many contexts of the enterprise activities, the user specifies the category "Human resource management" which automatically shows up by typing in the special keyword "category" (much like keyword-search functionality in file explorers of common operating systems). Based on

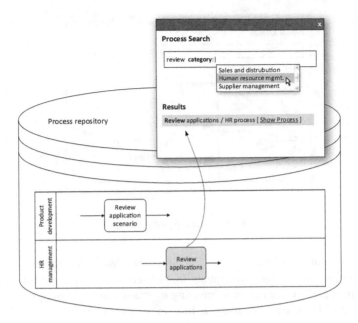

Fig. 2. Improved process retrieval

the automated annotation of processes, the activity "Review applications" is found that belongs to a process in the HR realm. *Hence with automated annotation, the retrieval of process knowledge on a semantic level can be improved.*

Process Analysis. Similar to process retrieval, current approaches for analyzing the contents of process models rely on keyword search or specialized query languages. Another way that is also common to analyze process models is to find similar process models, commonly referred to as *process model matching*. However, all these mechanisms have in common that an analysis is done in relation to what the user wants to know (which requires the user to know common terms in the business context) or what is available (when models are compared). However, in some situations of process analysis it may be favorable to introduce normative knowledge about which tasks *typically* occur in enterprises. With this, questions regarding the coverage of a process can be answered such as "Do we have a process for managing product quality?" which may be important for e.g. certain certification activities. Another example would be "In which area, we do not have yet specified processes?" or "Which of our processes are highly cross-cutting?" Fig. 3 illustrates how automated process model annotation may serve process analysis and comparison using normative knowledge.

At first, the user selects process models using a keyword search and adds models to the comparison (top). He or she subsequently inspects and compares the contents of the process model using a taxonomy of pre-defined business functions to guide this inspection (center). In more detail, the result of automatic annotation is displayed for each process model in a separate column. Each matching activity is displayed as a

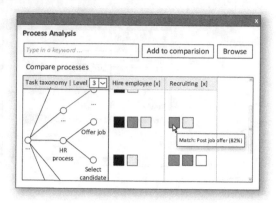

Fig. 3. Advanced process analysis

square that is saturated according to the matching score. Multiple squares are composed to a visualization that slightly resembles to well-known equalizer visualizations of audio-equipment. When the mouse is hovering over a square, matching score and other information can be shown such as a link to open the process model or other meta-information about the process. In order to zoom-in and -out, the user may also expand or reduce taxonomy levels (left).

Other visualizations that would be possible are histogram-like diagrams. *In this way, automated process model analysis that is enabled by exploiting annotation information is the basis for advanced analysis and visualization capabilities.*

3   State of the Art

Annotation in general has been discussed in the early stages of the Semantic Web movement [6]. In the following, annotation has also been explored in relation to enterprise modeling. For example, Boudjlida and Panetto describe annotation types in enterprise modeling [7]. The authors identify various semantic relations between an enterprise model and an element of an ontology and provide a schema for describing annotations. However, the authors also acknowledge that automation in annotation is largely missing: "However, an important feature is missing: it is the one that permits the automatic or the semi-automatic provision of the annotations." [7] Since no comprehensive overview of existing, manual annotation approaches for enterprise model is available so far, a structured literature analysis is conducted. With this, developers of automated annotation tools should be served with an overview that should inform and inspire the development of automated annotation procedures.

3.1 Selection and Analysis of Relevant Literature

For analyzing the literature, the literature data bases *EBSCO, Springer, ScienceDirect* and *Google Scholar* were examined. Different queries such as "process model" AND annotation or "model annotation" or "semantic annotation" AND annotation and variants of these queries were executed leading to 83 hits. The following inclusion and exclusion criteria were applied: Articles were excluded that use the term "annotation" to simply express that some additional information is written in the process model that has been generated automatically (i.e. to find semantic deficiencies). Further, works were excluded aiming at the semantic annotation of web services (e.g. by standards such as SAWSDL) or described by [8–12] or paper-based forms [13] since this is only slightly related to model annotation. Moreover, articles were excluded that describe high-level, general purpose annotation frameworks e.g. in the field of Semantic Web (annotation of web pages). Articles were included that sufficiently deal with business process modeling and that discuss annotation in sufficient detail. Regarding the latter aspect, this means not merely using/exploiting annotated process models that have been annotated somehow somewhere, but that are concerned with annotation itself.

In terms of completeness of the literature search, it can be assumed that most relevant papers have been identified since a high overlap between hits from databases and Google Scholar was found. Moreover, also all works in the area of process model annotation contained in the recent survey from [14] were retrieved. Hence it is likely that all important works were identified. For this reason, a forward- and backward search as requested e.g. from Webster and Watson [47] was not performed. Especially a backward search did not prove to be fruitful, since with this predominantly annotation tools of the semantic web community (such as OntoMat Annotizer etc., see [15, 16]) have been found that are not specific to process model annotation. If such approaches would be included, all the annotation work of the semantic web community (as an example list, see http://semanticweb.org/wiki/Tools) would be relevant. However, in the BPM community more focused approaches exist that leverage the process structure such as the works form Born et al. [17]. Hence it is more useful to more strictly look at the works from the BPM community that developed annotation techniques, which is done in the paper at hand.

3.2 General Overview on the State of the Art

In the following, the results of the literature analysis are presented (cf. Table 1). Relevant works are compared and reviewed by first giving a *Description of the overall approach*. Besides, a precise account on the notion of *Annotation concept* is given, that is, the specific approach the authors described, developed or implemented. In addition, approaches are compared in regard to whether they provide a (formal) definition of annotation (column *Def*) and the *Used technologies* such as e.g. lexical databases, string similarities etc. Moreover, approaches are compared in regard to two key characteristics. The first is their implemented or envisioned degree of automation (column *AU*). Symbol □ is used to indicate manual, ◼ for semi-automated and ▪ for

Table 1. Results of literature analysis

Paper	Description of the overall approach	Annotation concept	Def.	Used technology	AU	CO
[18]	The authors suggest a mapping strategy and present a tool developed to map BPEL4WS to OWL-S. With the help of the strategy and tool, BPEL4WS-processes can be translated to OWL-S process descriptions	The mapping relation of a BPEL4WS process to an OWL-S ontology. In addition, the relation between concepts from the OWL-S profile ontology to domain ontologies	–	No information	☐	no
[19]	Common modelling patterns are detected via an automatic semantic annotation of EPC process models. To automatically annotate the model elements, labels are decomposed using a lexical analysis and a pattern matching approach. If a suitable instance for annotation is missing in the ontology, then it will be created	Establishment of a semantic linkage from EPC functions and events to ontology instances	–	Lexicon (WordNet), term extraction, stemming	■	no
[17, 20, 21]	Execution-level business process modeling is supported that leverages a semantic annotation for process modeling and to automate process execution. Tool support aims at supporting the annotation by presenting the user only relevant annotation options. To do so, process	Establishment of a relation between model contents (e.g. actions, objects, states) and appropriate domain concepts or instances specified in an ontology	–	Term similarity and synonyms (no details provided). Analysis of the process context and structure	◨	yes

(*continued*)

Table 1. (*continued*)

Paper	Description of the overall approach	Annotation concept	Def.	Used technology	AU	CO
	structure and lifecycle information of the involved objects are considered					
[22]	The approach proposes to add security information to process models via annotation. The user is supported by suggestions provided by a "knowledge annotator"	A text attached to a BPMN element conforming to a specific syntax (annotation term, followed by a list of parameters)	–	Lexicon for synonym similarity (WordNet), path recognition	◨	no
[23–25]	Conceptual models are annotated to support e.g. benchmarking. Annotation support is based on the meta-modeling platform ADOxx and the integration of social network information to facilitate annotation is discussed	Adding properties of model elements or establishing relations to separate annotation models	–	No detailed information is provided	☐	no
[26, 27]	The approach aims at an easy creation of domain specific ontology and semantic annotations. The latter are supported via automated suggestions. The computation of the suggestions is based on the semantic similarity between BPMN element labels and ontology concepts	Establishment of a relation between an activity of a BPMN model and an ontology concept	–	Lexicon for word sense hierarchies (WordNet), various lexical analysis techniques	◨	no
[28]	With Process SEER, a tool for semantic	Tasks in process models are	–	Using ontologies and Natural	☐	no

(*continued*)

Table 1. (*continued*)

Paper	Description of the overall approach	Annotation concept	Def.	Used technology	AU	CO
	effect annotation of business process models has been developed. The tool requires analysts to describe the immediate effects of each task in natural language. These are then accumulated in an automated fashion	enriched by a structured description of their cumulative effects		Language Processing (NLP) is discussed to improve the approach		
[29–31]	Semantic annotations are discussed and an annotation framework for a range of applications is proposed such as systems Interoperability in a PLM environment	A semantic annotation relates an element of knowledge to one more ontology instances. It also captures an annotation relation type (e.g. subsumption) and the meta model element corresponding to the annotated element	✓	No information	☐	no
[32]	Organizational models are enriched through semantic annotation. A procedure to derive annotation suggestions is briefly sketched	A subject of annotation is related to an object by a predicate. Ontological annotation moreover means that the predicate and context are ontological terms and the object conforms to the ontological definition of the predicate	✓	Custom approach for ontology-based similarity calculation	☐	no
[33]	An ontological approach is developed to semantically	Linkage of the elements of a process model or	–	WordNet in conjunction with various syntactic,	◼	no

(*continued*)

Table 1. (*continued*)

Paper	Description of the overall approach	Annotation concept	Def.	Used technology	AU	CO
	annotate supply chain process models with a BPMN and SCOR ontology. The approach computes annotation suggestions that a user can select from a list	meta model with concepts from an ontology		linguistic and structural sim. measures		
[34–38]	A semantic annotation framework is proposed and applied in various settings such as for goal annotation or to increase the interoperability of process models. Automation is discussed predominantly in terms of translating a model to a pre-defined ontology based on the model constructs and meta-model information	Annotations of concrete process model elements are part of the more comprehensive PSAM (process semantic annotation model). Model annotation means to relate ontology concepts to model elements via pre-defined semantic relationships	✓	No detailed information is provided	☐	no
[16]	As part of the Pro-SEAT tool, semi-automatic annotation of goals for process models is implemented. Possible goal annotations can be deduced automatically based on the model annotation information	Relation between ontology concepts and model elements via pre-defined semantic relationships	✓	No detailed information is provided, apart from String match	◪	no
[5]	A model for semantically annotating business	CPSAM is a context-based process semantic	–	No information	☐	no

(*continued*)

Table 1. (*continued*)

Paper	Description of the overall approach	Annotation concept	Def.	Used technology	AU	CO
	process models is devised. The purpose of the model is to facilitate search, navigation and understandability of process models stored in repositories	annotation model for annotating business processes in a process model repository				
[39]	A framework and Wiki-based tool for the collaborative specification and annotation of business processes is provided. The tool provides a list of admissible annotations to the user	A relation between elements of a BPMN model and instances of a formal ontology	–	No information	◨	no
[40]	An approach for the automatic generation and annotation of capabilities based on the extraction of textual descriptions is developed	A semantic frame-based capability model for describing what an action (e.g. a task, or a service) achieves	–	Various NLP techniques from the CoreNLP library, WordNet	■	no
[41]	The sEPC ontology for EPC model serialization is presented and modelled in the WSML language. Competency questions serve to validate ontology development	Linking the sEPC-based process representation with elements from other ontologies	–	No automated approach	☐	no
[42]	An approach for the annotation of the artefacts including process models is developed	Annotation is understood as the specification of concrete values for a set of common properties, given by the metadata defined in the ontology	–	No automated approach	☐	no

(*continued*)

Table 1. (*continued*)

Paper	Description of the overall approach	Annotation concept	Def.	Used technology	AU	CO
[43]	Process models are annotated with their effects in order to apply a revision strategy that helps to obtain compliant process models from models that might be initially non-compliant	Descriptions of immediate effects of BPMN tasks provided in a formal form or derived from natural language (e.g. via Controlled Natural Languages - CNL)	–	No automated approach	☐	no
[44]	Artefacts such as process models are annotated to foster their reuse. The semantic annotation of processes is implemented using relations and concepts from a Business Ontology to describe processes or process fragments	The pairwise grouping of processes (or process fragments) with the elements "Business Goal", "Business Function", "Business Domain", "Business Role" and "Process Resource" via respective relations	–	No automated approach	☐	no
[45]	The semantic annotation of process models is introduced in order to provide for advanced querying of business process repositories	A correspondence between elements of a business process schema and concepts of a Business Reference Ontology. The relation is established in order to describe the meaning of process elements in terms of related actors, objects, and processes	–	No automated approach	☐	no
[46]	Annotations are introduced as a link between process	Annotations provide either formal definitions	(✓)	No automated approach	☐	no

(*continued*)

Table 1. (*continued*)

Paper	Description of the overall approach	Annotation concept	Def.	Used technology	AU	CO
	models and a reference ontology	of the entities involved in a process such as activities, actors, items (terminological annotations) or specify preconditions and effects for the activation of flow elements (functional annotations)				
[1]	An Ontology-based process representation is developed that is used to enrich model elements with machine processable semantics	Properties of ontology instances representing process elements that link to instances of classes from a domain ontology via defined properties	–	No automated approach	☐	no

automated approaches. The second key criteria is whether the approach accounts for the semantic context of a process model element (column *CO*), i.e. what previous activities lead to the activity or which activity are triggered by the activity. This criteria is important for the annotation of process models, since processes are essentially about the order of tasks executed in a business process. Consequently, the flow of activities is important for annotation. If for example an order is captured, checked and finally executed, it is highly unlikely that after order execution an activity such as "Confirm order" is relevant for annotation, even if it lexically matches an activity label such as "Confirm order fulfilment". So in essence, the criteria is about "knowing" the semantic context in which a process element occurs and considering this during automated annotation.

4 Conclusion and Outlook

In this study, general use cases that require an automated annotation approach have been presented. This underpins the relevance of such a research endeavor. Then a comprehensive overview on the state of the art in the literature was presented. A major result of this overview is that annotation is rarely automated. Even if it is suggested in the research works, no automation seems to be implemented. Also, rarely prototypes are shown. Regarding the semantics of annotation, context information is (apart from one work) almost never used. This is a surprising research gap that exists even today – after almost one decade of research on semantic technologies applied to BPM that started with simple process model annotation proposals. Therefore, a research opportunity lies in developing (semi-)automated annotation approaches in order to first leverage existing standards such as PCF (cf. the use cases in Sect. 2) and second to make use of the wealth of semantic technologies (e.g. for search and matching of models on the semantic level) when process models have automatically been annotated. All in all, this contribution may be a starting point for developing more sophisticated (semi-)automatic approaches capable of linking semi-formal process models with more formal knowledge representations. With this, new use cases are possible shifting the automated interpretation of process models to a new and more semantic level. This contribution should encourage research towards this goal.

References

1. Thomas, O., Fellmann, M.: Semantic process modeling - design and implementation of an ontology-based representation of business processes. Bus. Inf. Syst. Eng. **1**, 438–451 (2009)
2. Koschmider, A.: Ähnlichkeitsbasierte Modellierungsunterstützung für Geschäftsprozesse, PhD-Thesis (2007)
3. APQC: Process Classification Framework (PCF), Version 5.2.0 (2010)
4. Awad, A.: BPMN-Q: A language to query business processes. In: Reichert, M. et al. (eds.) Proceedings of EMISA 2007, St. Goar, Germany, October 8–9, pp. 115–128 (2007)
5. Mturi, E., Johannesson, P.: A context-based process semantic annotation model for a process model repository. Bus. Process Manag. J. **19**, 404–430 (2013)
6. Bechhofer, S., Carr, L., Goble, C., Kampa, S., Miles-Board, T.: The semantics of semantic annotation. In: Meersman, R., Tari, Z. (eds.) OTM 2002. LNCS, vol. 2519, pp. 1152–1167. Springer, Heidelberg (2002). doi:10.1007/3-540-36124-3_73
7. Boudjlida, N., Panetto, H.: Annotation of enterprise models for interoperability purposes. In: International Workshop on Advanced Information Systems for Enterprises, 2008, IWAISE 2008, pp. 11–17 (2008)
8. Hau, J., Lee, W., Newhouse, S.: Autonomic service adaptation in iceni using ontological annotation. In: Proceedings of the 4th International Workshop on Grid Computing. IEEE Computer Society, Washington, DC, USA (2003)
9. Ringelstein, C., Franz, T., Staab, S.: The process of semantic annotation of web services. In: Cardoso, J. (ed.) Semantic Web Services - Theory, Tools, and Applications. Idea Publishing Group, USA (2007)
10. Xu, C., Liang, P., Wang, T., Wang, Q., Sheu, P.C.Y.: Semantic web services annotation and composition based on ER model. In: 2010 IEEE International Conference. on Sensor Networks, Ubiquitous, and Trustworthy Computing (SUTC), pp. 413–420 (2010)
11. Aljoumaa, K., Assar, S., Souveyet, C.: Publishing intentional services using new annotation for WSDL. In: Proceedings of the 12th International Conference on Information Integration and Web-based Applications & Services, ACM, New York, NY, USA, pp. 881–884 (2010)

12. Nie, H., Li, S., Lu, X., Duan, H.: From healthcare messaging standard to semantic web service description: generating WSMO annotation from HL7 with mapping-based approach. In: 2013 IEEE International Conference on Services Computing (SCC), pp. 470–477 (2013)

13. Kim, S.W.: Form annotation framework for form-based process automation. In: Haller, A., Huang, G., Huang, Z., Paik, H.Y., Sheng, Q.Z. (eds.) WISE 2011-2012. LNCS, vol. 7652, pp. 307–320. Springer, Heidelberg (2013). doi:10.1007/978-3-642-38333-5_33

14. Liao, Y., Lezoche, M., Panetto, H., Boudjlida, N., Loures, E.R.: Semantic annotation for knowledge explicitation in a product lifecycle management context: a survey. Comput. Ind. **71**, 24–34 (2015)

15. Reeve, L., Han, H.: Survey of semantic annotation platforms. In: Proceedings of the 2005 ACM Symposium on Applied Computing, pp. 1634–1638. ACM (2005)

16. Lin, Y.: Semantic Annotation for Process Models: Facilitating Process Knowledge Management via Semantic Interoperability (2008)

17. Born, M., Hoffmann, J., Kaczmarek, T., Kowalkiewicz, M., Markovic, I., Scicluna, J., Weber, I., Zhou, X.: Supporting execution-level business process modeling with semantic technologies. In: Zhou, X., Yokota, H., Deng, K., Liu, Q. (eds.) DASFAA 2009. LNCS, vol. 5463, pp. 759–763. Springer, Heidelberg (2009). doi:10.1007/978-3-642-00887-0_67

18. Aslam, M.A., Auer, S., Shen, J., Herrmann, M.: Expressing business process models as OWL-S ontologies. In: Eder, J., Dustdar, S. (eds.) BPM 2006. LNCS, vol. 4103, pp. 400–415. Springer, Heidelberg (2006). doi:10.1007/11837862_38

19. Bögl, A., Schrefl, M., Pomberger, G., Weber, N.: Semantic annotation of EPC models in engineering domains to facilitate an automated identification of common modelling practices. In: Filipe, J., Cordeiro, J. (eds.) ICEIS 2008. LNBIP, vol. 19, pp. 155–171. Springer, Heidelberg (2009). doi:10.1007/978-3-642-00670-8_12

20. Born, M., Dörr, F., Weber, I.: User-friendly semantic annotation in business process modeling. In: Weske, M., Hacid, M.-S., Godart, C. (eds.) WISE 2007. LNCS, vol. 4832, pp. 260–271. Springer, Heidelberg (2007). doi:10.1007/978-3-540-77010-7_25

21. Born, M., Hoffmann, J., Kaczmarek, T., Kowalkiewicz, M., Markovic, I., Scicluna, J., Weber, I., Zhou, X.: Semantic annotation and composition of business processes with maestro. In: Bechhofer, S., Hauswirth, M., Hoffmann, J., Koubarakis, M. (eds.) ESWC 2008. LNCS, vol. 5021, pp. 772–776. Springer, Heidelberg (2008). doi:10.1007/978-3-540-68234-9_56

22. Ciuciu, I., Zhao, G., Mülle, J., Stackelberg, S., Vasquez, C., Haberecht, T., Meersman, R., Böhm, K.: Semantic support for security-annotated business process models. In: Halpin, T., Nurcan, S., Krogstie, J., Soffer, P., Proper, E., Schmidt, R., Bider, I. (eds.) BPMDS/EMMSAD -2011. LNBIP, vol. 81, pp. 284–298. Springer, Heidelberg (2011). doi:10.1007/978-3-642-21759-3_21

23. Fill, H.-G.: Using semantically annotated models for supporting business process benchmarking. In: Grabis, J., Kirikova, M. (eds.) BIR 2011. LNBIP, vol. 90, pp. 29–43. Springer, Heidelberg (2011). doi:10.1007/978-3-642-24511-4_3

24. Fill, H.G., Schremser, D., Karagiannis, D.: A generic approach for the semantic annotation of conceptual models using a service-oriented architecture. Int. J. Knowl. Manag. 9 (2013)

25. Fill, H.-G.: On the social network based semantic annotation of conceptual models. In: Buchmann, R., Kifor, C.V., Yu, J. (eds.) KSEM 2014. LNCS, vol. 8793, pp. 138–149. Springer, Cham (2014). doi:10.1007/978-3-319-12096-6_13

26. Francescomarino, C., Tonella, P.: Supporting ontology-based semantic annotation of business processes with automated suggestions. In: Halpin, T., Krogstie, J., Nurcan, S., Proper, E., Schmidt, R., Soffer, P., Ukor, R. (eds.) BPMDS/EMMSAD -2009. LNBIP, vol. 29, pp. 211–223. Springer, Heidelberg (2009). doi:10.1007/978-3-642-01862-6_18

27. Di Francescomarino, C., Tonella, P.: Supporting ontology-based semantic annotation of business processes with automated suggestions. Int. J. Inf. Syst. Model. Des. **1**, 59–84 (2010)

28. Hinge, K., Ghose, A., Koliadis, G.: Process SEER: A Tool for Semantic Effect Annotation of Business Process Models. Presented at the September 2009

29. Liao, Y., Lezoche, M., Panetto, H., Boudjlida, N.: Semantic annotation model definition for systems interoperability. In: Meersman, R., Dillon, T., Herrero, P. (eds.) OTM 2011. LNCS, vol. 7046, pp. 61–70. Springer, Heidelberg (2011). doi:10.1007/978-3-642-25126-9_14

30. Liao, Y., Lezoche, M., Loures, E., Panetto, H., Boudjlida, N.: Formalization of semantic annotation for systems interoperability in a PLM environment. In: Herrero, P., Panetto, H., Meersman, R., Dillon, T. (eds.) OTM 2012. LNCS, vol. 7567, pp. 207–218. Springer, Heidelberg (2012). doi:10.1007/978-3-642-33618-8_29

31. Liao, Y., Lezoche, M., Loures, E.R., Panetto, H., Boudjlida, N.: A semantic annotation framework to assist the knowledge interoperability along a product life cycle. Adv. Mater. Res. **945**, 424–429 (2014)

32. Vazquez, B., Martinez, A., Perini, A., Estrada, H., Morandini, M.: Enriching organizational models through semantic annotation. Procedia Technol. **7**, 297–304 (2013)

33. Wang, X., Li, N., Cai, H., Xu, B.: An ontological approach for semantic annotation of supply chain process models. In: Meersman, R., Dillon, T., Herrero, P. (eds.) OTM 2010. LNCS, vol. 6426, pp. 540–554. Springer, Heidelberg (2010). doi:10.1007/978-3-642-16934-2_40

34. Lin, Y., Ding, H.: Ontology-based semantic annotation for semantic interoperability of process models. In: Mohammadian, M. (ed.) Proceedings of CIMCA-IAWTIC'06 -, vol. 01, pp. 162–167. IEEE, Washington, DC, USA (2005)

35. Lin, Y., Strasunskas, D.: Ontology-based semantic annotation of process templates for reuse. In: Halpin, T., Krogstie, J., and Siau, K. (eds.) Proceeding of 10th CAiSE/IFIP8.1/EUNO International Workshop on Evaluation of Modeling Methods in System Analysis and Design (EMMSAD05), Porto, Portugal, June 2005 (2005)

36. Lin, Y., Strasunskas, D., Hakkarainen, S., Krogstie, J., Solvberg, A.: Semantic annotation framework to manage semantic heterogeneity of process models. In: Dubois, E., Pohl, K. (eds.) CAiSE 2006. LNCS, vol. 4001, pp. 433–446. Springer, Heidelberg (2006). doi:10.1007/11767138_29

37. Lin, Y., Sølvberg, A.: Goal annotation of process models for semantic enrichment of process knowledge. In: Krogstie, J., Opdahl, A., Sindre, G. (eds.) CAiSE 2007. LNCS, vol. 4495, pp. 355–369. Springer, Heidelberg (2007). doi:10.1007/978-3-540-72988-4_25

38. Lin, Y., Krogstie, J.: Semantic annotation of process models for facilitating process knowledge management. Int. J. Inf. Syst. Model. Des. **1**, 45–67 (2010)

39. Rospocher, M., Francescomarino, C., Ghidini, C., Serafini, L., Tonella, P.: Collaborative specification of semantically annotated business processes. In: Rinderle-Ma, S., Sadiq, S., Leymann, F. (eds.) BPM 2009. LNBIP, vol. 43, pp. 305–317. Springer, Heidelberg (2010). doi:10.1007/978-3-642-12186-9_29

40. Gao, F., Bhiri, S.: Capability annotation of actions based on their textual descriptions. In: WETICE Conference (WETICE), 2014 IEEE 23rd International, pp. 257–262 (2014)

41. Filipowska, A., Kaczmarek, M., Stein, S.: Semantically Annotated EPC within semantic business process management. In: Ardagna, D., Mecella, M., Yang, J. (eds.) BPM 2008. LNBIP, vol. 17, pp. 486–497. Springer, Heidelberg (2009). doi:10.1007/978-3-642-00328-8_49

42. Furdík, K., Mach, M., Sabol, T.: Towards semantic modelling of business processes for networked enterprises. In: Noia, T., Buccafurri, F. (eds.) EC-Web 2009. LNCS, vol. 5692, pp. 96–107. Springer, Heidelberg (2009). doi:10.1007/978-3-642-03964-5_10

43. Ghose, A., Koliadis, G.: Auditing business process compliance. In: Krämer, B.J., Lin, K.-J., Narasimhan, P. (eds.) ICSOC 2007. LNCS, vol. 4749, pp. 169–180. Springer, Heidelberg (2007). doi:10.1007/978-3-540-74974-5_14

44. Markovic, I., Pereira, A.C.: Towards a formal framework for reuse in business process modeling. In: Hofstede, A., Benatallah, B., Paik, H.-Y. (eds.) BPM 2007. LNCS, vol. 4928, pp. 484–495. Springer, Heidelberg (2008). doi:10.1007/978-3-540-78238-4_49

45. Missikoff, M., Proietti, M., Smith, F.: Querying semantically enriched business processes. In: Hameurlain, A., Liddle, Stephen W., Schewe, K.-D., Zhou, X. (eds.) DEXA 2011. LNCS, vol. 6861, pp. 294–302. Springer, Heidelberg (2011). doi:10.1007/978-3-642-23091-2_25

46. Smith, F., Proietti, M.: Behavioral reasoning on semantic business processes in a rule-based framework. In: Filipe, J., Fred, A. (eds.) ICAART 2013. CCIS, vol. 449, pp. 293–313. Springer, Heidelberg (2014). doi:10.1007/978-3-662-44440-5_18

47. Webster, J., Watson, R.T.: Analyzing the past to prepare for the Future. MIS Q. **26**, 13–23 (2002)

Multi-process Reporting and Analysis for Change Management and Performance Reviews

Mario Cortes-Cornax[✉] and Adrian Mos[✉]

Xerox Research Center Europe, 6 Chemin de Maupertuis, Meylan, France
{mario.cortes-cornax,adrian.mos}@xrce.xerox.com

Abstract. Business process design and governance are two important phases of Business Process Management (BPM). They are however usually performed using tools that tend to be too generic and technical for most business analysts. For instance, they promote Business Intelligence (BI) mechanisms to extract reports for the analysis of the executed processes, but they typically focus on one process definition at a time. This approach has shortcomings in organisations where there are large collections of processes that need to be managed consistently. In previous work, we proposed the generation of domain-specific studios, in order to enable analysts to design their processes in a much more intuitive way than with generic languages. This work is a logical continuation through the addition of domain-specific multi-process reporting and analysis. By defining analytics metrics in a domain-specific space, analysts are able to make business performance reviews and manage change in ways that apply directly and quickly to entire collections of process. The appropriateness and the feasibility of the approach are shown through a detailed use-case and a complete prototype implementation.

Keywords: Reporting · BPM · Change management · DSL · BI

1   Introduction

Business process design and governance are two critical components for Business Process Management (BPM) [19]. Both design and governance are tightly related: metrics defined in a design phase are then analysed using governance tools in order to improve the organization's processes. Today's BPM approaches are *too generic and technical* for fitting well with the business experts, limiting the involvement of business matter experts, in particular for design and governance activities [12]. In addition, *current approaches focus on the management of one process and its corresponding instances at a time, not taking into account organization's process collections that may share a number of activities.* Dijkman et al. [3] highlight the difficulties that face organizations to deal with large process collections. In these collections (or repositories), similarities between processes are detected. In particular, activities that meant to do the

© Springer International Publishing AG 2017
W. Abramowicz (Ed.): BIS 2017, LNBIP 288, pp. 91–105, 2017.
DOI: 10.1007/978-3-319-59336-4_7

same thing. In a recent BPM survey, van der Aalst [2] reported that "Business Intelligence (BI) tools focus on simple dashboards and reporting rather than clear-cut business process insights". The author also argued that BI tools rely on "to-be" processes and do not help the stakeholders to understand the "as-is" processes. Indeed, business analysts require dedicated means (i.e., specific type of task with implicit domain knowledge) to effectively model their business domain such as logistics, healthcare, transportation, etc. [17]. In such aforementioned domains, many processes could be defined, having a number of shared activities. Therefore, an intuitive, centralized and efficient way of designing and govern a collection of processes for a domain is necessary.

Domain-Specific Languages (DSLs) are an effective means to cope with application domains providing improvements in expressiveness and ease of use [10]. More specifically, Domain Specific Process Modeling Languages (DSPMLs) [8] can be used by business stakeholders to design their processes in a much more intuitive way than actual standards such as the Business Process Model and Notation [16] (BPMN 2.0). In previous work, a model-driven and generative approach is proposed to build domain-specific studios in order to adapt process design and monitoring to business matter experts [12,14]. In this paper, *we extend the previous work adding and analysing metrics in the domain specific perspective. The provided extension enables an understandable, centralized and efficient way to deal with multi-process analysis and reporting. The latter are critical for change management and business performance reviews.* Change management is a difficult problem when dealing with existing processes in an organization. Analysts would typically design the current processes in what is commonly called the "as-is" version. They would then propose various improvements, as a business proposition. The outcomes of these projected improvements are known as the "to-be" versions [18]. The business case is based on the perceived value of the migration between "as-is" process to the "to-be" process. Therefore, the tools that allow this analysis are crucial in securing such business deals. By integrating reporting tools in the domain-specific approach we obtain an important advantage over current tools because the business performance reviews can be applied to all the processes in a collection, and not just to one. Analysts will be able to define their metrics such as *cost* or *duration* for the different activity types. When these activity types are used in different processes, the metric's information will be centralized in the domain specification. This implies that the aforementioned analyst could *easily simulate the impact of a change of a concrete metric for a collection of processes.* For example, she/he could easily see the implications of automating a very time-consuming manual task in terms of overall cost over time in the process collection.

The appropriateness and the feasibility of our approach is shown through a use case and the integration of the reporting solution in a complete prototype implementation. The rest of the document is structured as follows. Section 2 describes a general overview of the approach. Section 3 presents an example that illustrates the solution. Section 4 details the different elements and their integration with the previous work. Section 5 presents more details about the prototype implementation. Section 6 focuses on related work and finally, Sect. 7 summarizes and put forwards future perspectives.

2 Overview of the Approach

This section gives an overview of the approach, which enables the possibility of dealing with metrics for activity types that can be reused in a collection of processes. As such, the solution can provide an overview of the impact of changes considering one or several metrics for an entire collection of processes. For instance, an analyst can investigate the change in process cost when certain types of manual tasks (such as entering data) are automated. The automated alternative of the task is proposed in a central domain definition and the impact is seen across all the processes that are affected. This contrasts with the current approaches of modifying the metrics in each individual process. For instance, by manually identifying the task and changing the type and characteristics to indicate automation [2]. As Fig. 1 shows, in our approach, similar activities can refer to a domain activity type. This means that they have equivalent behaviour. Therefore, when building up a report, the information from the aforementioned activity types can be extracted. This solution provides a wider view of the possible improvements concerning the organization's processes, as it considers the entire bunch of processes.

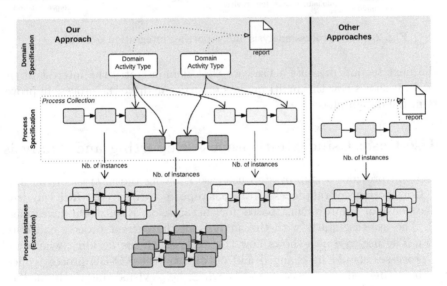

Fig. 1. The approach overview

Figure 2 gives an overview of the integration of the solution in the current framework. Each number in the figure corresponds to one key component of the domain-specific design method [13]. In the first step, the *Corporate Strategy Expert* defines the domain analytics metrics. By default, two metrics (duration and cost) are proposed, but new ones could be easily added, extending the domain description. In the second step, a *Technical Expert* creates a graphical template of the added metric. The latter, also creates a report template taking into account the new defined metrics. In our solution, a generic template is proposed, relying on the cost and duration of the activity types. Finally, the *Analyst* will eventually relate the process steps to the selected metrics and he/she will assign them a value. From the process studio, the analyst would be able to open a *Reporting View* that will present the report results. This report will automatically take into account the updates in the current process model.

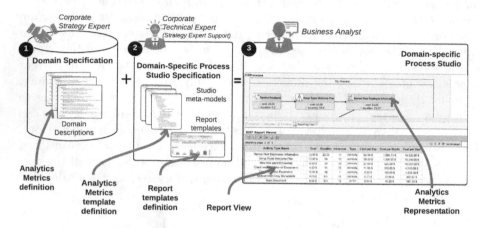

Fig. 2. Multi-process analysis and reporting integration overview

The next section presents a use-case that demonstrates the interest of the approach. It shows how this multi-process reports can drive the process improvements in the design phase.

3   Use-Case: Using Multi-process Reporting and Analysis

This section provides a step-by-step use-case showing the interest of the approach. Considering several processes, various reports are derived, which can drive the decisions for improvement based for instance on Return On Investment (ROI). The use-case starts with the analysis of the current process collection (AS-IS). The use-case also shows how the approach helps deciding where and when processes should be changed and optimized (the TO-BE process collection). Reporting relies on process models, which helps visualizing the immediate impact of reducing the duration and cost of an activity type in the entire process collection (e.g., automating a concrete step).

3.1   AS-IS Process Collection Analysis

Figure 3 shows a simple example of a process collection that contains three different processes. For a matter or visual fitting, the processes have no more than eight steps. All of them concerns the human resources domain (for instance, the submission of the new employee information or the set-up of a welcome meeting). Note that the first process is considered to be executed ten (10) times per day

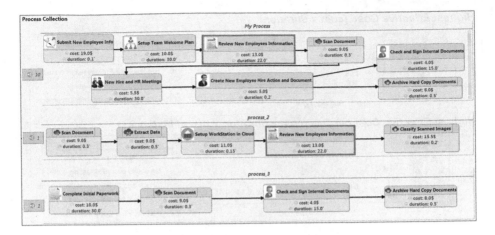

Fig. 3. Process collection (AS-IS)

(i.e., ten instances), while the others just one time. The number of instances is important to be considered in order to calculate the cost per period of time of a process. In the use-case, we focus on the squared steps, which are both linked to the same activity type holding the same name: "Review New Employee Information".

Process Name	Process id	Number of Instances per Day
My Process	process_1	10
process_2	process_2	1
process_3	process_3	1

Activity Type Name	Cost / hour	Duration (min)	Nb. Instances	Type	Cost per Day	Cost per Month	Cost per Year
Review New Employees Information	13.00 $	22	11	MANUAL	52.43 $	1,573.00 $	19,138.17 $
Setup Team Welcome Plan	10.00 $	30	10	MANUAL	50.00 $	1,500.00 $	18,250.00 $
New Hire and HR Meeting	5.50 $	30	10	MANUAL	27.50 $	825.00 $	10,037.50 $
Check and Sign Internal Documents	4.00 $	15	11	MANUAL	11.00 $	330.00 $	4,015.00 $
Complete Initial Paperwork	10.00 $	30	1	MANUAL	5.00 $	150.00 $	1,825.00 $

Fig. 4. Process collection report (AS-IS) - Activity types summary

Starting from a process collection as the one illustrated in Fig. 3, an analyst launches the reporting tool, which is integrated in the studio as an independent view. Three main reporting results are then generated:

- A **process collection overview** as the one illustrated in Fig. 4. This part provides an overall perspective of the different processes in the collection and a summary of the activity types' usage. The table sorts them by the *effective cost*, which is calculated multiplying the number of instances by the cost per hour and the duration (in minutes) (i.e., $nb.instances * duration * cost/60$).
- A number of **graphs that highlight the most costly activity types** (see Fig. 5). These graphs points out the most expensive activity types in terms of cost and duration in a visual way.

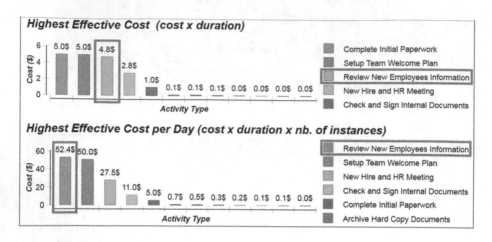

Fig. 5. Process collection report (AS-IS) - Highest effective costs' graphs

– A **daily, monthly and yearly report for each process of the collection.**
Figure 6 shows an example of a monthly report of the process called "My process".

Looking at the report artefacts of Figs. 4, 5 and 6, an analyst could easily conclude that one of the easiest and more effective ways to improve the overall process collection is to reduce the cost and duration of the "Review New Employee Information" activity type. Figure 7a shows its definition in the current domain, which indicates that there are no services linked to it. The next section presents the case where the analyst simulates the automation of the aforementioned activity type.

My Process

Activity Name	Id	Instances	Cost per hour	Duration (min)	Actual Cost (30d)	Type
Setup Team Welcome Plan	WelcomePlan	10	10.00 $	30	1,500.00 $	MANUAL
Review New Employees Information	ReviewInfo	10	13.00 $	22	1,430.00 $	MANUAL
New Hire and HR Meeting	HRMeeting	10	5.50 $	30	825.00 $	MANUAL
Check and Sign Internal Documents	SigningContract	10	4.00 $	15	300.00 $	MANUAL
Archive Hard Copy Documents	ArchiveHardCopyDocs	10	8.00 $	0.5	20.00 $	MANUAL
Scan Document	ScanDocument	10	9.00 $	0.3	13.50 $	AUTO
Submit New Employee Information	NewEmp	10	19.00 $	0.1	9.50 $	AUTO
Create New Employee Hire Action and Document	HireAction	10	5.00 $	0.2	5.00 $	AUTO
			TOTAL:	98.1	4,103.00 $	

Automated vs Manual Activities

auto — 3
manual — 5

Automated vs Manual Activities Actual Cost

4,075.00$ — manual
28.00$ — auto

Fig. 6. Process report (AS-IS) - Per month

3.2   The TO-BE Process Collection

This section presents the TO-BE Process Collection. The use-case shows the possible automation of the "Review New Employee Information" activity type (this could involve for instance an automatic service that analyses all the data about the employee; it could also be a mix between a reduced human involvement and an additional service). Figure 7b illustrates how the activity type is linked to a service. The time of the proper activity type is negligible but the cost and duration of the corresponding service is 6 s (0.1 min) and 1\$ per hour respectively. This information is shown as an aggregated value in the corresponding activity type.

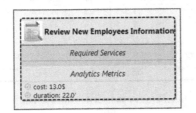

(a) Activity Type Definition AS-IS with no Associated Service (Manual)

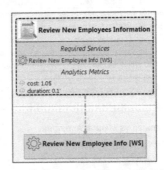

(b) Activity Type Definition TO-BE with Associated Service (Automated)

Fig. 7. Activity types definition

The change in the domain will impact all the steps in the process collection pointing to the updated activity type. Note that the modification impacts two processes in our use-case as Fig. 3 showed. However, *we could imagine dozens of processes updated with this punctual and centralized change.*

After the update of the process collection, the analyst could re-run the report tool. Figures 8 and 9 illustrate the impact of the change in the process collection. The analyst will conclude that the automation of the "Review New Employee Information" activity type will dramatically improve the performance of the entire organization. The comparison between the Figs. 6 and 8 indicates that this change could save up to 1,566.31\$ per month (1573 − 6.69) in the example when only 10 instances of the main process are executed per day. Of course in large organizations with many processes (such as banks) the changes could amount to millions in difference, and these reports can provide easy ways of calculating the various scenarios to justify investments in automation or other improvements.

Note that in the case when the reporting shows that the ROI is below the acceptable limit for the change, other alternatives could be explored, including automating other activity types, which for the same development cost have a better ROI. This may be because these activity types are used in processes that

Activity Type Name	Cost / hour	Duration (min)	Nb. Instances	Type	Cost per Day	Cost per Month	Cost per Year
Create New Employee Hire Action and Document	5.00 $	0.2	10	*AUTO*	0.17 $	5.00 $	60.83 $
Extract Data	9.00 $	0.5	1	*AUTO*	0.08 $	2.25 $	27.38 $
Classify Scanned Images	15.50 $	0.2	1	*AUTO*	0.05 $	1.55 $	18.86 $
Setup WorkStation in Cloud	11.00 $	0.15	1	*AUTO*	0.03 $	0.83 $	10.04 $
Review New Employees Information	1.00 $	0.1	11	*AUTO*	0.02 $	0.55 $	6.69 $

Fig. 8. Process collection report (TO-BE) - Activity types summary

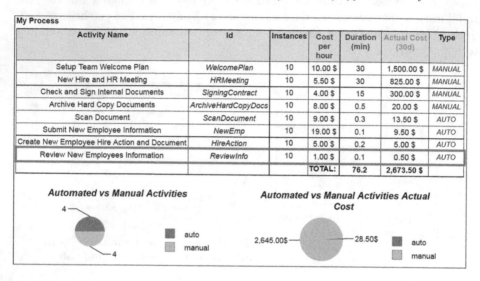

Fig. 9. Process report (TO-BE) - Per month

are more frequent. In our use-case, we choose one of the most frequent in terms of number of instances (see Fig. 5). Therefore, the impact of its automation appears to be of major interest.

This example shows how the reporting capability can help driving the strategic decisions from a high level of governance without getting into the details of individual process designs. Compared to related tools, *the changes are explored and investigated in the domain, not for every single process definition.*

4   Integrating the Analytics Metrics in the Domain

Our aim is to provide the analysts the means to be able to consistently and repetitively reuse metrics across business processes with all the advantages enjoyed by the domain-specific approach (such as sweeping changes, consistent connections between activities, reusable reports, etc.). The approach presented here proposes to *extend the previously documented domain meta-model* [14] *to include specific design analytic metrics.* This implies considering the latter as a first class citizen component to complete the definition the domain. As shown in the previous section, these metrics will support the creation of business performance reviews through reporting.

4.1   Terminology

The approach introduces the notions of *AnalyticsMetricType* and *AnalyticsMetricUsage*. This separation is necessary to distinguish between the static parts of the metric (the type) such as the units or symbol and the dynamic part (the usage), which corresponds to the actual value of the metric linked to an domain-specific activity type. In this section we briefly explain the two new concepts, in which we rely to provide the reporting functionalities:

Analytics Metric Type: refers to any type of measurement used to evaluate some quantifiable component of an activity or service performance such as duration or cost. It is defined as a governed object (has an id, a label, a description, an icon and a version).

Analytics Metric Usage: refers to the actual value of a metric in relation with a domain activity type or a domain service. Therefore, it relates a domain activity type or domain service with an *Analytic Metric Type* with a value.

The formal relation between these terms and the domain concepts are detailed via the (simplified) Domain meta-model (DomainMM) presented in the next section.

4.2   Domain Specification Extension

The DomainMM represents the business domain information for an enterprise, with regard to the specification of the activity types that are going to be reused in the business processes. Figure 10 illustrates a simplified version of the DomainMM where the new concepts are highlighted in grey. Note that terms such as the *Service Level Agreements* (SLA) and the *Data Objects* are not represented here as the focus is put on the meta-model-extension and the related constructs.

The meta-model contains the overall *DomainLibrary*, which has the individual business domains. Each *Domain* contains a *ActivityTypeLibrary*, a *ServiceLibrary*. The *ActivityTypeLibrary* holds the series of domain-specific activity types (*DSActivityType*). The *ServiceLibrary* contains *DSService* elements that refer to the actual SOA services required in the domain. The services can be abstract entities that are bounded later in the deployment process, as described in previous work on deployment [15]. The *DSActivityType* and the *DSService* are considered to be governed objects (*GovernedObject*). This abstraction has been added to the previous meta-model, which results from the work concerning domain-specific monitoring [12]. The monitoring infrastructure are out of the scope of the paper. Also note that with respect to the previous meta-model, the term *DSConcept* has evolved in *DSActivityType*. This evolution is the result of several exchanges with the BPM community as well as the aim of separating the behavioural term with the data-related term (i.e., domain specific activity type vs data objects).

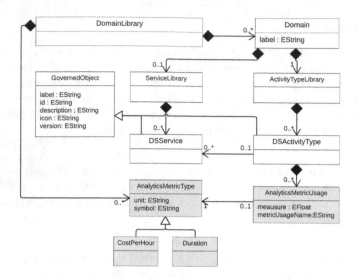

Fig. 10. DomainMM extended to incorporate the analytics metrics definition

As specified in a previous work [14], this meta-model is useful for several proposes: (1) to store the domain information in a central repository on the collaboration and distribution server; (2) to generate a domain editor (textual) that can be used stand-alone or embedded in a graphical editor as part of a diagram designer; (3) to make the connection with the behavioural view specifying how process steps are going to be represented; (4) to inform and update Service-Level Agreements (SLAs) for business concepts. This new solution allows the analysts to define the *Analytics Metrics* corresponding to each domain-specific activity type or domain service. This brings important advantages when considering changes in order to improve processes in a collection.

The mapping between activity-types with the process steps relies on a generic process meta-model and unique ID (UID) attributes as Fig. 11 shows. In our solution, the Mangrove meta-model [7] is used (here, we just focus on a subset). This meta-model constitutes a simplified representation of the main generic process concepts. It is significantly simpler than fully-fledged BPMN because its objective is simply to extract the essence of the structure of various business processes. Our hypothesis is that a descriptive level [20] (reduced amount of symbols but semantically enriched) is enough to define high-level domain-specific process models. The aim is to bring the DSPMLs to a common BPMN denominator, which can take advantage of the BPM suites (BPMS) investments while preserving its specificity and expressiveness. Figure 11 explicitly shows the distinction between de domain specification and the process specification. This separation is the key point to enable multi-process analysis and reporting. In addition, this decoupling permits to define particular metrics for a step, even if it refers to an activity type with its "default" metrics.

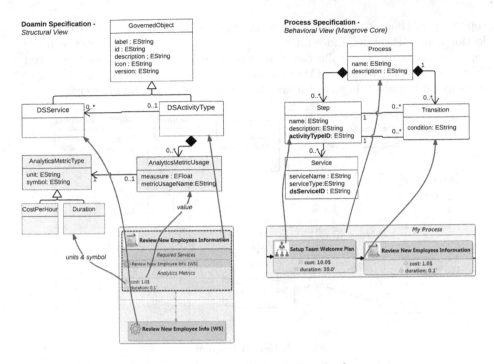

Fig. 11. Domain and process instantiations in the prototype

On the bottom of Fig. 11 two models instantiating the aforementioned meta-models are shown. The left-bottom one represents an activity-type in which the *cost* and the *duration* have been define. Thanks to the ID match between process steps and activity-types, the metrics can be graphically visualized for each step (model in the right-bottom). Note that as we said previously, these metrics are shared between all the steps that refer to the same activity-type.

5   Prototype Implementation

The prototype implementation relies on the Business Intelligence and Reporting Tools (BIRT) Project [4] for integrating the approach into the process studio. BIRT is *"an open source technology platform used to create data visualizations and reports that can be embedded into rich client and web applications"*. In our case, BIRT is embedded in the generated domain-specific process studio, which is an Eclipse-based rich client.

Figure 12 shows the reporting synchronization mechanism between the domain-specific process models and the reporting tool. The report template (report design) contains the definition (logic and style) of the tables and charts presented in the report. Also, it contains the scripting code (JavaScript in our case) that fills the data sets from the process models via a *ReportingService*. The BIRT repport's code loads the *ReportingService*, calling the implemented reporting methods to access the data corresponding to the current process collection semantic model. Indeed, this component is used as interface between the report and the process collection semantic model (instantiation of the linked DomainMM and MangroveMM containing the activity metrics). The data-sets correspond to data-tables, which are the entry point of the report's charts.

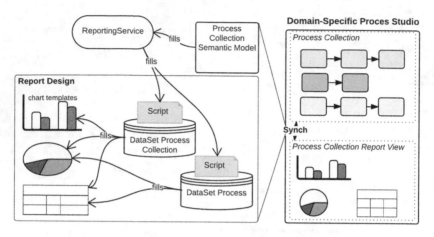

Fig. 12. Report template

Our entire prototype, comprising this new reporting feature, relies on a set of mature and open-source Eclipse technologies. We believe they are highly relevant for any BPM suite many of which are actually built using Eclipse. The screenshots shown in the previous sections give an overview of the prototyping work. We use the Eclipse Modelling Framework (EMF) [21] for the definition of the meta-models (to define the new elements *AnalyticsMetric* and *AnalyticsMetricsUsage*). The Eclipse Xtext framework [5] is used to generate tool support with a fully featured textual editor for domain descriptions relying on the DomainMM. The Ecore meta-models are the inputs for the Sirius domain-specific editor [6]. This tool allows an easy creation of the configurable graphical modeling studios (definition of the templates and the interpreted user interface). Mangrove Core [7] is used as process meta-model. The synchronization illustrated in Fig. 12 enables the data visualization when the report is executed, by simply clicking a button.

Some limitations are still observed in the prototype. Currently, the report template is highly dependant on the metric definition. A future extension of the work considers an automatic generation of the report. BIRT provides a rich API that enables a programmatic construction of the report template. Another limitation is that only two metrics are defined that are *Cost* and *Duration*. This separation enables an easier visual distinction in the studio palette. However, in order to define new metrics in the studio, the *Analyst* could instantiate the generic class *AnalyticMetricType*. Also, the framework permits an easy extension of the meta-model and the subsequent visual representation, that could be performed by the *Technical Expert*. Note that the visual definition of the graphical elements can also be easily changed with a one-click manipulation. Note also that today, we focus on the analysis of the steps referring to activity-types in a generic way. Even if the approach supports particular metrics for each step, this functionality has not been yet implemented in order to incorporate them to the rapport. Finally, richer reporting charts and information extraction has been considered. Indeed, we did not pushed the limits of BIRT, which is a very powerful tool. Nevertheless, we focused on showing the feasibility and interest of the approach.

6   Related Work

Dijman et al. [3] summarize and highlight the challenges of different academia researches around process collection's management. Most of these works focus on looking for similarities between processes in terms of labelling or structure but none of them in reporting mechanisms for collections. For instance, Leopold et al. [9] take similarities into account before adding a new process to a collection. Van der Aa et al. [1] deal with behavioural ambiguity in textual and label process descriptions proposing the so-called behavioural spaces, which support definition variability. In a similar direction, Weber et al. [22] propose a solution to identify refactoring opportunities in process repositories. Considering our domain-specific approach, this "ambiguity" problem is minimized, as the behavioural similarities are kept in a centralized, domain-specific activity type repository. However, the aforementioned works are complementary to our approach as they can be used as starting point to refactor (i.e., migrate it to a domain-specific approach) an already existing process collection. Then, we could apply our proposed process collection reporting technique. In this direction, families of process variants [11] could also be an entry-point of our approach.

If we look at the industry, some BPMN vendors such as Bizagy[1], PMSoft[2] or Bonita[3], offer sophisticated graphical tooling to create reports. Principally, they focus on execution metrics, which are stocked in a database. However, as we discussed before, as they do not consider the domain specific layer, they are not able to deal with a collection of processes. Their reports are therefore focused on single process definition and their corresponding instances.

Another differentiation between the previous tools is that our reports rely for the moment in the design phase and not in the execution phase. Focusing on the design phase facilitates the analyst to simulate possible changes and build reports without the need of the BPM infrastructure (that implies configuration and monitoring). In the future, we consider capturing the execution information plugging the report infrastructure to our monitoring infrastructure [12]. This will be useful to update different parameters that were designed in the initial phases with real data (for example, the number of executed instances of a process or the duration). These measures may be critical to properly define Service

---

[1] http://www.bizagi.com/.

[2] http://pmsoft.com/.

[3] http://www.bonitasoft.com/.

Level Agreements (SLAs). Tibco Nimbus[4] propose reporting features for design analysis as we do. As Nimbus does not have the concept of *DSActivityType* the reporting tool is limited to a single process again.

7   Conclusion

Existing reporting definition approaches are typically technology-specific and generic with respect to the business domain. They also focus on a single process definition at a time. This limits the ability of business matter experts to express their design intent and enact wide-ranging changes in multiple processes. The solution presented in this article provides the business stakeholders with means to analyse possible process problems and asses the improvement scenarios. It also allows to maintain the process metrics (such as cost and duration) at a high-level, with impact to the entire collection of business processes in a domain, if required.

The solution provides the means to decide how and when processes should be changed and optimized. For instance, this can help with an immediate impact assessment of automation of certain activity types. Therefore, the reporting capability presented here can help to drive the strategic decisions from a high level of governance without getting into the details of individual process designs. The changes that are explored are investigated in the domain definition and not for each individual process. Indeed, capturing domain-specific metrics provides three main advantages: (1) a centralized management of the analytics metrics that can be reused for different process definitions; (2) a better understanding of the process by business analysts since they can add precise information about expected cost and duration; and (3) the possibility to simulate outcome of scenarios and therefore help to take decisions about the necessary changes.

The solution is supported by tools that automate the creation of the reports based on the metrics defined in the domain. Today, our solution is focused on the BPM's design phase. For instance, we suppose that the number of instances is known at design time. In the future we could benefit from monitoring integration. This could help extracting insights by merging real measured data in analysis as well as to enhance business process analysis in real time as information becomes available. Also, we could extend the approach to relate the metrics definition to the actual performer's evaluation.

References

1. van der Aa, H., Leopold, H., Reijers, H.A.: Dealing with behavioral ambiguity in textual process descriptions. In: La Rosa, M., Loos, P., Pastor, O. (eds.) BPM 2016. LNCS, vol. 9850, pp. 271–288. Springer, Cham (2016). doi:10.1007/978-3-319-45348-4_16

---

[4] http://www.tibco.com/products/automation/business-process-management/nimbus.

2. van der Aalst, W.M.: Business process management: a comprehensive survey. ISRN Softw. Eng. **2013**, 37 (2013). Article ID 507984. http://dx.doi.org/10.1155/2013/507984
3. Dijkman, R.M., La Rosa, M., Reijers, H.A.: Managing large collections of business process models-current techniques and challenges. Comput. Ind. **63**(2), 91–97 (2012)
4. Eclipse-Fundation: Business Intelligence and Reporting Tools (BIRT) Project (2004). http://www.eclipse.org/birt/
5. Eclipse-Fundation: Xtext (2006). http://www.eclipse.org/Xtext/
6. Eclipse-Fundation: Sirius (2007). http://www.eclipse.org/sirius/
7. Eclipse-Fundation: Mangrove (2009). https://www.eclipse.org/mangrove/
8. Jablonski, S., Volz, B., Dornstauder, S.: Evolution of business process models and languages. In: 2nd International Conference on Business Process and Services Computing (BPSC), pp. 46–59. Citeseer (2009)
9. Leopold, H., Smirnov, S., Mendling, J.: On the refactoring of activity labels in business process models. Inf. Syst. **37**(5), 443–459 (2012)
10. Mernik, M., Heering, J., Sloane, A.M.: When and how to develop domain-specific languages. ACM Comput. Surv. (CSUR) **37**(4), 316–344 (2005)
11. Milani, F., Dumas, M., Ahmed, N., Matulevičius, R.: Modelling families of business process variants: a decomposition driven method. Inf. Syst. **56**, 55–72 (2016)
12. Mos, A.: Domain specific monitoring of business processes using concept probes. In: Toumani, F., et al. (eds.) ICSOC 2014. LNCS, vol. 8954, pp. 213–224. Springer, Cham (2015). doi:10.1007/978-3-319-22885-3_19
13. Mos, A., Cortes-Cornax, M.: Business matter experts do matter: a model-driven approach for domain specific process design and monitoring. In: La Rosa, M., Loos, P., Pastor, O. (eds.) BPM 2016. LNBIP, vol. 260, pp. 210–226. Springer, Cham (2016). doi:10.1007/978-3-319-45468-9_13
14. Mos, A., Cortes-Cornax, M.: Generating domain-specific process studios. In: 20th International Enterprise Distributed Object Computing Conference (EDOC), pp. 1–10. IEEE (2016)
15. Mos, A., Jacquin, T.: A platform-independent mechanism for deployment of business processes using abstract services. In: 17th International Enterprise Distributed Object Computing Conference Workshops (EDOCW), pp. 71–78. IEEE (2013)
16. OMG: Business process model and notation (BPMN) version 2.0 (2011). http://www.omg.org/spec/BPMN/2.0
17. Pinggera, J., Zugal, S., Weber, B., Fahland, D., Weidlich, M., Mendling, J., Reijers, H.A.: How the structuring of domain knowledge helps casual process modelers. In: Parsons, J., Saeki, M., Shoval, P., Woo, C., Wand, Y. (eds.) ER 2010. LNCS, vol. 6412, pp. 445–451. Springer, Heidelberg (2010). doi:10.1007/978-3-642-16373-9_33
18. Rosemann, M.: Potential pitfalls of process modeling: part a. Bus. Process Manage. J. **12**(2), 249–254 (2006)
19. Rosemann, M., Brocke, J.: The six core elements of business process management. In: Brocke, J., Rosemann, M. (eds.) Handbook on Business Process Management 1. IHIS, pp. 105–122. Springer, Heidelberg (2015). doi:10.1007/978-3-642-45100-3_5
20. Silver, B.: BPMN Method and Style: A Levels-Based Methodology for BPM Process Modeling and Improvement Using BPMN 2.0. Cody-Cassidy Press, USA (2009)
21. Steinberg, D., Budinsky, F., Merks, E., Paternostro, M.: EMF: Eclipse Modeling Framework. Pearson Education, USA (2008)
22. Weber, B., Reichert, M., Mendling, J., Reijers, H.A.: Refactoring large process model repositories. Comput. Ind. **62**(5), 467–486 (2011)

On Avoiding Erroneous Synchronization in BPMN Processes

Flavio Corradini, Fabrizio Fornari, Chiara Muzi$^{(\boxtimes)}$, Andrea Polini,
Barbara Re, and Francesco Tiezzi

University of Camerino, 62032 Camerino, Italy
{flavio.corradini,fabrizio.fornari,chiara.muzi,
andrea.polini,barbara.re,francesco.tiezzi}@unicam.it

Abstract. BPMN has acquired a clear predominance in the modeling
of organization processes. Since it is a fairly complex modeling language,
in some cases it is important to clarify the behavior of a modeled process,
especially when concurrency comes into play. We consider unsafe process
models with arbitrary topology, and we focus on the effects of concurrent
control flows activated within single process instances. We use text annotations to clarify the concurrent behavior, and tokens with identity to
regulate the synchronizations. We illustrate the benefits of our approach
by a simple, yet realistic, scenario about paper reviewing.

Keywords: Modeling · Concurrency · BPMN · Erroneous
synchronization.

1 Introduction

Concurrent issues in modeling and programming has been discussed for years [1].
With the growing number of distributed applications run by complex organizations, a proper management of concurrency became more and more important.
Focusing on enterprise architecture Zachman identified the different dimensions
to consider in order to reason on, and understand, the dynamics of a complex
organization [2]. Among the others process modeling describes how an organization structures its activities in order to achieve its goals [3]. Concurrency results
to be an issue with regards to the arrangement of these activities. Indeed, some of
them *"can be performed simultaneously by several autonomous workers that may
coordinate their work by means of communication"* [4]. Resolving concurrency
issues positively impacts on the organization performance.

To describe a process, de facto standard is BPMN 2.0 [5] provided by OMG.
It adopts a semi-formal approach combining a precisely specified syntax with
a token-based semantics given in natural language. A semi-formal description
is useful to allow different stakeholders to easily communicate and share ideas
so that BPMN can play the role of a bridge between business analysts and IT
developers [6].

© Springer International Publishing AG 2017
W. Abramowicz (Ed.): BIS 2017, LNBIP 288, pp. 106–119, 2017.
DOI: 10.1007/978-3-319-59336-4_8

Concerning the management of concurrency in BPMN, it is worth noticing that multiple process instances is explicitly addressed in the specification, while the effects of concurrent control flows within a single instance is underspecified. This can easily occur due to concurrent control flows initiated through the use of AND-split and OR-split gateways. The relevance of such an issue is pointed out also by studies stating that an increase in the level of concurrency for BPMN models implies an increase in modeling error probability [7,8]. Here, we focus on the management of concurrent behavior in a single process instance.

Imposing well-structured rules contributes to control and minimize the level of this form of concurrency by guaranteeing some correctness properties [9,10]. However, such restrictions are not easily applicable by all model designers. Indeed, on the one hand, designers with limited modeling experiences are prone to model spaghetti processes. On the other hand, more expert designers should be free to express their creativity in modeling the process according to the reality they feel [4]. In addition, not all process models with an arbitrary topology can be transformed into equivalent well-structured one [11,12]. Summing up, advantages of the structured process modeling style over the unstructured one (and vice versa) have been a topic of active debates for decades [4]. Hence, processes with arbitrary topology are still very common in practice.

In this work we do not impose any restriction on the usage of the modeling notation. We refer here to process models with an arbitrary topology including concurrent behavior, which may lead to the occurrence of erroneous synchronizations. Such a kind of processes generally include sequence flows that can be activated more than once at the same time, referred as *unsafe* processes. These processes are typically discarded by the modeling approaches proposed in the literature, as they are over suspected of carrying bugs. Unfortunately, this attitude significantly limits the use of concurrency in business process modeling, which is an important feature in modern systems and organizations. Instead, we believe that in these cases the designer could keep the 'offending' model and solve the issue by better clarifying the intended behavior. In fact, the problem typically is not in the model itself but it is due to the underspecification of the BPMN standard in dealing with concurrency issues within a single process instance.

Our work is thus mainly motivated by the need of achieving synchronization correctness in unstructured processes, which is still an open challenge. More specifically, the contribution of the paper is an advanced use of BPMN text annotations to enrich the model with information suitable to deal with concurrent execution of control flows. We also contribute by refining the process execution semantics by taking into account token identities to avoid erroneous synchronizations. The major benefit of our contribution is having the possibility to fully explore the modeling potentialities of BPMN notation in case of processes with arbitrary topology and concurrent behavior.

The rest of the paper is organized as follows. Section 2 introduces a motivating scenario, while Sect. 3 discusses on unsafe processes. Section 4 provides details on our methodology, and Sect. 5 reports the works found in literature that inspired our work. Finally, Sect. 6 closes the paper with some conclusions and future work.

2   A Motivating Scenario

To better clarify the issues we want to address, we introduce a scenario concerning the management of the paper reviewing process of a scientific conference. We use this scenario to motivate our approach, and throughout the paper to illustrate its technicalities. We rely therefore on a simplified version of the scenario, as in [3, Sect. 4.7.2].

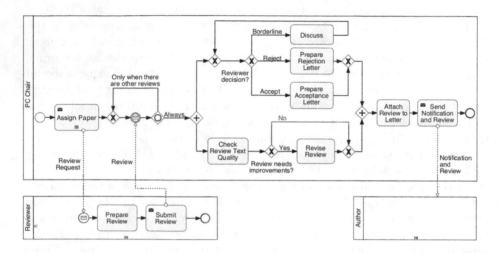

Fig. 1. Paper review process collaboration.

This is modeled in BPMN as the collaboration in Fig. 1. The participants are: **Program Committee (PC) Chair**, the organizer of the reviewing activities. For the sake of presentation, we assume that the considered conference has only one chair, whose behavior is represented by the process within the PC Chair pool; **Reviewer**, a person with knowledge in some of the conference topics. This role is modeled as a multi-instance pool. Each process instance describes the tasks that a reviewer has to accomplish to complete her/his assignment. For the sake of simplicity, we choose to assign only one paper to each reviewer; **Author**, who submitted a paper to the conference and acts on behalf of the other authors (contact author). This role is modeled as a multi-instance blackbox pool since details on the author behavior are not relevant to our purposes.

The reviewing process is started by the chair, who assigns (via a parallel multi-instance activity) each submitted paper to a reviewer. Then, the chair receives the reviews and evaluates them. In particular, as soon as a review is received, the chair starts its evaluation and is immediately ready to receive and process another review. This behavior is rendered in BPMN by means of a loop, realized via an OR split gateway and a XOR join gateway, whose single iteration consists of receiving a paper review and starting its evaluation. The evaluation of each review is modeled by the process fragment enclosed by the AND split gateway and the AND join one. Indeed, the evaluation proceeds along two concurrent control flows: *(bottom branch)* the chair checks the quality of the received review and, if necessary, he/she revises it to improve and *(top branch)*, according to the reviewer decision, the chair prepares the acceptance/rejection letter or, if the paper requires further discussion, the decision is postponed. In the last part of the process (after the AND join gateway), the chair attaches the review to the notification letter, and sends it to the contact author.

The model described so far represents in an intuitive and compact way the paper reviewing scenario. However, despite its simplicity, it hides some subtleties that may affect its correct enactment. For instance, it may happen that an author of a paper will receive a notification with attached the review of another paper. We describe below how this kind of situations may occur, by making use of the concept of *token*, thought of as a means to indicate the process elements that are active during the execution.

Let us consider the reception of the review for a paper, say *paper1*. This event produces a token that activates the OR split; assuming that other reviews are waited, the OR gateway produces in its own turn two tokens: one is used to reactivate the receiving message intermediate event, while the other to activate the evaluation of *paper1*'s review. This latter token is split into two tokens for activating the two evaluation branches described above. Then, a review for another paper, say *paper2*, is received and dealt with in a similar way. The evaluations of the two reviews proceed, hence, along two concurrent control flows. After some steps, we may have the current situation: *(i)* *paper1*'s review has been revised by the chair and a corresponding token reached the AND join gateway from the bottom incoming flow, while the other *paper1* token is still marking the Discuss task, as the paper received a borderline score; *(ii)* *paper2* received a reject score, thus, while the chair is still checking the review quality, a *paper2* token reached the AND join gateway from the top incoming flow.

In this situation, the two incoming flows of the AND join carry a token. Thus, according to the standard semantics of BPMN, the AND gateway triggers the flow through its outgoing sequence flow. In fact, the semantics does not distinguish tokens related to the evaluation of the *paper1*'s review from those related to the *paper2*'s one. This **erroneous synchronization** of tokens allows the process execution to continue with the notification task, using the revised review of *paper1* and the rejection letter of *paper2*.

To address this problem, we advocate the use of tokens with identity. This enables the AND join gateway to distinguish the two incoming flows, hence avoiding the erroneous synchronization. In fact, only tokens with the same identity, i.e. referring to the same paper review, synchronize. When synchronization cannot take place, the incoming tokens just wait for the arrival of 'brother' tokens.

Notably, in order to have the situation described above, during the execution of the considered process more than one token must concurrently transit along the same sequence flow. In the reviewing scenario this happens each time a review is assigned as result of the *OR Join* behavior specification. Moreover, the other condition leading to situations of erroneous synchronization is the presence of *concurrent control flows*, where the generated multiple tokens are split and then have to be synchronized. In our scenario, we have that the concurrent control flows correspond to the two evaluation activities performed by the chair. We present in Sect. 4 our approach to avoid erroneous synchronizations to take place when the above conditions are met.

Other possible solutions have been proposed in order to overcome the concurrency issues addressed by our approach. With reference to the introduced reviewing scenario, a first solution proposes to model the PC Chair by means of two processes: one that assigns papers, collects reviews and instantiates the other (multi-instance) process, whose instances separately deal with the evaluation of paper reviews. As no interaction can take place among these instances, erroneous synchronizations cannot emerge. A second solution suggests to put in sequence the various evaluation activities performed by the chair. This avoids concurrent flows and, hence, the possibility of erroneous synchronizations. Compared with our solution, where the chair behavior is modeled as a single process instance, the first alternative does not fit well with the reality, as the behavior of a single human person is split into two separate processes, one of which is multi-instance. Missing to represent concurrency aspects can be dangerous when the model is intended to model activities to be automated by information systems. The second one, instead, imposes to put in sequence a set of activities that originally were parallel. Most of all, the two alternative solutions require an alteration of the original structure, as well as of the semantics, of the designed process. This requires the designer to be expert enough to identify the concurrency issue in his model and, then, to solve it by properly restructuring the model. Moreover, these are ad-hoc application-specific solutions. Instead, our approach provides a general solution to the problem, without altering the structure of the process. In fact, we acts on the level of abstraction, which is lowered in order to distinguish token identities.

3   On Unsafe Processes

Unsafe processes emerge only when the control flow is organized in such a way that tokens can be dynamically generated during the process instance execution. In this section, we clarify how multiple tokens are generated, and how to recognize processes that do that and hence may be subject to erroneous synchronization problems.

First, we set the scene by introducing the necessary background notions. The first key concept is that of *token*. The BPMN specification states that "a token is a theoretical concept that is used as an aid to define the behavior of a process that is being performed" [5, Sect. 7.1.1]. A token is commonly generated by a start event, traverses the sequence flows of the process and passes through its elements (enabling in this way their execution), and is consumed by an end event when the execution terminates. Besides, tokens can be generated and consumed by gateways. The distribution of tokens in the process elements, called *marking*, defines a state of the process, as it indicates which activities are enabled and which sequence flows have been selected. The *process execution* is therefore defined in terms of marking evolution (i.e., changes of state).

Now, by relying on the above notions, we define when a process is *unsafe*.

Definition 1 (Unsafeness). *A process is* unsafe *if and only if during its execution it can reach a marking where more than one token marks the same sequence flow.*

Fig. 2. Token generator structures (bounded number of tokens).

Fig. 3. Token generator structures (unbounded number of tokens).

Intuitively, a process is unsafe if it contains either a process fragment capable of generating a bounded number of tokens (Fig. 2), or an unbounded number of tokens, by resorting to a loop (Fig. 3). Notably, it is evident that *well-structured* processes do not contain token generator fragments [10]. Looking at the structure of the process in Fig. 1 referring our running example, we can identify a pair of gateways, namely the first XOR and the OR, that form a fragment corresponding to the structure in Fig. 3. To establish if a process is unsafe, we can translate the BPMN model into a Petri Net [10] and resort to techniques for verifying safeness properties of Petri Nets. For an account of these techniques we refer to [13,14].

4   An Approach to Erroneous Synchronizations Avoidance

To manage unsafe BPMN models, we need a fine-grained view on the tokens flow within process instances. This allows us to distinguish tokens referring to different concurrent control flows; e.g., in our motivating scenario we want to distinguish the tokens referring to concurrent evaluations of different papers. We achieve this by relying on the use of *tokens with identity*. Such identity can evolve during the process execution, as the token can have different meanings in different parts of a process. Thus, the token can be identified by means of different (unique) identifiers, whose scope can be limited to the part of interest in the process. Such scope is application specific, hence it must be the designer in charge of explicitly specifying this information on the process model.

The proposed approach includes ingredients allowing to: (i) enrich BPMN models with additional information, via specific text annotations on sequence flows, called *check-in* and *check-out*, that enclose the part of process defining the scope of a token identifier; and (ii) refine the BPMN semantics by taking into account token identities to control synchronizations. In particular, such ingredients extend the modeling phase of a BPMN process with the following steps: **Step 1** - The designer controls if the designed process is unsafe (see Sect. 3). If the process is unsafe, the designer goes to step 2, otherwise the model can be safely implemented; **Step 2** - The designer introduces check-in and check-out annotations in the process; **Step 3** - The designer analyzes the process execution by means of a refined semantics that takes into account token identities, thus avoiding erroneous synchronizations between distinguished tokens. If the desired behavior is achieved, the model can be safely implemented; otherwise the designer either goes back to step 2 to revise the positions of check-in and check-out annotations, or redesigns the process model and reapplies the approach from step 1.

Our choice of using text annotations for defining the scopes of token identifiers, rather than introducing new modeling elements, is due to the intention of avoiding a syntactic extension of the BPMN notation. This allows us to easily apply our approach to existing BPMN models and, most of all, to use the whole plethora of tools already available for BPMN. These are indeed the usual benefits of approaches based on annotations, which nowadays are very common in the field of programming languages.

In the following, we describe the proposed use of annotations and how they permit refining the BPMN semantics. Then, we show how the approach works into practice.

4.1   Check-in and Check-out

Check-in and check-out annotations are used to explicitly specify the scope of token identifiers. In particular, **Check-in** represents a point of the process from where the identity of the traversing tokens is enriched with a fresh identifier; **Check-out** represents a point of the process where the identifiers created by the corresponding check-in are no longer needed and, hence, are removed from the identity of the traversing tokens.

Check-ins and check-outs are identified by their names, ranged over by n. Each check-out must be correlated with one check-in, i.e. there is a check-in in the process model with the same name; on the other hand, each check-in is correlated with zero or more check-outs. Graphically, see Fig. 4, a check-in (resp. check-out) is a standard BPMN text annotation with the peculiarity of being attached to a sequence flow and of enclosing a text of the form Check-in (n) (resp. Check-out (n)).

Fig. 4. Graphical notation of check-in and check-out annotations.

As already mentioned, a token can have different meanings in the process. This can be achieved by means of more check-ins. Notably, a check-in can occur inside a check-in/check-out block. Therefore, the identity of a token is defined as a set T of pairs of the form (n, id), where n is the name of a check-in traversed by the token and id is an identifier freshly[1] generated by the check-in n. When a token is generated by the activation of a start event, it is initialized with a default identity represented by the set $\{(init, 0)\}$, where $init$ is a reserved check-in name and 0 is an identifier. The identity of a token changes only when it traverses check-in or check-out points, while its flow during the process execution is regulated by the standard BPMN execution semantics unless when it meets a synchronization point (i.e., an AND or an OR join gateway). We explain below the resulting refined BPMN semantics.

When a token traverses a check-in point n, its identity is not altered if it already contains an identifier generated by n, otherwise the token identity is enriched with a new identifier pair. Formally, the token identity evolution determined by a check-in is defined by function $TraverseCheckIn$ that, given as input a check-in name and the identity set of an incoming token, it returns as output the identity of the outgoing token

$$TraverseCheckIn(n, T) = \begin{cases} T & \text{if } (n, id) \in T \\ T \cup \{(n, fresh(n))\} & \text{otherwise} \end{cases}$$

where $fresh(n)$ is a function that returns a fresh identifier for the check-in n (notably, this function can be straightforwardly implemented by relying on a counter local to each check-in). As an example, consider that during the execution of a process a token with identity $T_0 = \{(init, 0)\}$ passes through the check-in named $first$. The token identity set evolves to $T_1 = \{(init, 0), (first, 3)\}$, assuming that at the time of the check-in crossing $fresh(first)$ returns 3. Then, the token identity does not change until another check-in is reached. In particular, if the token then passes through the check-in $second$, the identity set becomes $T_2 = \{(init, 0), (first, 3), (second, 7)\}$, as $fresh(second)$ returns 7. If the token passes through the same check-in more than once, nothing happens if the identifier produced by such check-in is still considered in the identity set. In the example, if the token passes again through check-in $first$, its identity set remains T_2.

Dually, when a token traverses a check-out point n, its identity is not altered if it does not contain an identifier generated by n, otherwise the corresponding identifier pair is removed from the identity set of the token. Formally, the token identity evolution determined by a check-out is defined by the following function.

$$TraverseCheckOut(n, T) = \begin{cases} T \setminus (n, id) & \text{if } (n, id) \in T \\ T & \text{otherwise} \end{cases}$$

---

[1] An identifier, generated by a check-in n, is called $fresh$ if it is different from all other identifiers previously generated by the check-in n.

Let us consider again the example previously discussed, where the token currently has identity T_2. Now, if during the execution of the process the checkout *second* is reached, then the identity set T_2 changes into T_1. Instead, if the token reaches a check-out named *first*, then the identity set T_2 becomes $T_3 = \{(init, 0), (second, 7)\}$, while nothing happens if the token reaches a check-out named *third*.

Finally, as already said, when a token with identity traverses any element of the process model different from a check-in, a check-out or a synchronization point, the effect on the token and on the process execution is the one prescribed by the standard semantics of BPMN. For example, if a token with identity set T_1 traverses an AND split gateway, a token with the same identity T_1 is produced for each outgoing sequence flow. Instead, when a token with identity traverses a synchronization point, the BPMN semantics synchronizes tokens with the same identity, i.e., tokens whose identity sets coincide. In this way, erroneous synchronizations, which mix up different concurrent control flows, are avoided. Notably, the synchronization requires a complete match of identities among tokens, which means that the identity sets must have the same pairs; thus, for example, $\{(init, 0), (third, 3)\}$ and $\{(init, 0), (first, 5)\}$ do not match with T_1. It is also worth noticing that, in case of synchronization of tokens whose identity is given by the default value $\{(init, 0)\}$, our refined semantics coincides with the one prescribed by the BPMN standard. In other words, our semantics is conservative with respect to the standard one, i.e., if no check-in and check-out annotations are introduced in the model then the two semantics coincide.

To sum up, once the BPMN model under design is enriched with check-ins and check-outs, during its execution we can observe the evolution of token identities. In this way we are able to track the behavior of the process considering the paths traversed by the tokens and, most of all, their synchronizations (ensured to be non-erroneous).

4.2   The Approach at Work

In this section we illustrate how our approach can be applied in practice. Figure 5 shows how check-ins and check-outs are used to specify in which part of the process, within the PC Chair pool, tokens represent the control flows of the paper evaluation. For the sake of presentation, we identified three relevant parts of the process named A, B and C. Moreover, to show the flow of each token, we mark the corresponding path in the process with token identities (curly brackets and the default identifier are omitted).

At the beginning of the execution, the token placed on the start event has the default identity, represented by the set $\{(init, 0)\}$. Then, the token enters into, and hence activates, Part A of the process, which is a token generator. Thus, for each received review, a new token identity has to be generated. To this aim, the designer introduced a check-in named n so that, as soon as a token traverses the check-in, its identity is enriched with the new identifier $(n, 1)$, denoting that the token is related to review of *paper*1. The OR split gateway then splits the token

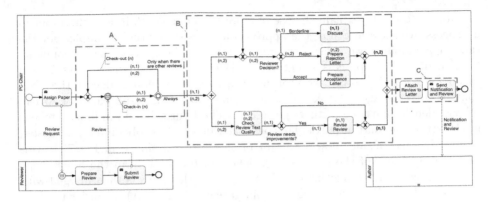

Fig. 5. Paper reviewing process collaboration simulation with tokens id.

into two tokens with the same identity $\{(init, 0), (n, 1)\}$. One of the generated tokens will go back into the loop, traversing the check-out point n and hence loosing its *paper*1 identity. This shows the usefulness of the check-out annotation in our approach: sometimes it is necessary that a token looses its identity as, e.g., it enters in a path where it is merely used as a control flow signal. In our case, the only purpose of the considered token is to activate a new iteration of the loop; in fact, without loosing its identity the token would fail in doing this. Instead, the other *paper*1 token will go into Part B of the process. This token will cross the AND split gateway and the evaluation of the *paper*1's review will start. From this point, the execution proceeds as described in Sect. 2, thus a new token with identity $\{(init, 0), (n, 2)\}$ enters in the game, and the marking represented by the tokens whose identity is written in bold in Fig. 5 is reached. Now, the AND join gateway has two incoming tokens, one per each incoming edge, and thus evaluates their synchronization. Anyway, according to the refined semantics, the synchronization does not take place, as the two incoming tokens have different identities. Therefore, they remain in the edges waiting for their brothers. In this way, the appropriate synchronization will take place, the tokens will go through Part C of the process, and a notification to the author with attached the review of the corresponding paper will be sent.

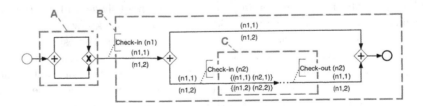

Fig. 6. Process structure combining check-ins and check-outs.

Now, in order to explain the use of more than one check-in and check-out in a process, we show also how the approach applies to another example. For the sake of readability, we show just the structure of the process, i.e. we omit its task elements. The process structure in Fig. 6 is divided in three parts: A is a token generator that produces a token for each of the outgoing edges of the AND split; B corresponds to a token identity scope[2]; and C is a sub-part of B where tokens identity must be further specialized. The figure shows the flow of tokens in the structure, and in particular how token identities evolve in case of nested scopes. Let us consider now a variant of this process, shown in Fig. 7. In this case the structure of the process is enriched with a path from the inner scope to an element of the enclosing one. In particular, this path is then merged with a path of the enclosing scope (AND join gateway in Part A). In order to allow the synchronization of tokens coming from these two paths it is necessary to remove the identifiers created by check-in $n2$. This is properly done via a second check-out $n2$. This example thus shows why we may need to associate two or more check-outs to a single check-in.

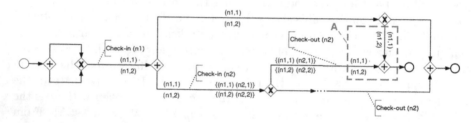

Fig. 7. A variant of the process structure in Fig. 6.

5   Related Works

Several techniques have been developed and applied to specify and reason on issues introduced by concurrency in software systems [15,16]. Concurrency is recognized as an important aspect of processes [4], in particular when processes have to be simulated and/or executed [17]. With reference to processes, three different kinds of concurrency have been highlighted [18]: concurrent processes, concurrent control flows inside a single process, and concurrent events/tasks.

The issues observed in managing concurrent control flows and related synchronizations have been already considered by some workflow patterns [19]. Among the others, the "And-join generalization" pattern corresponds to the general notion of AND-join where several execution paths are synchronized and merged together. The pattern supports situations, such as those non-safe, where one or more incoming branches may receive multiple tokens for the same process instance. The intended semantics for the pattern tends to be unclear in situations involving non-safe behavior. Our paper aims to contribute to close this

---

[2] Notably, no check-out is defined for $n1$, meaning that identifiers of the form $(n1, id)$ must be keep on token identities until the end of the execution.

gap. In the BPMN specification we can observe a similar issue. The BPMN standard uses the concept of token to facilitate the discussion about a process execution flow, however it does not impose conditions on how to keep track of tokens propagation.

Tokens with identity have been used in other research works. Nevertheless, they are mainly used to manage concurrent processes rather than, as we propose, to manage concurrent control flows. For instance Börger et al. discuss about the use of tokens identity represented as hierarchy sets for tracing the sequence flow of a process instance [20]. The notion of token identity has been also discussed with reference to the characterization of the OR-join behavior [21]. Also Colored Petri Nets, where tokens have identity (colors), have been used to represent concurrent processes [22]. Finally, token identifiers have been also used to control process execution in the interaction with database transactions enabled by the represented process [23].

Multiple instance management raises problems when passing from design time to run-time [24]. The problem of the run-time synchronization evaluation is also introduced with regard to the OR-join by Dumas et al. [25]. In this regard we believe that postponing the issues from the design to the implementation is not a general solution. On one side because it is well known the importance of early defect detection to avoid loss of time and money. On the other side because the implementation of BPMN processes needs a transformation to executable languages that can introduce further issues (i.e. those introduced by BPEL [26–28]).

6   Conclusions and Future Work

In this paper we presented an approach to solve issues caused by the inherent underspecification of synchronizations statements in BPMN models, and that can emerge when unsafe processes with an arbitrary topology and concurrent control flows are considered. To solve the issue we rely on the introduction of text annotations, which allows the model designer to clarify the intended behavior in terms of tokens with identity.

As a future work, we plan to investigate on possible strategies to automatize the placement of check-in and check-out annotations, which would help us to resolve issues regarding practical usage and scalability. Currently, this step is completely manual and requires some efforts from the model designer, who has to carefully arrange the annotations in the BPMN model. This could also help to evaluate our approach and make the proposed BPMN extension easy-to-use, useful and less prone to errors. Moreover, we plan to extend our BPMN formalisation in [29] with the check-in and check-out notion. Finally, we plan to develop a software tool exploiting the potentialities of the approach to automatically generate code that is free from synchronization issues from (annotated) BPMN models. This will also enable a systematic validation of the proposal.

References

1. Cleaveland, R., Smolka, S.A.: Strategic directions in concurrency research. ACM Comput. Surv. **28**(4), 607–625 (1996)
2. Zachman, J.A.: A framework for information systems architecture. IBM Syst. J. **26**(3), 276–292 (1987)
3. Weske, M.: Business Process Management: Concepts, Languages, Architectures. Springer (2007)
4. Polyvyanyy, A., Bussler, C.: The structured phase of concurrency. In: Bubenko, J., Krogstie, J., Pastor, O., Pernici, B., Rolland, C., Sølvberg, A. (eds.) Seminal Contributions to Information Systems Engineering, pp. 257–263. Springer, Heidelberg (2013)
5. OMG: Business Process Model and Notation (BPMN V 2.0). Technical report (2011)
6. Henderson, J.C., Venkatraman, N.: Strategic alignment: leveraging information technology for transforming organizations. IBM Syst. J. **32**(1), 4–16 (1993)
7. Mendling, J., Sanchez-Gonzalez, L., Garcia, F., La Rosa, M.: Thresholds for error probability measures of business process models. J. Syst. Softw. **85**(5), 1188–1197 (2012)
8. Moreno-Montes de Oca, I., Snoeck, M.: Pragmatic guidelines for business process modeling. Technical Report 2592983, KU Leuven, November 2014
9. Mendling, J., Reijers, H.A., van der Aalst, W.M.: Seven process modeling guidelines (7PMG). Inf. Softw. Technol. **52**(2), 127–136 (2010)
10. Dijkman, R.M., Dumas, M., Ouyang, C.: Semantics and analysis of business process models in BPMN. Inf. Softw. Technol. **50**(12), 1281–1294 (2008)
11. Polyvyanyy, A., García-Bañuelos, L., Dumas, M.: Structuring acyclic process models. Inf. Syst. **37**(6), 518–538 (2012)
12. Polyvyanyy, A., Garcia-Banuelos, L., Fahland, D., Weske, M.: Maximal structuring of acyclic process models. Comput. J. **57**(1), 12–35 (2014)
13. Murata, T.: Petri nets: properties, analysis and applications. Proc. IEEE **77**(4), 541–580 (1989)
14. Van Der Aalst, W.M.P.: Workflow verification: finding control-flow errors using petri-net-based techniques. In: Aalst, W., Desel, J., Oberweis, A. (eds.) Business Process Management. LNCS, vol. 1806, pp. 161–183. Springer, Heidelberg (2000). doi:10.1007/3-540-45594-9_11
15. Ramchandani, C.: Analysis of asynchronous concurrent systems by timed petri nets. Massachusetts Institute of Technology, Cambridge (1974)
16. Clarke, E.M., Emerson, E.A., Sistla, A.P.: Automatic verification of finite-state concurrent systems using temporal logic specifications. ACM Trans. Program. Lang. Syst. **8**(2), 244–263 (1986)
17. Vasilecas, O., Smaižys, A., Rima, A.: Business process modelling and simulation: hybrid method for concurrency aspect modelling. J. Mod. Comput. **1**(3–4), 228–243 (2013)
18. Sörensen, O.: Semantics of Joins in cyclic BPMN Workflows. Ph.D. thesis, Christian-Albrechts-University Kiel, Department of Computer Science (2009)
19. Russell, N., Ter Hofstede, A.H., Mulyar, N.: Workflow controlflow patterns: a revised view. Technical Report BPM-06-22, BPMcenter.org (2006)
20. Börger, E., Thalheim, B.: A method for verifiable and validatable business process modeling. In: Börger, E., Cisternino, A. (eds.) Advances in Software Engineering. LNCS, vol. 5316, pp. 59–115. Springer, Heidelberg (2008). doi:10.1007/978-3-540-89762-0_3

21. Thalheim, B., Sorensen, O., Borger, E.: On defining the behavior of OR-joins in business process models. J. UCS **15**(1), 3–32 (2009)
22. van Hee, K.M., Sidorova, N., van der Werf, J.M.: Business process modeling using petri nets. Trans. Petri Nets Other Models Concurrency VII, 116–161. Springer (2013)
23. Van Hee, K.M., Sidorova, N., Voorhoeve, M., others: Generation of database transactions with petri nets. Fundamenta Informaticae **93**(1–3), 171–184 (2009)
24. Barros, A.P., Grosskopf, A.: Multiple instance management for workflow process models. Google Patents US Patent 8,424,011, April 2013
25. Dumas, M.G., Grosskopf, A., Hettel, T., Wynn, M.T.: Evaluation of synchronization gateways in process models. Google Patents US Patent 8,418,178, April 2013
26. Recker, J.C., Mendling, J.: On the translation between BPMN and BPEL: conceptual mismatch between process modeling languages. In: CAISE, pp. 521–532 (2006)
27. Weidlich, M., Decker, G., Großkopf, A., Weske, M.: BPEL to BPMN: the myth of a straight-forward mapping. In: Meersman, R., Tari, Z. (eds.) OTM 2008. LNCS, vol. 5331, pp. 265–282. Springer, Heidelberg (2008). doi:10.1007/978-3-540-88871-0_19
28. Lapadula, A., Pugliese, R., Tiezzi, F.: Using formal methods to develop WS-BPEL applications. Sci. Comput. Program. **77**(3), 189–213 (2012)
29. Corradini, F., Polini, A., Re, B., Tiezzi, F.: An operational semantics of BPMN collaboration. In: Braga, C., Ölveczky, P.C. (eds.) FACS 2015. LNCS, vol. 9539, pp. 161–180. Springer, Cham (2016). doi:10.1007/978-3-319-28934-2_9

IFTT: Software Application Interfaces Regression Testing

Michał Kowalczewski[1(✉)], Michał Krawczyk[1], Elżbieta Lewańska[2], and Witold Abramowicz[2]

[1] INT4, ul. Przelot 10, 60-408 Poznań, Poland
{michal.kowalczewski,michal.krawczyk}@int4.com
[2] Poznań University of Economics and Business, al. Niepodległości 10, 61-875 Poznań, Poland
{e.lewanska,w.abramowicz}@kie.ue.poznan.pl

Abstract. As modern enterprises use a number of different information systems, the problem of integrating them becomes crucial. One of the most-used patterns for integration is through application interfaces. However, information systems change constantly in order to respond to the new informational needs of enterprises. The regression testing goal is to assure that systems work correctly after each introduced change. Because the regression testing process is both time and resource consuming, support tools are needed. Since the nature of the changes in a system might vary, the testing techniques and supporting tools must be specific to the system and testing scope. The paper describes the IFTT tool developed specifically for regression testing of software application interfaces.

Keywords: Regression testing · Integration testing · Application interfaces

1 Introduction

Modern enterprises use many different information systems that are deployed internally or run by their co-operators. The number of different systems used by a single company goes from a few to several dozen, thus making the integration process crucial for all the software components and testing. According to studies, up to 50% of software maintenance costs are related to testing [1]. Moreover, up to 80% of those costs are caused by regression testing activities.

Although there is a lot of research on testing software in general and specifically on regression testing, Engstrom and Runeson [1] identify a gap between the research on the subject and the way such tests are conducted in practice. While the research focuses mainly on selection and prioritisation of test cases, the general industry approach is to re-test all use cases that are available and consider the selection of test cases as not critical [19].

By using proper regression testing support tools, enterprises claim to reduce efforts and costs related to this activity. For example, reduction of costs related to the regression testing after introducing automated testing software was up to 80%, reduction of time – up from 70% [2]. The same survey shows that with the support of testing tools, more critical functionalities have been tested (80–90% instead of 10% before) and even 80% of non-critical functionality.

© Springer International Publishing AG 2017
W. Abramowicz (Ed.): BIS 2017, LNBIP 288, pp. 120–131, 2017.
DOI: 10.1007/978-3-319-59336-4_9

Engstrom and Runeson [1] conducted a survey to identify real-life problems with regression testing. The results showed that test case selection is one of the biggest issues in regression testing. Respondents stated that test case prioritisation and selection are very difficult to conduct. They also wished for clear regression testing guidelines. About 25% of respondents were dissatisfied with the test coverage while 40% assessed it as "neutral". This shows that dealing with test coverage is quite a challenge in organisations. Another important weakness of regression testing is test case design and maintenance. One third of the respondents claimed lack of resources for this task and were dissatisfied with this part of the testing process. Respondents wished for the possibility to reuse test cases from the development process during the regression testing. Authors identified problems with both automated and manual testing approaches. The first results in errors and causes problems, while the latter is time and resource consuming. It is worth highlighting that about 30% of respondents perform regression testing manually.

Improvements in regression testing processes are possible only if the support tool is well-fitted to the enterprise infrastructure, type of changes in the software, particular system component, and information needs. The IFTT tool was developed specifically for regression testing of software application interfaces between SAP systems and other software (third-party solutions or other SAP systems) but the IFTT's approach for performing regression testing is generic and it is planned to support non SAP systems in the future as well.

Software application interface is a set of rules and methods description that defines the way computer programs can communicate with each other.

Software application interfaces regression testing purpose is to check whether the communication works properly after any changes in the source or target system. It is conducted by using a set of **test cases**, i.e., executable steps validating the expected result.

Regression testing for software application interfaces differs from testing of other software components due to the following aspects. Firstly, interface serves as a connecting point for two or more independent systems but at the same time, it is not an integral part of any of it. Thus, it is very rare to run a regression test of an interface while testing the systems. Instead, a dedicated testing run is required. Secondly, interfaces should not depend on any of the connected systems. Thus, a separate set of test cases must be developed for regression testing. Finally, while systems connected by the interface focus on data processing and creation, an interface itself focuses solely on exchanging data. Therefore, even though the same application data is being tested, there might be need to check different aspects and utilise other regression testing techniques for software application interfaces.

The structure of the paper is as follows: Sect. 2 discusses related work, focused on both integration of software applications in the enterprises and regression testing. Section 3 describes the IFTT approach to testing of software application integration. Section 4 presents case studies and evaluation results. The paper is summarised with conclusions and future work.

2 Related Work

Software application interfaces are widely used in the area of Enterprise Application Integration as they allow different modules and systems to exchange data and communicate with each other. Thus, in the following section authors discuss different aspects of integration through interfaces in this area. In the Sect. 2.2, a short discussion on regression testing is provided.

2.1 Enterprise Application Integration

The goal of Enterprise Application Integration (EAI) is to create computer systems' architecture that supports integrating a number of applications and systems used in the enterprise. The EAI framework allows different applications to share data and communicate with each other. Those applications may be installed on different operating systems, use different database technologies, or involve different data semantics. Systems might be integrated via a database (with direct access to a database for all systems, or indirectly using the duplication technique) or via Web services (directly or indirectly) [3]. Each integration method has advantages and disadvantages and requires a different approach for testing (in particular, for regression testing).

Although EAI is a fairly well-researched topic, there are not many publications focused on the problem of application interfaces integration regression testing. This issue is vital for the software developers who work on EAI in practice. The relevant literature is also difficult to find due to the incoherence in the definition of "interface". This term relates to both the graphical user interface (GUI) and application program interface (API). GUI testing is out of the scope of the paper because most third-party integration issues are not reflected in a GUI. The data between systems is exchanged through application interfaces and IFTT focuses on assuring that the exchange process is free of errors after any corrections to the existing code.

Another problem with the literature on the subject is that authors frequently do not indicate if they investigated integration between systems as a whole, or if they focused only on integration through application interfaces. The difference is quite important because while integrating third-party systems through interfaces, an enterprise is not allowed to make any changes in external systems, does not have access to their source code, and works only in the interface layer. Thus, the scope of regression tests, available data, and testing scenarios are different.

Although there are some industry publications on the topic, e.g., [17], they are published for a specific programming environment and are not always transferable to third-party systems. There are a lot of very specific solutions for testing integration between system modules that are well-encapsulated and communicate through interfaces [4] or integration between VoiP systems [7]. Unfortunately, such solutions usually focus on a specific stage of regression testing (in the exemplary cases: on the test cases generation, which is just one of the aspects of the regression testing).

Hura and Dimmich [5] present another approach to automatic EAI testing. Test cases are defined in the unified modelling language (UML) and the proposed solution automatically compares system logs with the UML definitions. This approach requires the

UML model for the software application interfaces. Most of the companies never create them during the initial implementation and later on the effort is too big to justify this approach, therefore, even though using UML diagrams for software interface testing could help from pragmatical standpoint, it's too difficult and time-consuming to have them done. Moreover, initial research results suggested that UML models for software application interfaces were fairly simple and in most cases did not cover all aspects required to be checked during the regression tests.

2.2 Regression Testing

The main purpose of regression testing is to assure that previously developed and tested software (denote computer program as P and its set of specification as S) performs correctly after being changed or integrated with another application (P' – modified program with modified set of specification S'). Regression testing checks if P' meets the specification S' in the same way P met S. The testing process is performed using the test suite (T or modified set T'), which is a set of individual test cases (t). Regression testing covers functional and non-functional system requirements. Continuous integration of IT systems in organisations makes regression testing especially important [8]. The most common scenario for regression testing is to re-run previously developed test cases (T) and check to see if the results are the same as they were before the system changed. However, this approach is not always possible because changes made to the system might change its behaviour. Moreover, it is time and cost consuming. Thus, the following regression testing techniques have been developed [8]:

- Test case prioritisation (TCP) – order the test cases so specific testing criteria can be met faster (e.g., rate of fault detection). This technique seeks the optimal test case permutation. The test suite is not reduced, but the assumption is that testing might be terminated before all cases are checked. Thus, in this case T is equal to T'.
- Regression test selection (RTS) – a subset of test cases T' is selected from T in order to decrease costs (i.e., only those cases that are important or related to the certain requirement are selected to re-run). This technique is also referred to as Test Case Selection [9]. The downside of this technique is it might omit some cases which could have revealed faults in P'.
- Test suite reduction (TSR) or test suit minimisation (TSM) – reduces T into subset T' (also denoted as T_{min}), which achieves the same objectives as T (test cases are technically excluded from further runs). Yoo and Hartman [9] distinguish minimisation (permanent changes to the test set) and reduction (temporary changes).

The above-mentioned testing techniques focus only on existing test suites. However, Orso [8] points out that this is usually not enough because changes made to the system alter its functionality. In such situation, new test cases must be created. A technique called test suite augmentation (TSA) focuses on identifying areas where new test cases are needed, and on supporting users to generate them.

A regression testing purpose can be progressive (used when changes in P' are made due to the changes in specification S') or corrective (used when P' should be tested against S because specifications did not change).

Test cases are also classified into several categories, based on their usage [9]. Three types of test cases that already exist in T are: reusable (covers only unchanged parts of the program, i.e., the parts that have not changed between P and P'); retestable (covers parts of the program that have been changed); and obsolete (no longer valid for the testing purposes). There are also two classes of newly-developed test cases: new-structural (covers structural changes of P') and new-specification (covers changes caused by S').

Most of the available regression testing tools are focused on a specific regression testing technique, e.g., ORTS [6] specifically uses RTS for testing java-based applications. Some tools support test case creation rather than the testing process itself, e.g., eXpress [10] generates regression test cases based on the path-exploration-base test generation (PBTG) technique. BERT [11] uses the concept of behavioural regression testing, i.e., compares two versions of the same program and looks for a difference in its behaviour. Finally, TOSCA Test Suite [18] supports both test case development and regression testing automation and is designed for SAP systems testing. An interesting approach is discussed by Elbaum et al. [15] – the authors present a solution used in Google for regression testing in continuous integration development environments.

However, the authors of the paper were not able to find any solution designed particularly for regression testing of software application interfaces where, as noted in Sect. 1, other testing techniques may be more relevant as this kind of appliance differs significantly from other testing scenarios because it is focused on integration aspects between different solutions. There are, however, some tools for supporting the quality assurance of APIs (e.g., REMI [12]) or its compliance with specifications (e.g., ACART [13]).

3 Approach for Software Integration Regression Testing

The IFTT (Interfaces Testing Tool), developed by Int4, is a regression testing software currently dedicated for SAP software application interfaces. It fully covers software application interfaces testing across the complete SAP landscape: middleware platforms (SAP PI/PO) and backend system software application interface logic (custom code and enhancements) together with system customising. The motivation for developing the IFTT was twofold. Mainly, there was a lack of a tool specifically designed for software interfaces regression testing on the market. And a number of problems related to regression testing, described in Sect. 2, occurred in the area of software application interfaces testing. Thus, the main goals formulated for the IFTT were: to allow automation of regression test processes by speeding up the creation of new test cases, and to provide a comprehensive and fully automatic assertion results. This section presents the solution that allows to achieve the defined goals.

3.1 IFTT Architecture and Data Flow

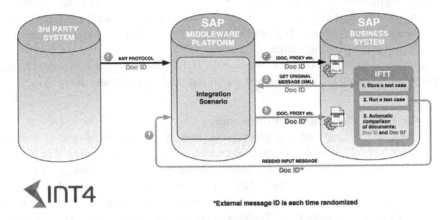

Fig. 1. IFTT architecture and data flow diagram

The IFTT is an ABAP add-on deployed on the SAP application server of a business system. It comes together with its own database tables. The tables are used to store interface definitions (i.e., IFTT configuration) and the test cases themselves (i.e., IFTT test case repository). The IFTT can be connected to many middleware platforms with the use of adapters. In most cases, the IFTT is used with the SAP Process Orchestration (SAP PO, former SAP PI) middleware.

Typical EAI communication follows these steps:

1. Sender business system creates the outbound messages.
2. Middleware platform routes the messages between sender and receiver, applying conversions and mappings (step 1 and 2 on the Fig. 1).
3. Receiver business system (in this case SAP Business System) runs the integration logic to process business documents on the basis of the inbound message.

The data flow to enable testing of such interface in the IFTT is following:

1. In order to perform regression tests, the adapter gets the original inbound messages from the middleware platform from IFTT add-on on the backend system (step 3 on Fig. 1).
2. Consequently, at the next step when the test case is being executed by the IFTT, the adapter will resend the same messages (changing only the message identifier) back to the middleware platform (step 4 on Fig. 1). In the last step, the regular interface processing is triggered and the message is processes in the backend system (step 5 on Fig. 1) again.
3. Finally, the IFTT creates a comparison showing in detail how the documents have been processed originally and during the test run.

The whole process from the user's perspective is described in detail in the following sections.

3.2 Interfaces Regression Testing Process

The IFTT application interfaces testing is intended for message-driven data exchange. It operates on the final business document (a set of consistent database table entries within the system's database). It provides testing of all underlying layers like routing, mappings and system configuration.

The regression testing process flow using the IFTT consists of the following phases: test planning, test data provisioning, test execution, and results comparison.

3.2.1 Test Data Planning and Provisioning

Usually, due to the time and cost restrictions, one of the regression test techniques (described in Sect. 2) is used to reduce the number of available test cases. At the same time, development of new test cases is usually resource consuming, thus new cases are created only to cover crucial functionalities. With the IFTT the authors took a different approach – "intelligent" brute force method – which allows creating many test cases with a minimal time consumption. The process is semi-automatic and the tester needs only to select a range of validated business documents created by interface during the previous runs. Later on, the IFTT will reprocess the old business documents and expect the same result after posting them so there's no need to create additional test cases manually.

During day-to-day work with customers, authors discovered that testers spend most of their time trying to find and replicate the steps made by business users to re-create the stories and business documents created by them (similar conclusions are described in [1]). Thus, one of the goals for the IFTT development was to enable easy and quick single-test creation with the use of existing business documents instead of creating them. Using the existing business document for testing follows the waterfall SAP implementation methodology. The implementation of an application interface requires phases that include: blueprinting, developing, and unit testing. After that, official user acceptance tests (UATs) are performed where the business users approve the implementation of the interface and sign off on it. This is generally how interface development moves from the development environment to the production environment.

Changes are allowed in every stage of this process, creating a need for regression testing. Usually, regression testing starts with the UAT. In case of a change request implementation or bug resolution, even if the interface is still not moved to production, there is a need to prove that the remaining functionality behaves in exactly the same way as it did when it was previously approved by business users.

Hence, there is a need to build a test case repository for software application interfaces. The IFTT allows existing business documents created in external systems by users during the UAT phase to be used as a test cases, so test case provisioning does not require any additional help or time of sender applications or business experts. As the IFTT is tailored for software application integration only, there is no need to create step-by-step recordings in sender and receiver systems to mirror all the steps normally performed by the users to create the business documents.

All test cases (i.e., business documents) are stored in a central repository and are grouped by various criteria. The IFTT test case is based on an action that triggers an inbound message for a particular business document. Typical actions include creating or updating the business document.

Second unique approach of the IFTT for test provisioning is related to creating assertions and validation rules for each test case. Other regression testing tools required validation rules to be specified during each test case creation as a part of the data provisioning step. The IFTT operates in a completely different way: the test results are based on a full comparison of the final business documents on the database layer (the original reference document and the one created during test execution). Thus, there is no need to define validation rules during the test provisioning phase, which drastically simplifies the test case creation process. The only thing which needs to be set up once, are the database table names where the business documents are stored.

A typical process run for test case provisioning includes:

Step 1. To create a test case from a reference document, the tester identifies a middle-ware message GUID (Globally Unique Identifier) that created the business document. Together with the message GUID, the tester must specify the IFTT business object, i.e., a single business document (for example a sales invoice, purchase invoice, purchase order etc.).
Step 2. The IFTT calls the middleware platform (for example SAP PO) and retrieves the original input message received by the platform from the sender application.
Step 3. The original message is written to the IFTT database and associated with the final business document number and the business object type.

This simple procedure for creating test cases allows the test creator to avoid going into any sender system to record each test case and eliminates the need to create any assertions in the receiver system, but most importantly allows creating hundreds of test cases in a matter of hours which was not possible previously (by manual work or with the support of other tools without any heavy development inside them, known to authors).

3.2.2   Automated Test Execution

Another crucial aspect of regression testing is test execution time. In cases when screen recording and scripting tools are being used in order to replicate the interface run, increased execution time requires additional hardware and makes the process relatively slow. With the IFTT, the sender's initial message is being used and sent directly to the middleware platform when it's being reprocessed and then is delivered to the backend application system. All messages sent from the IFTT use the same unified approach and according to the first referential customer the execution time was improved 5 times compared to the previous regression testing solution, based on screen recording and scripting tools.

A typical technical process of a test run is described below.

Step 1. The IFTT reads the original message from the test case repository.
Step 2. Based on the configuration of the business object associated to the test case, the original business document identifier is replaced in the message with a

random value that will be later tracked. The system assures the integrity by generating a unique number from a special number range that is not used by original sender system.

Step 3.   The new message is sent by the IFTT adapter to the middleware platform.

Step 4.   The middleware platform processes the new message exactly as it would be received from the sender system. The platform recognises that the message comes from the sender system. The middleware applies the same routing and transformation rules as the real flow and passes the message created in step 3 to the SAP business system.

Step 5.   The SAP business system processes the message as it would come from the sender system and posts a new business document with the identifier generated in step 2. Despite the fact that steps 4 and 5 are triggered by the IFTT, the execution is completely autonomic and not affected by the IFTT. The regular flow is executed to process the interface message and post the final business document.

Step 6.   The final automated comparison report is created (see Sect. 3.3).

3.3   Reporting Functions

Typically, there are two approaches to validate the results of an interface test case run. In the first approach, the message status can be checked as successful or not successful processing. This approach, however, does not validate any potential changes in the business document itself, which would result in the same message status. In the second approach, the scripting techniques open the business document generated from the test run and compare it with predefined values. The main issue with this approach, apart from the long execution time of this process, is that test creators need to specify all fields which need to be verified. The IFTT works a bit differently. Each new business object generated from the incoming message reads the IFTT configuration of a business object type which also contains the validation rules.

Validation rules are used to compare the referenced documents to the one created during each test execution. The rules are used to confirm the business documents are created as expected. They reflect the way the business documents are written in the application system database. Technically speaking, the IFTT stores the names of database tables and the relationships between them. For instance, the configuration of ECC[1] financial document will be represented as set of BKPF, BSEG, and BSET tables[2]. After specifying the tables, it is possible to configure the database fields to be included in the regression test report, but by default the IFTT expects that all fields from all business-object-related tables will be validated. Therefore, during the test case execution, the IFTT extracts the database entries from all linked tables and compares the defined fields for referenced and newly created documents. This approach allows testing in shorter execution time.

---

[1]   ECC stands for SAP ERP Central Component.

[2]   SAP ECC database tables: BKPF - FI(nance) document headers, BSEG - FI document items and BSET - taxes.

As the business object validation rules must be setup only once during the IFTT implementation, they can be prepared very comprehensively and accurately. Moreover, they contain all possible assertions and the IFTT can find errors in fields never meant to be checked by the test case creator, which provides an additional level of confidence in the regression test results and increase regression test coverage.

The advantage of using validation rules defined on the basis of database tables structure can be explained by the following example: a tester wants to add two test cases for verifying the booking of the incoming sales invoices. The target of the first test case is the validation of tax data and the second is to check the general ledger account determination for a particular business scenario. With the IFTT's business objects approach, both targets will be validated for these two documents automatically by using the validation rule defined for the invoice document. Such a rule covers all fields from database related to the document. Thus, there is no need to take this into consideration during test case provisioning.

4   IFTT Case Studies and Evaluation

The conceptual work on the IFTT started in 2011. To date, the tool has been successfully implemented at three global customers where it has significantly reduced the testing time as well as improved the production system quality and reliability.

The first use case of the IFTT is a global SAP ECC implementation for an international group that operates over 100 assets around the world and supports more than 60 offices. The use of the IFTT started after implementing and deploying the global template in the first location. During rollouts to other countries, a lot of changes were required to the existing application interfaces in order to support local systems and legal requirements. Each location (country) for which the system was implemented, has built its own test case repository and is continuously validating if the introduced changes cause any negative impact on live interfaces.

With the use of the IFTT, the customer is able to perform a run of 4000 test cases in two-and-a-half hours which is significantly shorter than when using their previous support tool (depending on the scenario, the test cases execution time is up to 5 times shorter than with other solutions used by the customer).

Two other IFTT implementations were utilised in a different scenario. The two existing SAP implementations relied heavily on data provided by the interfaces. The interfaces were implemented incorrectly, which resulted in low data quality and high maintenance costs after a few years of operating. Both clients decided to perform the improvement projects and to re-implement the interfaces.

The business involvement in those projects was low because none of the existing business requirements changed and were already specified. However, there was still a strong need to perform highly efficient regression testing to prove that the new interfaces implementation produces correct results. The project sponsors could not expect a strong business involvement (as there were no new business requirements) and the IFTT was a great tool to prove that the system behaves correctly after the re-work. Both clients decided to enable parallel runs for the time being. The parallel run is a re-configuration of the test system to receive the same data as production system. Thanks to the IFTT, the clients were able to compare thousands of business documents posted in the old and new way and catch all inconsistences before starting the new interfaces on the production systems. A typical daily report generated during the project was comparing 20,000 business documents in less than 10 min.

5 Summary

The paper describes problems related to the regression testing of software application interfaces and proposes the potential solution, i.e., the Interface Testing Tool (IFTT). The main advantages of the IFTT come directly from the fact that it is a solution designed specifically for software application interfaces regression testing. Differences between regression testing of information systems in general and regression testing of interfaces are described in Sect. 1. The IFTT uses a novel approach to the preparation of test cases. Instead of focusing on techniques for test case set reduction, it offers a solution that allows to conduct an "intelligent" brute-force testing by semi-automatically creating hundreds of test cases. Moreover, the test cases execution is fully automatic and since it focuses only on interface testing, it does not require to re-do all business functions in the sender and receiver business system. Instead, the IFTT compares entries in the database which is a quick and efficient solution. As explained in Sect. 3.2.2, the complex test results evaluation is based on original business documents, called references, and their representation in the database. It allows users to make the testing process flexible and independent, and to avoid the laborious recording of actions performed in the receiver application to collect all expected values for fields and screens that need to be verified.

The tool has been successfully implemented for three clients' systems and the clients were able to: run automated regression testing with much more test cases than with previous testing tools and with a very limited budget on human resources (due to two IFTT's advantages for test data provisioning), and, according to their statement, do it more quickly.

Future works on the tool will focus on enabling the performing of regression testing of outbound interfaces in a similar, easy to customise way and on further automation of the current solution.

References

1. Engström, E., Runeson, P.: A Qualitative survey of regression testing practices. In: Ali Babar, M., Vierimaa, M., Oivo, M. (eds.) PROFES 2010. LNCS, vol. 6156, pp. 3–16. Springer, Heidelberg (2010). doi:10.1007/978-3-642-13792-1_3
2. Wefersm, M.: Best practice/next practice: regression testing of SAP-centric business processes. Active Global Support, SAP AG (2011). http://scn.sap.com/docs/DOC-14714
3. Kaneshima, E., Vaccare Braga, R.T.: Patterns for enterprise application integration. In: Proceedings of the 9th Latin-American Conference on Pattern Languages of Programming (SugarLoafPLoP 2012), Article 2, 16 p. ACM, New York (2012)

4. Yuan, H., Xie, T.: Substra: a framework for automatic generation of integration tests. In: Proceedings of the 2006 International Workshop on Automation of Software Test (AST 2006), pp. 64–70. ACM, New York (2006)
5. Hura, D., Dimmich, M.: A method facilitating integration testing of embedded software. In: Proceedings of the Ninth International Workshop on Dynamic Analysis (WODA 2011), pp. 7–11. ACM, New York (2011)
6. Huang, S., Zhu, J., Ni, Y.: ORTS: a tool for optimized regression testing selection. In: Proceedings of the 24th ACM SIGPLAN Conference Companion on Object Oriented Programming Systems Languages and Applications (OOPSLA 2009), pp. 803–804. ACM, New York (2009)
7. Hao, R., Lee, D., Sinha, R.K., Griffeth, N.: Integrated system interoperability testing with applications to VoIP. IEEE/ACM Trans. Netw. **12**(5), 823–836 (2004)
8. Orso, A., Rothermel, G.: Software testing: a research travelogue (2000–2014). In: Proceedings of the on Future of Software Engineering (FOSE 2014), pp. 117–132. ACM, New York (2014)
9. Yoo, S., Harman, M.: Regression testing minimization, selection and prioritization: a survey. Softw. Test. Verif. Reliab. 1–7 (2007). doi:10.1002/000. Published online in Wiley InterScience
10. Taneja, K., Xie, T., Tillmann, N., de Halleux, J.: eXpress: guided path exploration for efficient regression test generation. In: Proceedings of the 2011 International Symposium on Software Testing and Analysis (ISSTA 2011), pp. 1–11. ACM, New York (2011)
11. Jin, W., Orso, A., Xie, T.: . BERT: a tool for behavioral regression testing. In: Proceedings of the Eighteenth ACM SIGSOFT International Symposium on Foundations of Software Engineering (FSE 2010), pp. 361–362. ACM, New York (2010)
12. Kim, M., Nam, J., Yeon, J., Choi, S., Kim, S.: REMI: defect prediction for efficient API testing. In: Proceedings of the 2015 10th Joint Meeting on Foundations of Software Engineering (ESEC/FSE 2015), pp. 990–993. ACM, New York (2015)
13. Yilmaz, L., Kent, D.: ACART: an API compliance and analysis report tool for discovering reference design traceability. In: Proceedings of the 49th Annual Southeast Regional Conference (ACM-SE 2011), pp. 243–248. ACM, New York (2011)
14. Haftmann, F., Kossmann, D., Lo, E.: A framework for efficient regression tests on database applications. VLDB J. **16**(1), 145–164 (2007)
15. Elbaum, S., Rothermel, G., Penix, J.: Techniques for improving regression testing in continuous integration development environments. In: Proceedings of the 22nd ACM SIGSOFT International Symposium on Foundations of Software Engineering (FSE 2014), pp. 235–245. ACM, New York (2014)
16. Haraty, R.A., Mansour, N., Daou, B.: Regression testing of database applications. In: Proceedings of the 2001 ACM Symposium on Applied Computing (SAC 2001), pp. 285–289. ACM, New York (2001)
17. Documentation for ASP.NET. http://docs.asp.net/en/latest/testing/integration-testing.html. Accessed 13 Apr 2016
18. Platz, W.: 2010 TOSCA@SAP. TOSCA Testsuite in the SAP environment. Tricentis Whitepaper. www.tricentis.com. Accessed 13 Apr 2016
19. Onoma, A.K., Tsai, W.T., Poonawala, M.H., Suganuma, H.: Regression testing in an industrial environment: progress is attained by looking backward. Association for computing machinery. Commun. ACM **41**(5), 81–86 (1998)

ICT Project Management

Influence of Task Interdependence on Teamwork Quality and Project Performance

Kondwani F. Kuthyola[1], Julie Yu-Chih Liu[1(✉)], and Gary Klein[2]

[1] Department of Information Management, Yuan Ze University, 320 Chung-Li, Taiwan
`s1036241@mail.yzu.edu.tw`, `imyuchih@saturn.yzu.edu.tw`
[2] College of Business and Administration, University of Colorado, Colorado Springs, USA
`gklein@uccs.edu`

Abstract. Although task interdependence is regarded as a key factor in determining individual performance, empirical evidence on the relation between interdependence and project performance is limited. This work investigates how task interdependence influences project performance. Specifically, we empirically examine the relationship between task interdependence, teamwork quality, and project management performance using a questionnaire survey of 300 software personnel. The analysis results show the mediating effects of different aspects of teamwork quality on the relationship between task interdependence and project performance. Therefore, when assigning tasks with high interdependence for an agile process, managers shall provide team members the platform for facilitating their teamwork behaviors.

Keywords: Teamwork quality · Interdependence · Task interdependence · Project performance

1 Introduction

Interdependence is a common trait of a software project team. Interdependence refers to the extent to which the outcomes of individuals are related to the efforts of others [1]. Team members often cooperate to accomplish the desired outcomes of the team. Task interdependence arises when individuals need to cooperate and support each other to accomplish tasks [2]. Team members are often assigned to tasks with a high level of interdependence [3], especially for certain software development process, such as an agile process.

The impact of interdependence on project performance has received considerable scholarly attention and brought two different perspectives. One viewpoint indicates that interdependence is related to project performance as individuals can accomplish tasks more efficiently in a team [4]. Interdependence promotes team members to provide one another with information, advice, assistance and resources [1]. Another perspective claims the potential negative impact of interdependence through raising the chance of conflict. Along this line, team members have less responsibility, and more disagreement on the tasks that need to be accomplished with other team members. The inconsistent findings suggest a need to develop a deeper understanding of how interdependence

© Springer International Publishing AG 2017
W. Abramowicz (Ed.): BIS 2017, LNBIP 288, pp. 135–148, 2017.
DOI: 10.1007/978-3-319-59336-4_10

affects the performance of software project teams [5]. Until now, the evidence provides a limited understanding of the influence task interdependence has on the performance of project management.

Several studies indicate that teamwork quality represents the nature of interactions among individuals in their work [6]. Marks, Mathieu and Zaccaro [7] reported that the interactions are important for team members to achieve an effective outcome. To extend our knowledge to task interdependence, this study extends the prior research by considering different aspects of teamwork behaviors. The research questions proposed are: Which teamwork behaviors will task interdependence promote? Do different teamwork behaviors have different mediating effects on task interdependence to project management performance? To answer these questions, we conduct a survey of IS professionals. More specifically, this survey includes five teamwork behaviors, communication, coordination, cohesion, mutual support, and learning. The result provides a more precise evaluation of the mediating role of each single variable but also allows for integrating behaviors to relate task interdependence to performance.

2   Background

Task interdependence. Task interdependence represents that tasks in a team are designed to enable individuals to rely on other team members' efforts for completing the duty [5]. The tasks that individuals perform in a team may vary from being independent at one extreme to being interdependent at the other. A highly interdependent task requires team members to interact more so as to be able to accomplish their project work [8]. Team members are task interdependent when they must share resources, information or knowledge so as to accomplish their desired outcomes [9]. Particularly, task interdependence increases when individuals require greater assistance from each other to execute their project tasks [10]. For example, in most IS project teams, tasks are often complex and interdependent in nature. On the other hand, research also indicated the tasks of high interdependence enhance the opportunity individuals to discuss better ways to execute their tasks which will likely result in the formation of social relationships between members [5]. Task interdependence is an important design feature for teams. When task interdependence is high, members will have more opportunity to communicate about their work and have a better chance to form social relationships [11]. Also, with high task interdependence, there is a greater need to share information to clarify project assignments, establish effective work performance strategies, make better decisions and obtain feedback [12].

Teamwork quality. Teamwork quality represents the nature of task-related and social interactions among members of a software project team [13]. Particularly, members interact to set their project objectives, coordinate individual efforts, and encourage each other in their work. Teamwork quality includes several dimensions, such as communication, coordination, cohesion, mutual support, and effort [6].

Communication. Communication refers to the process by which information is clearly and accurately transmitted between individuals in a team in a logical manner and with a proper description of terms [14]. Previous studies indicate that communication provides the means to enable members to be able to share information that is critical to the successful execution of their project tasks. Individuals may also engage in communication for brainstorming new ideas, reviewing the progress of their project work and receiving feedback on their performance from their project manager. Project team's communication is described regarding timeliness, formalization, structure, and openness of the information exchange [13]. Moreover, communication is an important ingredient to fostering better teamwork as members have more chance to share their ideas about their project work which results in better problem solving. Individuals' communication could provide more information to other members and reduce the level of uncertainty [15].

Coordination. Coordination refers to the interactions that occur in the team aimed at organizing resources and individuals' expertise interdependencies [16]. The process of developing software on a large scale requires high coordination as the process is carried out by many individuals and teams, after which, the final product is compiled [17]. Individual members must have a common view of the functionality of the software they develop, such as, how it should be organized and how it should fit in with the other software modules already developed [18]. Furthermore, software modules need to integrate and interoperate properly, creating interdependencies among the tasks and the members [17]. In many cases, however, team members often have different opinions about the design and functionality of the modules they develop [18]. Also, the users of the system are likely to demand new requirements that had not been envisioned by the team during the initial design phase. These coordination problems increase when software projects are very large and beyond the ability of an individual to understand in detail [18].

Cohesion. Cohesion refers to individual members' liking of and commitment to their team, fellow teammates and the tasks they execute [19]. An individual with a strong desire for belonging or a high need for affiliation will likely be more motivated to stay in a team [20]. Thus, as cohesion increases, members are more concerned with their sense of belonging and are thus motivated to contribute to the team's well-being, to promote its objectives and increase their participation in its tasks [19]. The social relationships that form in teams are more satisfying to members [21]. As such, cohesion might be enhanced when individuals recognize the team to be a means for acquiring satisfying social relationships. On the other hand, cohesion helps teams to have better interactions with each other and to coordinate their efforts effectively [20]. Particularly, individuals in cohesive teams interact more in their work, develop better ways of resolving task conflicts and assign tasks effectively. Teams with lower cohesion will have difficulty in resolving their task conflicts which may result in the creation of additional problems [7]. Through cohesion, individuals coordinate their efforts, interact more to accomplish work and develop better problem solutions when executing interdependent tasks [22].

Mutual support. Mutual support refers to the extent to which individuals in a team offer support to each other when needed during the execution of tasks [23, 24]. A team of mutual support gives more opportunity to individuals to obtain help from each other [23]. Members of a mutually supportive team encourage fellow teammates to develop and contribute their ideas [25]. In line with this view, individuals in such teams will likely develop better solutions in their problem-solving. Previous studies acknowledge that individuals in teams with high mutual support may help each other in different ways such as assisting or filling in for someone falling behind schedule to complete a task, guiding a fellow teammate to correct mistakes in his or her work and providing resources to each [26]. By doing so, team members will have more opportunity to learn better ways of carrying out their work.

Learning. Learning is viewed as a process, as such, members carry out learning activities to obtain and process information needed to carry out tasks [14]. Learning is an important means by which expertise is distributed and utilized in a team [20]. Learning in teams begins when individuals interact with each other about different aspects of their work. As a result of interacting with members having different expertise, an individual in a team may increase his or her learning ability by being exposed to new perspectives and a wealth of better ideas [27]. Learning in this way helps an individual to gain new knowledge that he or she did not have previously. Team members may be involved in learning activities where they provide suggestions to one another, seek assistance from each other, share information in their tasks and discuss better ways of resolving their problems [28]. Through these activities, members may improve their understanding of a problem and uncover better ways of resolving previous shortcomings in the team's tasks.

3   Hypothesis Development

Empirical research has shown that task interdependence is positively related to communication in project teams [8, 11]. Particularly, when task interdependence increases, team members rely on each other to complete their tasks. The more that project work is designed so that members depend on one another for the accomplishment of their tasks, the better will be the interactions likely to take place between individuals in the team [5]. Along these lines, these interactions will most probably increase when team members are close to each other. For example, when individuals are near each other, such as located on the same floor of a building, they spend less effort to initiate interactions in their work.

Communication makes transmitted information available to all individuals in a team which increases the amount of information and, thus, minimizes the level of uncertainty such as tasks conflicts [15]. By reducing task conflicts, individuals will be able to understand and consider each other's opinions more. Team members who understand each other's opinions will be more capable of seeing the shortcomings in their views and integrate different perspectives, thus, leading to better decision-making [29]. When members make better decisions, they will likely develop effective problem solutions. As a consequence, the performance of the team will likely increase. This line of reasoning about the relationships between task interdependence, communication, and project performance leads to the following hypothesis:

H1: Communication mediates the effect of task interdependence on project performance.

Previous research suggests that teams with high task interdependence demonstrate effective coordination in their work [30, 31]. For example, when there is high task interdependence, individuals share resources, information, and advice so as to be able to accomplish their desired project outcomes [1, 10]. The need to share information may be high especially when the level of uncertainty, such as task complexity increases in the team's work. For example, with high levels of uncertainty involved with a task, members in interdependent teams who make decisions process much information during the execution of project work [32]. In such a context, these decision makers tend to develop knowledge about which team members are around, where and when, as appropriate for the task [17]. Also, having knowledge about individuals in the team could result in high coordination as members are more likely to develop an accurate understanding of who knows what and how they may respond to particular situations, thus, enabling them to plan their efforts [17].

The relationships between coordination and project performance have also been studied. A recent stream of research suggests that the ability to effectively coordinate with a team can be an important driver of project performance [20]. By coordinating their individual efforts, members can improve the collective mind so as to help the team develop a better understanding of tasks. It follows that having a better understanding of tasks will help members reduce process conflicts, such as being more aware of their roles and how resources are allocated, thus, leading to better accomplishment of project work [33]. Consequently, the performance of the team will likely increase. Thus, we propose the following hypothesis:

H2: Coordination mediates the effect of task interdependence on project performance.

When task interdependence is high, team members tend to have more interaction in their work, provide assistance to each other and share a lot of information [11]. The interactions that members have in a team will likely lead to the creation of better social relationships [34]. Interpersonal influences such as having good social relationships help to create coordinated membership attitudes and behaviors as well as facilitating better resolution of disagreements in the team [19]. Thus, by having coordinated attitudes such as, an individual's desire to remain in a team and behaviors related to member's participation in tasks [19], cohesion between teammates will likely be strengthened.

Research has supported a positive relationship between cohesion and project performance. For example, members in cohesive groups care about the success of each other, as their attainment of individual objectives is often linked to achieving those of the team [35]. When individuals care about each other's accomplishments, their team will likely be in more agreement in trying to fulfill its objectives. Particularly, teams in agreement tend to use their competencies with great efficiency as they know their members better and are committed to accomplishing project work. In this respect, by committing more to tasks, members will most probably generate better solutions in their problem-solving. Highly cohesive teams have been shown to create many quality solutions for a particular problem and accomplish work better regarding technical quality, cost, meeting assigned schedules and overall team performance [36]. Therefore, we propose the following hypothesis.

H3: Cohesion mediates the effect of task interdependence on project performance.

Highly interdependent tasks require team members to interact more in their work so as to achieve their desired outcomes [8]. It follows that by having frequent interactions, individuals in a team will likely agree on better ways to execute their tasks. Research shows that individuals who are agreeable are more friendly, trustworthy and tolerant with their fellow teammates [37]. Especially when trust is unconditional, such as individuals beginning to sense that they are not just coworkers but friends or teammates, feelings of owing each other are irrelevant, as shared values which include creating better interpersonal relationships and reciprocated emotions make people want to cooperate [37]. Indeed, individuals who cooperate tend to understand each other better, discuss their tasks and help one another in carrying out project work [38, 39]. Thus, by assisting each other to complete tasks, members of the team will be more mutually supportive to one another.

Mutual support promotes a sense of belonging in a team and enables members to develop better social relationships [40]. When members have good social relationships, they will likely help each other more to complete their work such as, resolving difficult tasks and providing encouragement to an individual to make he or she feel cared for [23]. The encouragement that one individual provides to another will make his or her contributions feel as valuable in the team. Especially when members feel that their contributions are being valued, they will likely be more committed to their team [41]. In this view, members will be more willing to put extra effort for the team by increasing their commitment to tasks and having better attitudes and behaviors such as maintaining a strong desire to remain in the group [41]. As a consequence, the performance of the team will most probably be enhanced. Based upon the above discussion, we propose the following hypothesis:

H4: Mutual support mediates the effect of task interdependence on project performance.

Task interdependence has been proven to increase interactions among individuals as well as helping and sharing of information in a team [42]. Thus, by interacting frequently, individuals in a team will likely agree on better ways to resolve their problems. Agreeable team members tend to be more useful, friendly, trustworthy and tolerant with their fellow teammates. Especially when two individuals begin to trust each other, they become more willing to share their resources or knowledge without having a concern that one will take advantage of the other [33]. By sharing knowledge, members can discuss and build different skills and perspectives with fellow teammates, and thus, adjust their methods to improve the way they carry out tasks [42], thereby increasing learning in the team.

Learning reduces the time spent for IT teams to complete tasks by providing quick feedback to problem identification, solution formulation, and selection of better alternatives [42]. As members provide more feedback to each other in a team, they will develop better approaches to executing their work. Furthermore, team members who create better approaches to carrying out work are likely to be satisfied with working in the team and more committed to it as a result of improved coordination of efforts and other benefits following from learning [43]. Therefore, satisfaction with working and commitment to the team will enable members to put more effort in their tasks. In this regard, putting more effort to the team's tasks will result in high project performance. Therefore, we propose:

H5: Learning mediates the effect of task interdependence on project performance.

Accordingly, we proposed the research model as shown in Fig. 1.

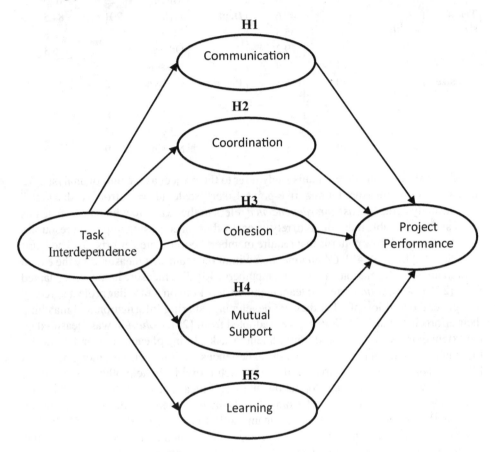

Fig. 1. Proposed research model

4   Methods

The proposed research model is tested empirically with data collected through a survey of 238 IS staffs in IS project teams from several sectors in Taiwan. Table 1 summarizes the demographic characteristics of the sample. Non-response bias was tested by comparing the early and late respondent groups on key variables [44], and no significant differences were found.

Table 1. Sample demographics (n = 238)

Var.s	Categories	#	%	Var	Cat	#	%
Gender	Male	170	71.4	**Position**	IS man	38	16.0
	Female	67	28.2		IS dev	195	81.9
	Missing	1	1.0		Missing	25	10.5
Team size	<6	118	49.6	**Dept**	IT/IS	201	84.5
	6 – 10	84	35.3		Man	21	8.8
	>10	22	9.2		Others	15	6.3
	Missing	14	5.9		Missing	1	1.0
Prj. size	<7 members	40	16.8	**Prj. dur**	<=1 year	113	47.5
	8 – 15	62	26.1		1 – 2 years	58	24.4
	16 – 25	30	12.6		>2 years	48	20.2
	>26	19	8.0		Missing	19	8.0

All constructs considered in this study refer to the project as the unit of analysis. All indicators were measured on a five-point Likert scale (1 = "strongly disagree," 5 = "strongly agree"). *Task interdependence* refers to the extent to which tasks in teams are designed to enable individuals to rely on each other for accessing important resources and creating workflow patterns that require members' coordinated efforts [5]. The items were adapted from [4, 45]. *Communication* within the team was measured by the extent of exchange of information among team members [46]. The measure items were adapted from [25]. *Coordination* refers to team members working on a familiar project agreeing to a general definition of what they are developing, sharing information and matching their efforts in the tasks [15]. Items were adopted from [25]. *Cohesion* was measured by the extent of individuals' shared commitment to tasks, liking of each other and the desire to remain in the team that arises from the experiences and interactions among members [47]. *Mutual Support* refers to the extent in which individuals help others in executing their tasks when required, develop fellow members' ideas and contributions to the team [46]. The instrument is adapted from [25]. *Learning* refers to the extent to which members acquired knowledge in the team [48]. Items were adapted [25]. *Project Performance* refers to the outcome of project management. The measurement items employed in this study were adapted from [49]. We control the effects of team size, project duration and team tenure in the structural model analyses. This study used PLS-Graph Version 3.0 [50] to verify the measurement model and structural model.

5 Results and Discussion

Table 2 shows all indicators, where loadings were all significant and ranged from 0.72 to 0.92 (except K2 with loading 0.66), the ITC ranged from 0.46 to 0.81 (>0.3 recommended), and the Cronbach's α estimates all ranged from 0.66 to 0.90 respectively. are at or above 0.7 except three items, still indicating the acceptable level of convergent validity.

Table 2. Measurement model – confirmatory factor analysis results

Construct/indicators
Task Interdependence (CR: 0.81; AVE: 0.58; CA: 0.66)
K1: Other members provide me some hard to find information
K2: I cannot complete most of my work without help from others
K3: Within my team, tasks performed by team members are related to one another
Communication (CR: 0.91; AVE: 0.71; CA: 0.86)
COM4: Conflicts concerning the openness of the information flow
COM5: Happy with the timeliness in receiving information
COM6: Happy with the precision in receiving information
COM7: Happy with the usefulness in receiving information
Coordination (CR: 0.87; AVE: 0.68, CA: 0.77)
COOR1: The work done on subtasks within the project was closely synchronized
COOR2: There were clear and fully understood goals for subtasks within the team
COOR3: The goals for subtasks were accepted by all team members
Cohesion (CR: 0.92; AVE: 0.67; CA: 0.90)
COH1: It was important to be part of this project
COH2: Strongly attached to this project
COH3: The team worked in a friendly atmosphere
COH4: Our team stuck together
COH5: Felt responsible for maintaining and protecting the team
COH6: Felt proud to be part of the team
Mutual Support (CR: 0.91; AVE: 0.64; CA: 0.89)
MS1: Helped and supported each other as best as they could
MS2: Conflicts were easily and quickly resolved
MS3: Discussions and controversies were conducted constructively
MS4: Suggestions and contributions were respected
MS5: Suggestions and contributions were discussed and developed further
MS6: Able to reach consensus regarding important issues
Learning (CR: 0.93; AVE: 0.82; CA: 0.89)
PL1: Acquired useful knowledge through this project
PL2: This project is a technical success
PL3: Learned important lessons from this project
Project Performance (CR: 0.89; AVE: 0.59; CA:0.84)
PP1: Met project goals
PP2: Work completed
PP3: High quality of work
PP5: Adherence to budget
PP6: Task operations were carried out efficiently

*CR: Composite Reliability; CA: Cronbach's Alpha

The AVE of all constructs exceeded the recommended cutoff of 0.5 [51], confirming convergent validity. Although a threshold level of 0.7 for Cronbach's α is recommended,

0.6 has been accepted as well for reliability [52]. The composite reliabilities for the eight constructs range from 0.81 to 0.93 (>0.70 recommended), ensuring adequate internal consistency. Discriminant validity was also examined by checking if the correlation between any two constructs was less than the square root of AVE by the items measuring the constructs. The results demonstrate that the measurement model discriminated adequately between the constructs.

Multiple regression analyses were conducted to examine the hypothesis. The three control variables were entered in the first step, and then independent variable. The mediating variables were entered at the final step. The results from multiple regression analysis conducted to test the study hypotheses are presented in Table 3. The variance inflation factors (VIF) ranged from 1.156 to 2.445 (<10 recommended) indicate no sizable multicollinearity among independent variables and control variables.

Table 3. Regression results with project performance dependent variable

Variables	Model 1	Model 2	Model 3
Control			
Team size	0.025	0.047	0.068
Project duration	0.058	0.076	0.065
Team tenure	−0.149	−0.067	−0.102
Independent			
Task interdependence		0.188*	−0.112
Mediator			
Teamwork quality proc			
Communication			0.110
Coordination			0.166*
Cohesion			0.282**
Mutual support			0.070
Learning			0.153*
R^2	0.013	0.039	0.355
$\triangle R^2$	0.013	0.026	0.329

*P < 0.05 **P < 0.01 °P < 0.001

Table 3 shows that the coefficients of the relationship between task interdependence and project performance decreased from ($\beta = 0.188$, p < 0.01) in model 2 to ($\beta = -0.112$, n.s.) in model 3. Particularly, among the dimensions of teamwork quality, cohesion ($\beta = 0.282$) is the most significant predictor of project performance, followed by coordination ($\beta = 0.166$) and learning ($\beta = 0.153$). In Model 2, task interdependence is significantly related to project performance. As shown in Model 3, only coordination, cohesion and learning are positively related to project performance, but communication and mutual support are not. These results indicate that teamwork quality dimensions such as coordination, cohesion, and learning mediate the relationship between task interdependence and project performance.

Table 4 summarizes the hypotheses test results of this study. From these results, we can conclude that the teamwork quality processes such as coordination, cohesion, and learning, but not communication and mutual support are viable mediators between task interdependence and final project performance.

Table 4. Summary of hypothesis tests

Hypothesis	Mediator	Path coefficient		Supported
		TI to TQ	TQ to PP	
H2	Communication	0.217^{**}	0.110	No
H3	Coordination	0.255^{**}	0.166^{*}	Yes
H4	Cohesion	0.259^{**}	0.282^{**}	Yes
H5	Mutual support	0.286^{**}	0.070	No
H6	Learning	0.299^{**}	0.153^{*}	Yes

Note: TI: Task Interdependence; TQ: Team Quality; PP: Project performance
$^{*}P < 0.05.$ $^{**}P < 0.01.$

6   Discussion and Conclusion

This study attempted to offer a contribution to the management literature by exploring the effects of interdependence on project performance in software development teams. First, the results provide strong support for the argument that task interdependence is significantly associated with teamwork quality. The evidence for the positive relationship between task interdependence and teamwork quality is an important finding, and it is consistent with research by [8]. Second, teamwork quality was found to mediate the relationship between task interdependence and project performance. This finding may close the gap between theoretical and empirical practices by revealing a mediating mechanism for the relationship between task interdependence and project performance. Third, the finding that coordination, cohesion, and learning were positively related to project performance is consistent with the results from previous project team studies [13, 25, 53]. Counter to expectations, neither communication nor mutual support were related to project performance. The unexpected finding that communication and mutual support demonstrated a negative relationship with software project performance rather than the hypothesized positive relationships is interesting and warrants further investigation. Fourth, the relationship between teamwork quality and project performance becomes stronger as task interdependence increases. However, our study's finding indicates that teamwork quality provides a partial explanation for the results and underscores the need to search for additional explanations of the relationship between task interdependence and software project performance in future research. This study adds to the literature by providing evidence that task interdependence very much facilitates the positive effects of teamwork quality in software development teams.

The findings of this study suggest that it is unimaginative for managers who are interested in enhancing the performance of software development teams to focus on interdependence without considering teamwork quality. The results presented here highlight the importance of teamwork quality processes as actualizations of interdependence and as mechanisms that transmit the effects of interdependence to project performance. This study illustrates that interdependence affects project performance in a variety of ways and that some of the ways in which interdependence is believed to affect performance may have been overstated in previous studies. Through studying the concept of interdependence, this study adds new insights to the growing literature on teamwork quality in software project teams, presents concepts and measures that may be adopted by future scholars, and suggests directions for future research. It also provides a pathway for managers to establish relevant courses of action to motivate software project teams to engage in productive teamwork quality.

References

1. Van Der Vegt, G., Emans, B., Van De Vliert, E.: Effects of interdependencies in project teams. J. Soc. Psychol. **139**(2), 202–214 (1999)
2. Stewart, G.L., Barrick, M.R.: Team structure and performance: assessing the mediating role of intrateam process and the moderating role of task type. Acad. Manage. J. **43**(2), 135–148 (2000)
3. Vegt, G.S., Emans, B.J., Vliert, E.: Patterns of interdependence in work teams: a two-level investigation of the relations with job and team satisfaction. Pers. Psychol. **54**(1), 51–69 (2001)
4. Campion, M.A., Medsker, G.J., Higgs, A.C.: Relations between work group characteristics and effectiveness: implications for designing effective work groups. Pers. Psychol. **46**(4), 823–850 (1993)
5. Courtright, S.H., Thurgood, G.R., Stewart, G.L., Pierotti, A.J.: Structural interdependence in teams: an integrative framework and meta-analysis. J. Appl. Psychol. **100**(6), 1825–1846 (2015)
6. Dayan, M., Di Benedetto, C.A.: Antecedents and consequences of teamwork quality in new product development projects: an empirical investigation. Eur. J. Innov. Manage. **12**(1), 129–155 (2009)
7. Marks, M.A., Mathieu, J.E., Zaccaro, S.J.: A temporally based framework and taxonomy of team processes. Acad. Manage. Rev. **26**(3), 356–376 (2001)
8. Wageman, R., Gordon, F.M.: As the twig is bent: how group values shape emergent task interdependence in groups. Organ. Sci. **16**(6), 687–700 (2005)
9. Pinjani, P., Palvia, P.: Trust and knowledge sharing in diverse global virtual teams. Inf. Manage. **50**(4), 144–153 (2013)
10. Van der Vegt, G., Van de Vliert, E.: Intragroup interdependence and effectiveness: review and proposed directions for theory and practice. J. Manag. Psychol. **17**(1), 50–67 (2002)
11. Wageman, R.: Interdependence and group effectiveness. Adm. Sci. Q. **40**, 145–180 (1995)
12. Sharma, R., Yetton, P.: The contingent effects of management support and task interdependence on successful information systems implementation. MIS Q. **27**, 533–556 (2003)
13. Suprapto, M., Bakker, H.L., Mooi, H.G.: Relational factors in owner–contractor collaboration: the mediating role of teamworking. Int. J. Proj. Manage. **33**(6), 1347–1363 (2015)
14. Akgün, A.E., Lynn, G.S., Keskin, H., Dogan, D.: Team learning in IT implementation projects: antecedents and consequences. Int. J. Inf. Manage. **34**(1), 37–47 (2014)
15. Hsu, J.S.-C., Shih, S.-P., Chiang, J.C., Liu, J.Y.-C.: The impact of transactive memory systems on IS development teams' coordination, communication, and performance. Int. J. Proj. Manage. **30**(3), 329–340 (2012)

16. Faraj, S., Sproull, L.: Coordinating expertise in software development teams. Manage. Sci. **46**(12), 1554–1568 (2000)
17. Espinosa, J.A., Slaughter, S.A., Kraut, R.E., Herbsleb, J.D.: Team knowledge and coordination in geographically distributed software development. J. Manage. Inf. Syst. **24**(1), 135–169 (2007)
18. Kraut, R.E., Streeter, L.A.: Coordination in software development. Commun. ACM **38**(3), 69–81 (1995)
19. Friedkin, N.E.: Social cohesion. Ann. Rev. Sociol. **30**, 409–425 (2004)
20. Kozlowski, S.W., Ilgen, D.R.: Enhancing the effectiveness of work groups and teams. Psychol. Sci. Public Interest **7**(3), 77–124 (2006)
21. Carron, A.V., Brawley, L.R.: Cohesion conceptual and measurement issues. Small Group Res. **31**(1), 89–106 (2000)
22. Gully, S.M., Devine, D.J., Whitney, D.J.: A meta-analysis of cohesion and performance effects of level of analysis and task interdependence. Small Group Res. **43**(6), 702–725 (2012)
23. Aubé, C., Rousseau, V.: Team goal commitment and team effectiveness: the role of task interdependence and supportive behaviors. Group Dyn. Theor. Res. Pract. **9**(3), 189 (2005)
24. Cha, J., Kim, Y., Lee, J.-Y., Bachrach, D.G.: Transformational leadership and inter-team collaboration exploring the mediating role of teamwork quality and moderating role of team size. Group Organ. Manage. **40**, 715–743 (2015). doi:10.1177/1059601114568244
25. Hoegl, M., Gemuenden, H.G.: Teamwork quality and the success of innovative projects: a theoretical concept and empirical evidence. Organ. Sci. **12**(4), 435–449 (2001)
26. Rousseau, V., Aubé, C., Savoie, A.: Teamwork behaviors a review and an integration of frameworks. Small Group Res. **37**(5), 540–570 (2006)
27. Van Der Vegt, G.S., Bunderson, J.S.: Learning and performance in multidisciplinary teams: the importance of collective team identification. Acad. Manage. J. **48**(3), 532–547 (2005)
28. Edmondson, A.: Psychological safety and learning behavior in work teams. Adm. Sci. Q. **44**(2), 350–383 (1999)
29. Hirst, G., Mann, L.: A model of R&D leadership and team communication: the relationship with project performance. R&D Manage. **34**(2), 147–160 (2004)
30. Rico, R., Sánchez-Manzanares, M., Gil, F., Gibson, C.: Team implicit coordination processes: a team knowledge–based approach. Acad. Manage. Rev. **33**(1), 163–184 (2008)
31. Swaab, R.I., Schaerer, M., Anicich, E.M., Ronay, R., Galinsky, A.D.: The too-much-talent effect team interdependence determines when more talent is too much or not enough. Psychol. Sci. **25**(8), 1581–1591 (2014)
32. Sherman, J.D., Keller, R.T.: Suboptimal assessment of interunit task interdependence: modes of integration and information processing for coordination performance. Organ. Sci. **22**(1), 245–261 (2011)
33. Huang, C.-C.: Knowledge sharing and group cohesiveness on performance: an empirical study of technology R&D teams in Taiwan. Technovation **29**(11), 786–797 (2009)
34. Keller, R.T.: Cross-functional project groups in research and new product development: diversity, communications, job stress, and outcomes. Acad. Manage. J. **44**(3), 547–555 (2001)
35. Zaccaro, S.J., Rittman, A.L., Marks, M.A.: Team leadership. Leadersh. Q. **12**(4), 451–483 (2002)
36. Jones, M.C., Harrison, A.W.: IS project team performance: an empirical assessment. Inf. Manage. **31**(2), 57–65 (1996)
37. Barrick, M.R., Stewart, G.L., Neubert, M.J., Mount, M.K.: Relating member ability and personality to work-team processes and team effectiveness. J. Appl. Psychol. **83**(3), 377 (1998)

38. Pinto, M.B., Pinto, J.K., Prescott, J.E.: Antecedents and consequences of project team cross-functional cooperation. Manage. Sci. **39**(10), 1281–1297 (1993)
39. Ng, K.Y., Van Dyne, L.: Antecedents and performance consequences of helping behavior in work groups a multilevel analysis. Group Organ. Manage. **30**(5), 514–540 (2005)
40. Maton, K.I.: Social support, organizational characteristics, psychological well-being, and group appraisal in three self-help group populations. Am. J. Commun. Psychol. **16**(1), 53–77 (1988)
41. Sheng, C.-W., Tian, Y.-F., Chen, M.-C.: Relationships among teamwork behavior, trust, perceived team support, and team commitment. Soc. Behav. Pers. Int. J. **38**(10), 1297–1305 (2010)
42. Bachrach, D.G., Powell, B.C., Collins, B.J., Richey, R.G.: Effects of task interdependence on the relationship between helping behavior and group performance. J. Appl. Psychol. **91**(6), 1396 (2006)
43. Zellmer-Bruhn, M., Gibson, C.: Multinational organization context: implications for team learning and performance. Acad. Manage. J. **49**(3), 501–518 (2006)
44. Armstrong, J.S., Overton, T.S.: Estimating nonresponse bias in mail surveys. J. Mark. Res. **14**, 396–402 (1977)
45. Van Der Vegt, G., Emans, B., Van De Vliert, E.: Team members' affective responses to patterns of intragroup interdependence and job complexity. J. Manage. **26**(4), 633–655 (2000)
46. Parolia, N., Jiang, J.J., Klein, G., Sheu, T.S.: The contribution of resource interdependence to IT program performance: a social interdependence perspective. Int. J. Proj. Manage. **29**(3), 313–324 (2011)
47. Barrick, M.R., Bradley, B.H., Kristof-Brown, A.L., Colbert, A.E.: The moderating role of top management team interdependence: implications for real teams and working groups. Acad. Manage. J. **50**(3), 544–557 (2007)
48. Liang, T.-P., Jiang, J., Klein, G.S., Liu, J.Y.-C.: Software quality as influenced by informational diversity, task conflict, and learning in project teams. IEEE Trans. Eng. Manage. **57**(3), 477–487 (2010)
49. Henderson, J.C., Lee, S.: Managing I/S design teams: a control theories perspective. Manage. Sci. **38**(6), 757–777 (1992)
50. Chin, W.W., Marcolin, B.L., Newsted, P.R.: A partial least squares latent variable modeling approach for measuring interaction effects: results from a Monte Carlo simulation study and an electronic-mail emotion/adoption study. Inf. Syst. Res. **14**(2), 189–217 (2003)
51. Fornell, C., Larcker, D.F.: Evaluating structural equation models with unobservable variables and measurement error. J. Mark. Res. **18**, 39–50 (1981)
52. Hair, J.F., Anderson, R.E., Tatham, R.L., Black, W.C.: Multivariate Data Analysis, 5th edn. Prentice Hall International, New York (1998)
53. Chun-Yu, T.: The Effects of the Task Interdependence and Team Work Quality on Project Performance. Yuan Ze University, pp. 1–51 (2011)

On the Use of ISO/IEC Standards to Address Data Quality Aspects in Big Data Analytics Cloud Services

Jonathan Roy[✉], Hebatalla Terfas, and Witold Suryn

École de technologie supérieure, Montreal, QC, Canada
{jonathan.roy.1,hebatalla-s-h.terfas}@ens.etsmtl.ca,
witold.suryn@etsmtl.ca

Abstract. With data volumes constantly growing, cloud computing provides a model for Big Data Analytics where solutions can benefit from rapid elasticity and scalability. This model changes the level of control that cloud service customers have on their data. Understanding how data is handled by cloud service providers is therefore critical in achieving data quality objectives. This paper presents an analysis on the applicability of ISO/IEC standards to Big Data Analytics cloud services, focusing on data quality. Based on results, we provide observations, identify challenges, and offer recommendations on the application of standards and future development.

Keywords: Data quality · Big Data · Cloud computing · Quality models · SLA

1 Introduction

Big Data Analytics is concerned with how high data volumes of high velocity from a high variety of data types and sources is assembled and mined to derive insights and obtain knowledge. Its workflow is comprised of four phases: data sources and their storage, data management, modeling, and analysis and visualization of results [1]. With data volumes constantly growing, cloud computing provides an interesting model for Big Data Analytics, where solutions can be hosted on the cloud to benefit from its key characteristics, such as rapid elasticity and scalability, resource pooling, and measured services. When taking into account cloud deployment models (i.e., private, public, and hybrid) [2], the following options are typically evaluated when deploying Big Data Analytics workflows on the cloud, that is, that data and models are private, data are public and models are private, data and models are public, and data are private and models are public [1,3]. Given that public cloud services offer the lowest level of personal data control to the customer, we focused on public cloud deployments of Big Data Analytics workflows where data and models are located in the public domain. With a low level of control, software quality engineers working for

© Springer International Publishing AG 2017
W. Abramowicz (Ed.): BIS 2017, LNBIP 288, pp. 149–164, 2017.
DOI: 10.1007/978-3-319-59336-4_11

the cloud service customer must understand and quantify the impact of cloud service quality on data quality objectives of solutions. While widely accepted ISO/IEC quality and cloud computing standards have been put into practice, only limited research has been conducted on their application in the context of Big Data Analytics cloud services. Therefore, the extent to which these standards support software quality engineers in the identification and definition of data quality requirements in this particular context has clearly not been fully investigated at this point in time.

To address this problem, in Sect. 2, we first select the standards for this analysis based on the software quality engineering (SQE) practical framework by Suryn [5]. Then, we identify and group under the term "data quality aspects" the different aspects influencing data quality in Big Data Analytics cloud services, that is, the characteristics specific to the Big Data Analytics process and workflow, the context of use, and the capabilities provided by the cloud service. Section 3 presents related research on the application of the selected standards to Big Data Analytics cloud services. The following research questions are then addressed for Big Data Analytics cloud services from the perspective of the cloud service customer:

RQ1: What are the *data quality aspects* investigated in scientific literature?
RQ2: To what extent do ISO/IEC standards support software quality engineers in the identification and definition of data quality requirements that address *data quality aspects* investigated in scientific literature?

By identifying the *data quality aspects* investigated in scientific literature, the findings related to RQ2 will guide the future development of standards as well as research in the field of SQE. Section 4 presents the proposed methodology by which to gather the *data quality aspects* in Big Data Analytics cloud services and the method to conduct the applicability analysis of standards is presented. In Sect. 5, we present the results, which include observations, challenges, and recommendations. Section 6 provides our final conclusions.

2   Data Quality Aspects in Big Data Analytics Cloud Services

In order to help support software quality engineers with the identification and definition of quality requirements, the SQE practical framework by Suryn [5] presents a *"continuous, systematic, disciplined, quantifiable approach to the development and maintenance of quality throughout the whole life cycle of software products and systems"*. It is the only "best practices" previously published framework of its type to date. As a result, it constitutes the SQE foundation of this study. The application of SQE requires the use of a quality model with the capacity to support both definitions of quality requirements as well as their evaluation. For this purpose, the SQE refers to widely accepted quality models of the ISO/IEC 25000 series, that is:

- ISO/IEC 25010 Systems and software engineering—Systems and software Quality Requirements and Evaluation (SQuaRE) [6]
- ISO/IEC 25012 Software engineering—Software product Quality Requirements and Evaluation (SQuaRE)—Data quality model [7]

Although ISO/IEC 25000 quality models are applicable to both computer systems and software products, the conceptual shift from conventional computing environments to cloud computing environments has led to the recent publication of standards series as well as a framework that aims to provide support in the context of cloud computing, that is:

- ISO/IEC 17788 Information technology — Cloud computing — Overview and vocabulary [2]
- ISO/IEC 17789 Information technology — Cloud Computing — Reference architecture [8]
- ISO/IEC 19086-1 Information technology — Cloud computing — Service level agreement (SLA) framework — Part 1: Overview and concepts [9]
- ISO/IEC 19086-2 Information technology — Cloud computing — Service level agreement (SLA) framework — Part 2: Metric Model [10]
- ISO/IEC 19086-3 Information technology — Cloud computing — Service level agreement (SLA) framework — Part 3: Core conformance requirements [11]
- ISO/IEC 19086-4 Information technology — Cloud computing — Service level agreement (SLA) framework — Part 4: Security and privacy [12]
- ISO/IEC 27018 Information technology — Security techniques — Code of practice for protection of personally identifiable information (PII) protection in public clouds acting as PII processors [13]
- ISO/IEC 27017 Information technology — Security techniques — Code of practice for information security controls based on ISO/IEC 27002 for cloud services [14]

It is important to note that ISO/IEC 19086-2 [10], ISO/IEC 19086-3 [11], and ISO/IEC 19086-4 [12] have not yet been published and are still under development. Nevertheless, they have been made available for research purposes.

Therefore, the quality models as well as the recently published and under development standards for cloud environments are part of this analysis, with the exception of ISO/IEC 17789 [8] and ISO/IEC 19086-2 [10]. Given that the focus of this analysis is on guidelines that aim to support software quality engineers with the identification and definition of data quality requirements, reference architectures and metrics discussed in these standards are not within the scope of this analysis.

As previously mentioned, Big Data Analytics is concerned with how high data volumes of high velocity from a high variety of data types and sources is assembled and mined to derive insights and obtain knowledge. Its workflow is comprised of four phases: data sources and their storage, data management, modeling, and result analysis and visualization [1]. From these characteristics, which are specific to the Big Data Analytics workflow, software quality engineers must identify and define the quality in use requirements in the context of use. Based on ISO/IEC 25010 [6], dynamic quality requirements derive from the quality in use requirements and static quality requirements derive from dynamic quality requirements. Therefore, dynamic and static quality requirements must also be identified and defined. This relationship model proposed by ISO/IEC experts was later validated empirically [15]. As a complement to the product quality model, ISO/IEC 25012 [7] comprises data quality characteristics from two points of view, that is, inherent and system dependent, for which some data quality characteristics share both points of view. From the inherent point of view, data quality refers to the data itself (e.g., consistency) [7]. From the system dependent point of view, data quality depends on and is achieved by the capabilities provided by the computer system [7]. Consequently, in cases whereby Big Data Analytics workflows are deployed on the cloud, inherent data quality remains under the control of the owner of the data. As it relates to system dependent data quality, it depends on and is achieved by the capabilities provided by the cloud service. As a result, the control of system dependent data quality is shared between the cloud service customer and provider. Moreover, for data quality characteristics sharing both points of view, the inherent data quality is also influenced by the capabilities provided by the cloud service. Therefore, data quality in Big Data Analytics cloud services is influenced by the following aspects:

– The characteristics specific to the Big Data Analytics process and workflow
– The context of use
– The capabilities provided by the cloud service (e.g., the architectural tactics used to achieve quality objectives)

In this paper, the term "data quality aspects" groups all aforementioned aspects and will also be referenced in this study to facilitate readership.

3   Related Work

3.1   Application of ISO/IEC Selected Standards

Villalpando et al. [17,18] focused on a specific application and quality characteristic and proposed a method integrating ISO/IEC 25010 [6] quality models and ISO/IEC 25023 [19] quality measures for the performance analysis of Big Data applications. Although the experiments were limited to the Hadoop[1] framework and the MapReduce [20] programming model, the authors provided a concrete example on the use of ISO/IEC 25010 [6] quality models for quality evaluation of Big Data cloud services.

Merino et al. [21] proposed a data quality in use model using ISO/IEC 25012 [7] as a foundation and tailored it to the assessment of quality in use of data from Big Data solutions. The objective of this model is to support the evaluation of quality in use of input data for Big Data analysis. The proposed data quality

---

[1] http://hadoop.apache.org.

in use model is validated against the technological and managerial challenges of Big Data identified under the assumption that once the business rules that sets the constraints over data are identified, the related dynamic properties will be covered by the ISO/IEC 25012 [7] data quality characteristics and associative data quality measures.

Laranjeiro et al. [22], conducted a literature review on Big Data and data quality and identified frequently used data quality dimensions, which included the ISO/IEC 25012 [7] data quality model characteristics and points of view (i.e., inherent and system dependent). Data quality issues are also gathered and mapped to data quality dimensions. This approach is presented as a first step toward the measurement of data quality.

Kemp [23], describes how ISO/IEC 27018 [13] is succeeding in addressing the gap between the rapid development of cloud services and the slow legislative changes in data protection. For example, Kemp reports that one year after the publication of the standard, Microsoft Corporation was the first to undertake and use ISO/IEC 27018 [13] in several layers of its cloud services. Kemp states that the standard can be applied across cloud services delivery models and layers. In addition, according to Kemp, as cloud service providers are not systematically aware of the PII nature of data sent by the cloud service customer, ISO/IEC 27018 [13] can be applied, taking into account its applicability whether the data is or is not in itself PII information.

Mitchell [24] provides a review of several standards applicable in the context of cloud service providers that aim for certification of security and privacy compliance, such as the ISO/IEC 27000 standards series that include ISO/IEC 27017 [14] and ISO/IEC 27018 [13]. Mitchell also highlights the growing need for data de-identification techniques and the challenges related to their application. Consequently, as stated by Mitchell, the standard committee responsible for ISO/IEC 27018 [13] agreed to create a new standard (i.e., N15297) [46] that covers data de-identification techniques.

Panth et al. [25], present data security compliance to the ISO/IEC 27001 family (Information security management systems) [26], ISO/IEC 27017 [14], and ISO/IEC 27018 [13] as countermeasures to data security threats and vulnerabilities.

Liu et al. [27] highlight issues in cloud security governance and the lack of a cloud governance framework. According to Liu, these issues are being partially addressed by emerging standards, such as ISO/IEC 27017 [4].

Although the research provides examples for the application of standards to Big Data and cloud services, the extent to which the standards support the identification and definition of data quality requirements in Big Data Analytics cloud services has not been evaluated and remains unknown. Moreover, given that ISO/IEC 19086-1 [9], ISO/IEC 19086-3 [11], and ISO/IEC 19086-4 [12] have only been published recently, our literature review has revealed no available research on their application in the context of Big Data Analytics cloud services.

4 Methodology

4.1 Data Quality Aspects

To address RQ1, we conducted a broad search on Big Data, cloud computing, and quality in scientific databases. Articles were then screened to extract those that specifically referred to *data quality aspects* in the context of Big Data Analytics cloud services. The *data quality aspects* from the perspective of the cloud service customer were then extracted from selected articles. Results are provided in Tables 2, 4, and 6.

4.2 Applicability Analysis of Standards

Given that software quality engineers may infer *data quality aspects* from standard guidelines that can be used for the identification and definition of data quality requirements, we propose a content analysis in order to address RQ2. In content analysis, the activity of recording inferences based on specific rules is referred as "coding". This activity can be performed by computers or humans. When humans are performing coding, they are referred as "coders".

In this research, two coders (SQE researchers in this case) analyzed the content of the selected standards by applying the following coding rule for each standard and *data quality aspect*:

– If a guideline can be used to identify and define the data quality requirements associated with the *data quality aspect*, the guideline information and section should be recorded.

4.3 Reliability of Analysis

In order to draw conclusions from content analysis, its reliability must be demonstrated [29]. Intercoder reliability, or, more specifically, intercoder agreement in the context of content analysis, is widely used to measure the extent to which independent analysts make the same coding decisions in evaluating a characteristic of an artifact. In this analysis, agreement means that coders agree that at least one guideline of the analyzed standard can be used or that no guidelines can be used to identify and define the data quality requirements associated with the *data quality aspect*.

According to content analysis researchers, the Krippendorff's alpha coefficient [29] is the recommended measure for intercoder agreement given that it addresses the weaknesses of measures, such as percent agreement, the S score by Bennett et al., Scott's pi, Cohen's kappa coefficient, Fleiss' kappa, and Cronbach's alpha [29]. In order to compute the Krippendorff's alpha coefficient, Tables 3, 5, and 7 are used as reliability datasets. From coding results of the applicability analysis of standards, for each coder (C) and *data quality aspect*

(D) 1 indicates that at least one guideline of the standard can be used and 0 indicates that no guidelines of the standard can be used to identify and define data quality requirements associated with the *data quality aspect*. Intercoder agreements must be evaluated for each variable of the sample as recommended by researchers. In this analysis, the variable is defined as potentially usable guideline(s) of a specific standard. As it pertains to the sample, it is the complete set of selected standards for the analysis. Consequently, the Krippendorff's alpha coefficient was evaluated for each standard included in this analysis, and the results are presented in Table 8. The Krippendorff's alpha coefficient general form is as follows [4]:

$$\alpha = 1 - \frac{D_o}{D_e} \tag{1}$$

wherein D_o is the disagreement observed [4]:

$$D_o = \frac{1}{n} \sum_c \sum_k o_{ck \ metric} \delta_{ck}^2 \tag{2}$$

wherein D_e is the disagreement resulting from chance [4]:

$$D_e = \frac{1}{n(n-1)} \sum_c \sum_k n_c \cdot n_{k \ metric} \delta_{ck}^2 \tag{3}$$

The arguments o_{ck}, n_c, n_k and n, are the frequencies of values as showed in the coincidence matrix in Table 1(a) [4].

To perform the evaluation, we used computational steps applicable to the reliability data generated by this content analysis, that is, binary data using two coders and no missing data [4]. As an example, Table 1(b) presents the coincidences matrix that accounts for all values contained in the ISO/IEC 19086-1 reliability data, that is, Tables 3, 5 and 7 combined. Units are entered twice, once as $c - k$ pairs and once as $k - c$ pairs [4] where $n_0 = 18$ is the number of 0 s, $n_1 = 34$ is the number of 1 s, and $n = 2N = 52$ is the total number of paired values.

Table 1. Coincidence matrix

(a) Binary

	0	1	
0	o_{00}	o_{01}	n_0
1	o_{10}	o_{11}	n_1
	n_0	n_1	$n = 2N$

(b) ISO/IEC 19086-1

	0	1	
0	16	2	18
1	2	32	34
	18	34	52

Table 2. Characteristics specific to the Big Data Analytics process and workflow

Data quality aspects	
D1	Sensitive data identified and omitted from the analysis and data store [30]
D2	Sensitive data included in the analysis and data store after data anonymization [30]
D3	Sensitive data included in the analysis and data store after encryption [30]
D4	Capability to operate on encrypted data limiting the movement of data to be decrypted [30]
D5	Data arrival can require processing at different speeds (e.g. batch, continuous and real-time) [1]
D6	Data and techniques used for analytics are *"emergent and not necessarily defined upfront"* [31]
D7	Data governance of structured and unstructured data sources [31]
D8	*"Timely processing of data for building and scoring"* [1]
D9	*"Isolation of analytical artifacts"* [32]
D10	Hard deadline analysis tasks [33]
D11	Avoid expensive data movements by offloading computation to sources [33]
D12	Analysis accuracy negatively affected by an increase in data volume and in a variety of data sources and their origins [34]
D13	Privacy protection during data and knowledge mining [34]
D14	Data mining with sufficient user interaction [34]
D15	Data visualization tools considering data quality and representation to facilitate exploration [35]
D16	Potential effects of data volume, velocity, variety, and veracity on users' cognitive loads [31]. Cognitive load: is the amount of working memory required when solving a problem [36]

For ISO/IEC 19086-1, based on the coincidence matrix:

$$\alpha = 1 - (n-1)\frac{o_{01}}{n_0 \cdot n_1} = 1 - (52-1)\frac{2}{18 \cdot 34} = 0.833 \qquad (4)$$

As a reference, the Krippendorff's alpha coefficient considers 0.800 as an acceptable intercoder agreement with only tentative and cautious conclusions to be drawn with results between 0.667 and 0.800 [28].

Table 3. ISO/IEC quality and cloud computing standards applicability analysis to characteristics specific to the Big Data Analytics process and workflow (coding results)

| | 25010 | | | | 25012 | | 17788 | | 19086- | | | | | | 27018 | | 27017 | |
| | Product quality | | Quality in use | | | | | | 1 | | 3 | | 4 | | | | | |
Data quality aspects	C1	C2	C1	C2	C1	C2	C1	C2	C1	C2	C1	C2	C1	C2	C1	C2	C1	C2
D1	1	1	0	0	1	1	1	1	1	1	1	1	1	1	1	1	1	1
D2	0	0	0	0	0	0	0	0	0	1	0	0	1	1	0	1	1	1
D3	1	1	0	0	1	1	0	0	0	0	0	0	1	1	1	1	1	1
D4	0	0	0	0	0	0	0	0	0	0	0	0	1	1	1	0	1	0
D5	1	1	0	0	1	1	1	1	1	1	1	1	0	0	0	0	1	1
D6	1	1	0	0	1	1	1	1	1	1	1	1	1	1	1	1	1	1
D7	1	0	0	0	1	1	1	1	1	1	1	1	0	0	0	0	0	0
D8	1	1	0	0	0	0	1	1	1	1	0	1	0	0	0	0	1	1
D9	1	1	0	0	0	0	1	1	0	0	0	0	1	1	0	0	1	1
D10	1	1	0	0	0	0	1	1	1	1	0	1	0	0	0	0	1	1
D11	1	0	0	0	1	0	1	0	1	1	0	1	0	0	0	0	0	0
D12	1	1	0	0	1	1	0	0	0	0	0	0	0	0	0	0	0	0
D13	1	1	0	0	1	1	1	1	1	1	0	0	1	1	1	1	1	1
D14	1	1	1	1	0	0	0	0	0	0	0	0	0	0	0	0	0	0
D15	1	1	1	1	0	0	0	0	0	0	0	0	0	0	0	0	0	0
D16	0	0	1	1	0	0	0	0	0	0	0	0	0	0	0	0	0	0

Table 4. Context of use (Big Data Analytics solutions requiring profiling)

Data quality aspect	
D17	Privacy protection using data de-identification and pseudonymization procedures to reduce the risk of direct identification of the subject to whom the personal data belongs [37]

Table 5. ISO/IEC quality and cloud computing standards applicability analysis to the context of use (coding results)

| | 25010 | | | | 25012 | | 17788 | | 19086- | | | | | | 27018 | | 27017 | |
| | Product quality | | Quality in use | | | | | | 1 | | 3 | | 4 | | | | | |
Data Quality aspects	C1	C2	C1	C2	C1	C2	C1	C2	C1	C2	C1	C2	C1	C2	C1	C2	C1	C2
D17	0	0	0	0	0	0	0	0	0	0	0	0	0	0	0	0	0	0

Table 6. Capabilities provided by the cloud service

Data quality aspects	
D18	*"Data can be modified only by authorized parties"* [38]
D19	Capability provided to the cloud service user to verify if data is maintained [38]
D20	Capability provided to the cloud service user to obtain a proof of data integrity [39]
D21	Data availability achieved through geo-distribution of data replicated across data centers [30]
D22	Data stored off premises by a third party [30]
D23	Limited control over where the data is physically located [30]
D24	*"Data physically located in a particular country and subject to local rules and regulations"* [30]
D25	Partition tolerance in a distributed system [40]. CAP theorem (i.e. consistency, availability, partition tolerance).
D26	Complete data loss resulting from direct DoS attacks [40]

Table 7. ISO/IEC quality and cloud computing standards applicability analysis to capabilities provided by the cloud service (coding results)

Data Quality aspects	25010 Product quality C1	C2	Quality in use C1	C2	25012 C1	C2	17788 C1	C2	19086- 1 C1	C2	3 C1	C2	4 C1	C2	27018 C1	C2	27017 C1	C2
D18	1	1	0	0	1	1	1	1	1	0	0	0	1	1	1	1	1	1
D19	0	0	0	0	0	0	1	1	1	1	1	1	1	0	1	1	1	0
D20	1	1	0	0	1	1	1	1	1	1	0	1	0	0	1	1	1	1
D21	1	1	0	0	1	1	1	1	1	1	1	1	0	0	1	1	1	1
D22	1	1	0	0	0	1	1	1	1	1	1	1	1	1	1	1	1	1
D23	0	0	0	0	0	0	0	0	1	1	1	1	0	0	1	1	1	1
D24	0	0	0	0	1	1	1	1	1	1	1	1	0	0	1	1	1	1
D25	1	1	0	0	1	1	1	1	1	1	1	1	0	0	0	0	0	0
D26	1	1	0	0	1	1	1	1	1	1	1	1	0	0	1	1	1	1

5   Results

5.1   Reliability of the Analysis

Reliability results as presented in Table 8 show that the Krippendorff's alpha coefficient was greater than 0.800 for all analyzed standards with the exception of ISO/IEC 19086-3 where the Krippendorff's alpha coefficient was between 0.667 and 0.800. This indicates that observations and conclusions can be drawn from the applicability analysis of standards with only tentative and cautious conclusions in the case of ISO/IEC 19086-3.

Table 8. Results from Krippendorff's alpha coefficients

	25010		25012	17788	19086-			27018	27017
	Product quality	Quality in use			1	3	4		
Alpha	0.823	1	0.848	0.919	0.833	0.698	0.919	0.849	0.833

Table 9. Categorization of *data quality aspects*

Data Quality aspects	**Inherent**	System dependent
D1	x	x
D3	x	x
D6	x	x
D12	x	
D13	x	x
D15	x	x
D18	x	x
D19		x
D20	x	x
D21		x
D22	x	x
D24	x	x
D25		x
D26		x

The following sections provide the observations, challenges, and recommendations drawn from both, the *data quality aspects* extracted from the scientific literature, and coding results related to the applicability analysis of standards. They are categorized in accordance with the *data quality aspects* as defined in Sect. 2. The categorization of *data quality aspects* presented in Table 9 is based on ISO/IEC 25012 [7] guideline information and section recorded by coders in accordance with the coding instructions in Sect. 4.

5.2 Characteristics Specific to the Big Data Analytics Process and Workflow

Observation: The data and techniques used in Big Data Analytics processes are modeled during its application and have therefore not been necessarily defined upfront [31,41] (see Table 2; D6).

Challenge: This dynamic characteristic of the Big Data Analytics process combined with the frequency at which it can be executed challenges the application of standards used for quality assessments and evaluations of: (1) the data and techniques used in the process, and (2) derived insights and knowledge artifacts resulting from the process itself.

Recommendation: In order to ensure confidentiality and privacy, the quality assessment and evaluation of aspects (1) and (2) mentioned above may require automation in order to achieve the quality in use objectives of the Big Data Analytics solution. Moreover, the quality model proposed in [21] could be used for the quality assessment and evaluation of input data for Big Data analysis.

Observation: The processing of encrypted information (see Table 2; D4) is not covered by analyzed standards.

Challenge: Although ISO/IEC 19086-4 [12] refers to data encryption (see Table 3; D4), it only considers data at rest and in transit. Based on this challenge, we conducted a search on the processing of encrypted data in scientific databases. In addition, we also conducted a search on its application in the industry. This search revealed that the capability to process encrypted data, such as in [42], is gaining momentum in the industry and offerings are starting to emerge. For instance, Microsoft Azure and IBM are now offering the capability to process analytical queries over encrypted data.

Recommendation: The processing of encrypted data should be applied when required given that solutions are now emerging in the industry. Moreover, ISO/IEC standards should include guidelines for their application.

5.3   Context of Use

Observation: As seen from the analysis (see Table 5; D17), data de-identification using pseudonymization procedures are not covered by any of the analyzed standards.

Challenge: Big Data Analytics solutions that require profiling, that is, the identification of a person (e.g., marketing and healthcare) under data anonymization (see Table 2; D2), have limitations [37]. On the one hand, the capability provided to process data (i.e., to assemble and mine data) from which to determine or predict actions or events is linked to data security and privacy issues. On the other hand (see Table 2; D13), it is not always possible to make data anonymous while simultaneously retaining the information necessary for processing purposes [37]. In this case, as demonstrated in [37], data de-identification using pseudonymization procedures reduces the risk of direct identification of the subject to whom the personal data belongs. Pseudonymization procedures *"...allows for the removal of an association with a data subject. It differs from anonymization (anonymous) in that it allows for data to be linked to the same person across multiple data records or information systems without revealing the identity of the person ...it can be performed with or without the possibility of re-identifying the subject of the data (reversible or irreversible pseudonymization)"* [43].

Based on this challenge, we conducted a search on pseudonymization within the ISO/IEC standard database. The search revealed several standards covering pseudonymization procedures, that is, ISO/TS 25237 Health informatics—Pseudonymization [44], ISO/IEC 29100 Information technology—Security techniques Privacy framework [16], and ISO/IEC 15944-8 Information technology—Business operational view—Part 8: Identification of privacy protection requirements as external constraints on business transactions [45].

However, as highlighted in [24], challenges related to the application of data de-identification techniques, such as pseudonymization procedures, resulted in the development of a new standard (i.e., N15297) [46] that aims to address these challenges.

Recommendation: For Big Data Analytics solutions that require profiling, standards under development that aim to address the challenges of the application of data de-identification should be used once published. For example, this standard could support software quality engineers in addressing privacy protection (see Table 2; D13) during data and knowledge mining procedures [34].

5.4   Capabilities Provided by the Cloud Service

Observation: As seen from this analysis (see Table 9), 13 out of the 26 *data quality aspects* extracted from scientific literature have been categorized as system dependent or both system dependent and inherent, based on ISO/IEC 25012 coding results.

Challenge: This confirms that, in cases whereby Big Data Analytics workflows are deployed on the cloud, an informed agreement between the cloud service customer and provider is crucial in order to achieve data quality objectives. Given that the ISO/IEC 19086 cloud computing SLA framework (which aims to establish a common understanding between cloud service customers and providers for the definition of cloud SLAs) was published only recently, our literature review has revealed no available research to date on its application in the context of Big Data Analytics cloud services.

Recommendation: Further research on its application is therefore required in order to identify potential challenges and provide support on SLA definitions to software quality engineers.

6   Conclusions

At this point in time, results of the content analysis have not been validated with other methodologies. Also, the results only reflect the *data quality aspects* in Big Data Analytics cloud services that were taken from scientific literature.

Nevertheless, results from this study will help software quality engineers working for cloud service customer organizations identify and define data quality requirements in Big Data Analytics cloud services. In addition, the recommendations we provided are meant to help software quality engineers to foresee the challenges ahead and to guide the future development of standards as well as research in the field of SQE.

References

1. Assuno, M.D., et al.: Big data computing and clouds: trends and future directions. J. Parallel Distrib. Comput. **79**, 3–15 (2015)
2. ISO/IEC 17788: Information technology – Cloud computing – Overview and vocabulary (2014)
3. Krishna, P., Varma, K.: Cloud analytics: A path towards next generation affordable BI. White paper, Infosys (2012)
4. Krippendorff, K.: Computing Krippendorff's alpha reliability. Departmental papers (ASC), p. 43 (2007)
5. Suryn, W.: Software Quality Engineering: A Practitioner's Approach. Wiley, Hoboken (2013)
6. ISO/IEC 25010: Systems and software engineering – Systems and software Quality Requirements and Evaluation (SQuaRE) – System and software quality models (2010)
7. ISO/IEC 25012: Software engineering – Software product Quality Requirements and Evaluation (SQuaRE) – Data quality model (2008)
8. ISO/IEC 17789: Information technology – Cloud Computing – Reference Architecture (CCRA) (2014)
9. ISO/IEC 19086-1: Information technology – Cloud computing – Service Level Agreement (SLA) framework – Part 1: Overview and concepts (2015)
10. ISO/IEC 19086-2: Information technology – Cloud computing – Service level agreement (SLA) framework – Part 2: Metric Model
11. ISO/IEC 19086-3: Information technology – Cloud computing – Service level agreement (SLA) framework – Part 3: Core conformance requirements
12. ISO/IEC 19086-4: Information technology – Cloud computing – Service level agreement (SLA) framework – Part 4: Security and privacy
13. ISO/IEC 27018: Information technology – Security techniques – Code of practice for PII protection in public clouds acting as PII processors (2014)
14. ISO/IEC 27017: Information technology – Security techniques – Code of practice for information security controls based on ISO/IEC 27002 for cloud services (2015)
15. Cheikhi, L., Abran, A.: Investigation of the relationships between the software quality models of the ISO 9126 standard: an empirical study using the Taguchi method. Softw. Qual. Prof. **14**(2), 22–34 (2012)
16. ISO/IEC 29100: Information technology – Security techniques – Privacy framework (2011)
17. Villalpando, L.E.B., April, A., Abran, A.: Performance analysis model for big data applications in cloud computing. J. Cloud Comput. **3**(1), 1–20 (2014)
18. Bautista, L., Abran, A., April, A.: Design of a performance measurement framework for cloud computing. J. Softw. Eng. Appl. **5**(2), 69–75 (2012)
19. ISO/IEC 25023: Systems and software engineering – Systems and software Quality Requirements and Evaluation (SQuaRE) – Measurement of system and software product quality (2014)
20. Dean, J., Ghemawat, S.: MapReduce: simplified data processing on large clusters. Commun. ACM **51**(1), 107–113 (2008)
21. Merino, J., et al.: A data quality in use model for big data. Future Gener. Comput. Syst. **63**, 123–130 (2015)
22. Laranjeiro, N., Soydemir, S.N., Bernardino, J.: A survey on data quality: classifying poor data. In: 2015 IEEE 21st Pacific Rim International Symposium on Dependable Computing (PRDC). IEEE (2015)

23. Kemp, R.: ISO 27018 and personal information in the cloud: first year scorecard. Comput. Law Secur. Rev. **31**(4), 553–555 (2015)
24. Mitchell, C.: Privacy, compliance and the cloud. In: Zhu, S.Y., Hill, R., Trovati, M. (eds.) Guide to Security Assurance for Cloud Computing, pp. 3–14. Springer, Heidelberg (2015)
25. Panth, D., Mehta, D., Shelgaonkar, R.: A survey on security mechanisms of leading cloud service providers. Int. J. Comput. Appl. **98**, 24 (2014)
26. ISO/IEC 27001: Information technology – Security techniques – Information security management systems – Requirements (2013)
27. Liu, Y., et al.: A survey of security and privacy challenges in cloud computing: solutions and future directions. J. Comput. Sci. Eng. **9**(3), 119–133 (2015)
28. Krippendorff, K.: Content Analysis: An Introduction to its Methodology. Sage, Thousand Oaks (2004)
29. Hayes, A.F., Krippendorff, K.: Answering the call for a standard reliability measure for coding data. Commun. Methods Measures **1**(1), 77–89 (2007)
30. Abadi, D.J.: Data management in the cloud: limitations and opportunities. IEEE Data Eng. Bull. **32**(1), 3–12 (2009)
31. Abbasi, A., Sarker, S., Chiang, R.: Big data research in information systems: toward an inclusive research agenda. J. Assoc. Inf. Syst. **17**(2), 3 (2016)
32. Sun, X., et al.: A cost-effective approach to delivering analytics as a service. In: 2012 IEEE 19th International Conference on Web Services (ICWS). IEEE (2012)
33. Kambatla, K., et al.: Trends in big data analytics. J. Parallel Distrib. Comput. **74**(7), 2561–2573 (2014)
34. Che, D., Safran, M., Peng, Z.: From big data to big data mining: challenges, issues, and opportunities. In: Hong, B., Meng, X., Chen, L., Winiwarter, W., Song, W. (eds.) DASFAA 2013. LNCS, vol. 7827, pp. 1–15. Springer, Heidelberg (2013). doi:10.1007/978-3-642-40270-8_1
35. Davey, J., Mansmann, F., Kohlhammer, J., Keim, D.: Visual analytics: towards intelligent interactive internet and security solutions. In: Álvarez, F., Cleary, F., Daras, P., Domingue, J., Galis, A., Garcia, A., Gavras, A., Karnourskos, S., Krco, S., Li, M.-S., Lotz, V., Müller, H., Salvadori, E., Sassen, A.-M., Schaffers, H., Stiller, B., Tselentis, G., Turkama, P., Zahariadis, T. (eds.) FIA 2012. LNCS, vol. 7281, pp. 93–104. Springer, Heidelberg (2012). doi:10.1007/978-3-642-30241-1_9
36. Sweller, J.: Cognitive load during problem solving: effects on learning. Cogn. Sci. **12**(2), 257–285 (1988)
37. Bolognini, L., Bistolfi, C.: Pseudonymization and impacts of big (personalanonymous) data processing in the transition from the directive 9546EC to the new EU general data protection regulation. Comput. Law Secur. Rev. **33**, 171–181 (2017)
38. Hashem, I.A.T., et al.: The rise of "big data" on cloud computing: review and open research issues. Inf. Syst. **47**, 98–115 (2015)
39. Kumar, R.S., Saxena, A.: Data integrity proofs in cloud storage. In: 2011 Third International Conference on Communication Systems and Networks (COMSNETS 2011). IEEE (2011)
40. Khan, N., et al.: Big data: survey, technologies, opportunities, and challenges. Sci. World J. **34**, 518–522 (2014)
41. Fayyad, U., Piatetsky-Shapiro, G., Smyth, P.: The KDD process for extracting useful knowledge from volumes of data. Commun. ACM **39**(11), 27–34 (1996)
42. Tu, S., et al.: Processing analytical queries over encrypted data. Proc. VLDB Endowment **6**, 289–300 (2013). VLDB Endowment
43. International Organization for Standardization: "Pseudonymization" – new ISO specification supports privacy protection in health informatics (2009)

44. ISO/TS 25237: Health informatics – Pseudonymization (2008)
45. ISO/IEC 15944-8: Information technology – Business operational view – Part 8: Identification of privacy protection requirements as external constraints on business transactions (2012)
46. ISO/IEC N15297: Proposal for a new work item on Privacy enhancing data de-identification techniques, June 2015

Potentials of Digitization in the Tourism Industry – Empirical Results from German Experts

Ralf-Christian Härting[1(✉)], Christopher Reichstein[1], Nina Härtle[2], and Jürgen Stiefl[1]

[1] Business Administration, Aalen University of Applied Sciences, Aalen, Germany
{ralf.haerting, christopher.reichstein, juergen.stiefl}@hs-aalen.de
[2] Competence Center for Information Systems, Aalen University of Applied Sciences, Aalen, Germany
nina.haertle@kmu-aalen.de

Abstract. The paper deals with the topic digitization in the tourism industry. An empirical study was conducted in Germany based on a theoretical foundation. The aim of this study is to find out how far the digitization has already changed the tourism industry and what is still going to change in order to find potential benefits of digitization in the tourism industry. The results of the structural equation model approach show six main driver (sales increase, classic booking, sharing economy, personalized offers, social media and customer reviews) that have a significant impact on the potential of digitization in the tourism industry.

Keywords: Potentials · Digitization · Tourism industry · German experts · Quantitative study · Empirical results

1 Introduction

"The number, speed and adoption of technological innovations grow exponentially. [...] This influences almost all aspects of our lives: the way we communicate, the way we produce energy and the way we distribute. [...] The physical world becomes digitalized. [...] We are in the middle of the third-industrial revolution and we move into the time frame of the Internet of Things. We have become familiar with the internet of communication [1]."

Digitization offers many new opportunities that can be exploited by providers in the tourism industry. At the same time, competition is being intensified and companies have to keep pace with digitization in order to remain on the same level. Without any question, "digitization can be viewed as the motor of transformation for the tourism industry in the age of the internet economy [2]."

In fact, more than 57% of all travel reservations each year are carried out with the internet and internet travel booking revenue has grown by more than 73% over the past

© Springer International Publishing AG 2017
W. Abramowicz (Ed.): BIS 2017, LNBIP 288, pp. 165–178, 2017.
DOI: 10.1007/978-3-319-59336-4_12

5 years, while 65% of tourist book hotels reservations for the same day are made on a mobile device [3]. In the course of this development, travel providers need to tackle the challenges of individually addressing different stakeholders from country to region through knowledge, time, and a sophisticated strategy. Professional digitization strategies of travel providers such as classic travel agencies require resources such as specialists with IT expertise and a certain budget. In return, the introduction and implementation of a digitization project in the tourism sector might lead to considerable cost savings and productivity gains. Significant examples of this are the automation of work-flows, the increase in employee motivation as well as cost and time savings [4]. In connection with the described increase of importance of digitization, it is still unclear, which potential benefits and which possible barriers evolve for providers in the tourism industry with respect to this development [5].

The paper is structured as follows: In Sect. 2, the terms of digitization and tourism are defined, as it will be understood in this study. Section 3 deals with background information and the research design of our empirical study. Thereafter, the authors are going to describe research methods in Sect. 4 in order to understand among others how data were collected. The study results are presented in Sect. 5 followed by a conclusion of the paper in Sect. 6.

2 Digitization in the Tourism Industry

The term of digitization is used in different interpretations: on the one hand, digitization is the "conversion of analog information in any form to digital form with suitable electronic devices so that the information can be processed, stored, and transmitted through digital circuits, equipment, and networks [6]." On the other hand, digitization is the process, which is caused by the adoption of digital technologies and the application systems that build upon them. Digitization can be defined according to different levels of intensity: from the pure presentation and information (website), the sales channel function (e-commerce), business process integration (E-Business) to new business models with virtual products or services [2].

A common interpretation of digitization stands for intelligent business processes and the using of efficient and new technology concepts, such as Big Data, Cloud and Mobile Computing, Internet of Things or Social Software [7]. Most of the industrial sectors had undergone dramatic changes in earlier times, which also led to profound social changes. In the first industrial revolution, human labor power was replaced by water and later steam power [7]. The concepts for the organization of the mass production as developed by Taylor [8] or Ford's highly advanced assembly line solved the artisan-oriented production in many areas. The advent of electronic control systems from the 1970s also led to serious changes in industry. The term "Digitization" [9] identifies the phase of the upheaval which is just beginning, in which for the first time intelligent actions of men are replaced by machine actions.

The "intelligent" output of new digital approaches is embedded in networks of intelligent products, commonly referred to as Internet of Things [9]. Especially the ability to collect data and communicate is the basis for a number of new concepts. Companies can learn more about the use of their services in a much more detailed manner than before, and can then improve their products. This information is supplemented by information from social networks. On this basis it is possible to develop new functions, processes and business models using extended analytical possibilities created by digitization technologies such as Big Data [10, 11].

Tourism includes national and international travel, the temporary stay (location) of strangers on a destination as well as the organization of the travel preparation and travel preparation at the hometown. Tourism is the totality of all phenomena and relationships resulting from travel and stay, where the place of residence is neither residential nor working place [12]. In this context, digitization offers promising potential in the tourism industry regarding both the supplier and customer perspective. Therefrom, all business processes before, during and after a journey are affected. These processes include the application and preparation of travel offers, the digital implementation, post-processing or customer recovery [13, 14]. In addition to the digital transformation of processes, the digitization offers opportunities for new business models in the tourism industry [15]. Some examples for these business models are virtual journeys, sharing services for accommodation and transportation or data driven product configuration.

In the following, digitization in the tourism sector will be investigated. Firstly, in terms of its impact on the individual phases of the transaction cost model. Secondly, in terms of its potentials to optimize the approach, agreement, implementation and control of all different phases within the digitization process.

3   Background and Research Design

In order to examine the potential of digitization in the tourism industry, a structural equation model (SEM) has been developed based on the current literature [16], as shown in the following graph (Fig. 1).

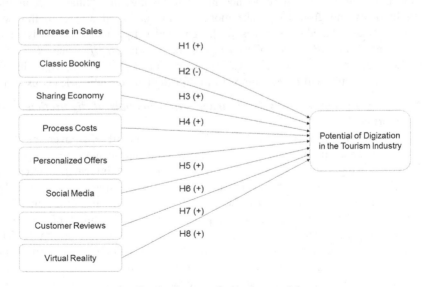

Fig. 1. Structural equation model

With regard to the research design above, eight determinants were found. Through digitization, many processes in the company become more effective and thus more cost-efficient. This results in a large potential sales volume because "the use of the internet technology enables the transition and distribution of information to be quicker, better, and cheaper regardless of geographical and time limitation [17]."

Digitization offers opportunities for a wide range of processes such as brand building, customer acquisition and retention, product development as well as quality assurance [18]. Especially for companies in the service sector like in the tourism industry, new technologies of digitization offer a great potential to achieve higher sales figures which leads us to H1:

H1: An increase in sales positively influences the potentials of digitization in the tourism industry

In the future, the entire organization will take place digitally in the tourism industry. This will affect the entire process, i.e. before the trip, during and after the trip. Digitization offers new opportunities and information to both providers and customers [19]. Also, "the importance of information technology [...] in tourism has increased tremendously over the past years" [17]. Therefore, "new technologies, rapid changes in the business environment, industry markets and consumer needs continue to challenge tourism destination organizations in fundamental ways [20] ". Travel bookings in a classical sense within a travel agency do not take advantage of the overall potentials of digitization which leads us to the following hypothesis:

H2: Classical booking negatively influences the potentials of digitization in the tourism industry

There are many opportunities for the tourism industry provided by digital markets [21]. Thereby, new business models were created that represent a great competition for classic suppliers (i.e. travel agencies). Particular business models arise from the idea of a Sharing Economy which can be defined in the following way: "Sharing economies allow individuals and groups to make money from underused assets. In this way, physical assets are shared as services [22]." Examples for business models of the Sharing Economy are *AirBnb* and *Couchsurfing* for hospitality, *Uber* and *Sidecar* for transportation or *Spotify* and *SoundCloud* for entertainment [22]. The new possibilities of a Sharing Economy will be encouraged through digitization which leads us to H3:

H3: A Sharing Economy positively influences the potentials of digitization in the tourism industry

In addition, we assume that digitization offers substantial benefits over traditional means of communication in order to optimize processes and to increase competitiveness from following points of view [23]: reducing costs of information exchange; increasing speed of information transfer and retrieval; increasing customer involvement; controlling of transactions; increasing flexibility using the marketing mix.

"The combination of enhancements in processing and flexibility of processing capability allows organizations to use their resources more wisely and profitably [24]." "The economic impacts stem from lower marketplace transaction costs and production that is more efficient [25]." Therefore, we assume that one of main potentials of digitization in the tourism industry is to reduce transaction, and therefore process costs:

H4: A reduction in process costs positively influences the potentials of digitization in the tourism industry

Regarding hypothesis five, researcher found out that the individualization of products and services for customers plays an increasingly important role nowadays [13]. Digitization has a positive effect on individualization because it can be used for targeted advertising and offerings [13]. "In near future, it is expected that tourism suppliers will easily be able to contact travelers […] and personalized offers through […] wearable devices (e.g. 3-D glasses) [26] ". As the demand of personalized services extremely grows [27], we hypothesize:

H5: Personalized offers positively influence the potentials of digitization in the tourism industry

Hypothesis six regards the importance of social media for advertising. Consumers stated to trust in social media and word-of-mouth, which includes recommendations from friends and family members, more than any other form of advertising [28]. Social media marketing is a relatively new form of advertising and highly accepted by consumers of the tourism industry. The impact of different social media channels such as Facebook, YouTube, Google and Twitter on the tourism industry is document by recent studies [29]. Regarding the German market, the social media channels Facebook, Instagram, Twitter, Google+, LinkedIn and Xing are having the most users [30] in this context. As a result, it is increasingly important for companies in tourism to use social media in order to communicate with their customers. Moreover, companies can use Big Data from social media to analyze customers. Therefore, we assume a high impact of social on digitization which leads us to:

H6: Social Media positively influence the potentials of digitization in the tourism industry

Past studies have shown that successful marketing begins with understanding how consumers think [14]. Therefore, customer reviews are good for advertising products or services because personalized advertising is presented from the customer's perspective. Currently, there is a supply surplus in all markets as well as in the tourism industry. Customer reviews are becoming increasingly important because they "are a valuable information source that can affect customers' pre-purchase evaluations and decisions [31]." Nothing is more helpful for generating offers and selling products and services than actual customer reviews and ratings of products [32]. The rapid expansion of E-Commerce means that more and more products are being sold online. In order to increase customer satisfaction, it has become a common practice for online merchants to allow their customers to express opinions about the products they have purchased [33]. On rating portals and on social media channels, travelers can share their travel experiences with others [34]. This increases both the transparency of the provider side and, on the other hand, it helps the provider to generate new customers. A wealth of information is needed within the tourism industry to understand customer needs [35]. Hence, is important to understand changes in technologies as well as consumer behavior that impact the distribution and accessibility of information for traveler [36]. As a result, social media can improve the process of generating offers, raise the transparency on the provider side and facilitating the acquisition of new customers bearing customer reviews in mind that leads us to H7:

H7: Customer reviews positively influence the potentials of digitization in the tourism industry

Virtual Reality has great potential for the tourism industry as it can be used in the tourism industry to present an illusion for the customer [37]. "The 3D virtual world provides opportunities for destination marketing organizations to communicate with targeted markets by offering a rich environment for potential visitors to explore tourism destinations [38]." Travel provider "could use immersive virtual reality technology to integrate sensory experience into their communication strategies, utilizing experience-based internet marketing to support the tourist's information search and decision-making process [38] ". Virtual Reality offers great possibilities to the tourism industry, so we hypothesize:

H8: The application of Virtual Reality promoting offers positively influences the potentials of digitization in the tourism industry

The literature review is summarized in the following table (Table 1). We selected 25 journal articles of which we checked the goodness by using the internationally accepted ranking relevant to business research [39], the SCImago Journal & Country Rank as well as Core Conference Rank.

Table 1. Summary of literature review

Determinant	Author
Sales increase	[17, 18]
Classic booking	[17, 19, 20]
Sharing economy	[21, 22]
Process costs	[23–25]
Personalized offers	[13, 26, 27]
Social media	[28–30]
Customer reviews	[14, 31–36]
Virtual reality	[37, 38]

In order to examine the impact of the specified determinants on the potential of digitization in the tourism industry, the online responses of the participants on a Likert scale [40] ranged from very high to very low (1: very high, 2: high, 3: neutral, 4: low, 5: very low). Besides, hypothesis two was supported by the question whether the firms of our interviewed experts offer digital support within business processes in their company (Fig. 3). The question of whether companies generate personalized offers (Fig. 4) has been used to reaffirm hypothesis five. The response ranged from less than one year to more than 5 years. All questions were conceived equivalent to general quantitative study guidelines [41].

4   Research Methods and Data Collection

Corresponding to our research model, we examine our hypotheses using a quantitative research approach using a web-based online survey [42] in Germany. In addition to qualitative research methods such as interviews, quantitative research methods such as questionnaires are used to assess the adaptation of a theoretical model to empirical data [42]. The survey was carried out in Germany via the open source software LimeSurvey [43]. Firstly, a preliminary study of the study was conducted in April 2016 to ensure a high quality of research standards. Based on the results the questionnaire has been improved. The main study started in May 2016 and ended in November 2016. We collected a sample of n = 301 responses from tourism experts.

After data cleaning, we received a final sample of n = 118 for tourism experts in Germany. 57% of the surveyed tourism experts are active in a company with less than 50 employees. In a medium-sized company with a number of employees between 51–250 employees, 28% of the interviewees work. 15% of the tourism experts are working in a company which employs more than 250 people. The importance of Virtual Reality and its role in the tourism industry in future is shown in Fig. 2:

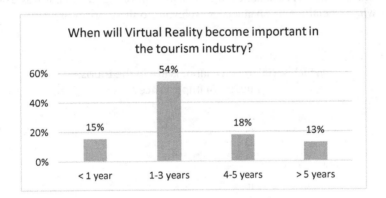

Fig. 2. Importance of virtual reality

54% of the surveyed tourism experts, voted for in to 1–3 years. 18% of the tourism experts are of the mind that Virtual Reality will play a role in 4–5 years. 15% of the tourism experts think in less than a year, Virtual Reality will play a big role. 13% will not see any impact of Virtual Reality on the tourism industry over the next five years. The following figure shows the percentage of companies providing digital support.

The study investigated whether the company of tourism experts offer their customers digital support in business processes (Fig. 3). This example involves payments such as *PayPal, NFC* or the boarding pass on the Smartphone. 63% of the companies in which the tourism experts are involved, provide digital support. Still, 37% against, have not taken note of these digital processes. A connection could be existing with the size of the company. Smaller companies do not have the possibilities to restructure processes so quickly due to lack of budget and expertise. In addition, it was examined whether the companies are creating personalized offers (Fig. 4) on the basis of personal

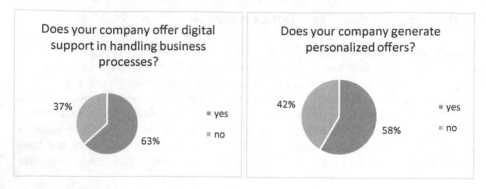

Fig. 3. Digital support   **Fig. 4.** Personalized offers

data, data of third parties, demographic data, statistical profiles or company-related data. 58% of the companies offer personalized offers. 37% of the companies in which the tourism experts work offer only standardized offers. Again, there might be a context with the company size. As social media is a major part of digitization, it was examined in Fig. 5 which social media channels are relevant to the tourism industry.

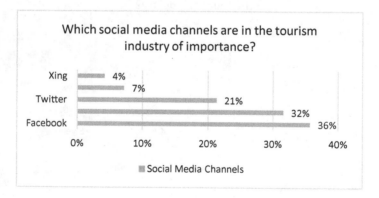

Fig. 5. Social media channels

36% of the questioned tourism experts answered, that Facebook is the most important. YouTube followed closely with 32%. Twitter received 21% of the votes. Google+ and Xing were found to be unimportant with 7% and 4% of the votes. Since Facebook is by far the largest platform, it is not surprising that Facebook is called the main platform. Because YouTube as a video platform can display content from the tourism industry very well, this platform also plays a role. Other social media channels that were named were: Instagram, WhatsApp, Pinterest, Periscope, Snapchat, Tumblr, LinkedIn, Flickr and TripAdvisor.

A structural equation modelling approach (SEM) was used to analyze the theoretical causal model with empirical data. In general, the relationship between different variables can be visualized using SEM [44]. SEM can be defined as a second generation of multivariate analysis to get a deeper insight in the analysis of the different (data) relations (in difference to e.g. cluster analysis or linear regression) [45]. A measurement and a structural equation model are two parts of the SEM [46]. The measurement model validates the latent variables and the structural equation model analyze the relationships between the research model and latent variables [46]. There are a number of different SEM approaches like AMOS, LISREL or Smart PLS [44]. We used Smart PLS 3.2 to analyze the data due to its robustness and data requirements [47, 48]. Furthermore, Smart PLS is frequently used in the information systems community [49]. In opposite to AMOS or LISREL the structural path significance is calculated via Bootstrapping [44]. Furthermore, single item sets are also allowed and often used in information systems research [50]. In case of single item sets, there is no need to calculate metrics like Cronbach's Alpha or AVE [50].

5   Results

After analyzing the empirical sample with a structural equation modelling approach, we received the following results (Fig. 6):

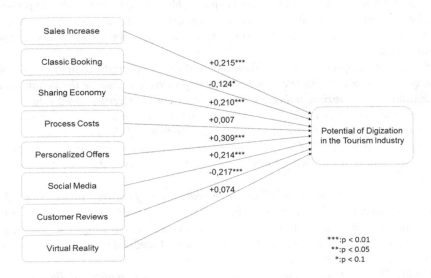

Fig. 6. Structural equation model with coefficients

Five out of eight path coefficient values are highly significant (p < 0.01). One path coefficient value with p < 0.1 is significant on a ten percent level. The coefficient of determination (R Square) is in a satisfying range (0.344 > 0.19) according to Chin [47].

The first hypothesis (*An increase in sales positively influences the potentials of digitization in the tourism industry*) examines the influence of sales on the potentials of digitization in the tourism industry. Based on our investigation, we can confirm H1 due to a positive value (+0.215) of the path coefficient on a high significance level (p < 0.01) of the structural equation model (Fig. 6). Hence, the potentials of digiti-zation are positively affected by increasing sales.

The second hypothesis (*Classical booking negatively influences the potentials of digitization in the tourism industry*) examines the impact of classical bookings (i.e. travel agencies) on the potentials of digitization in the tourism industry. H2 can be confirmed as well because of the negative path coefficient value (−0.124) on a ten percent significance level (p < 0.1). As a result, classical bookings do have a negative influence on the potentials of digitization in the tourism industry underlying the strong influence and possibilities of new applications of digitization in tourism markets.

The increase in competition between classical and new approaches by digitization (e.g. online booking websites) is examined using H3 (*A Sharing Economy positively influences the potentials of digitization in the tourism industry*) asking the respon-dents about the importance of business models of the Sharing Economy through dig-itization. The results show that business models of the Sharing Economy do have a positive (highly significant) impact (+0.210) on the potentials of digitization in the tourism industry. Therefore, hypothesis three can be confirmed, too. New business models such as those of the Sharing Economy will strengthen the competition between classic providers and new providers as it is one main driver of digitization benefits.

H4 (*A reduction in process costs positively influences the potentials of digitization in the tourism industry*) is not significant (p = 0.941 > 0.05). Hence, this hypothesis must be rejected. The possibility to reduce costs does not have a positive impact on potentials of digitization in the tourism industry with respect to different processes within companies.

The positive impact of the individualization of offers on digitization in the tourism industry is examined by H5 (*Personalized offers positively influence the potentials of digitization in the tourism industry*). The results show that personalized offers has a positive impact (+0.309) on the potentials of digitization in the tourism and emphasizes the importance of personalized offers through digitization. H5 shows a p-value of 0.000 at a significance level of 0.01. As a consequence, H5 can be confirmed. The possi-bilities for individual offerings obviously feature great potential for the tourism industry.

H6 (*Social Media positively influences the potentials of digitization in the tourism industry*) examines the influence of social media on the potential of digitization in the tourism industry. We can confirm H6 based on our results due to a high path coefficient (+0.214) on a high significance level. Therefore, social media do have an influence on the potentials of digitization in the tourism industry.

Finally, H7 (*Customer reviews positively influence the potentials of digitization in the tourism industry*) examines the impact of customer reviews on the process of generating offers, the transparency on the provider side and the acquisition of new customers. Analyzing the results of the structure equation model, H7 must be rejected due to a negative path coefficient value (−0.217). The result shows a highly significant p-value (0.007) wherefore it could be assumed that customer reviews are not a main driver of digitization. Instead, classical bookings within travel agencies and classical word-of-mouth marketing might have several advantages in this case. For instance, people might place one's trust in employees of travel agencies rather than reading manipulated information online about travelling. A further reason might be the fact that customers weight negative reviews more heavily than positive reviews [51].

H8 *(The application of Virtual Reality promoting offers positively influences the potentials of digitization in the tourism industry)* is not significant (p = 0.479 > 0.05) Therefore, this hypothesis must be rejected meaning that Virtual Reality currently plays no important role in the tourism industry.

Important values of the SEM are shown in Table 2. For single-item sets there is no need to calculate metrics like Cronbach's alpha [50]. Furthermore, single item sets are often used in research [50]. The potential of digitization in the tourism industry are shortened as PDT.

Table 2. SEM coefficients

Hypothesis	SEM-Path	Path coefficient	Significance (P values)
H1	Sales increase → PDT	+0.215	0.008
H2	Classic booking → PDT	−0.124	0.069
H3	Sharing economy → PDT	+0.210	0.010
H4	Process costs → PDT	+0.007	0.941
H5	Personalized offers → PDT	+0.309	0.000
H6	Social media → PDT	+0.214	0.010
H7	Customer reviews → PDT	−0.217	0.007
H8	Virtual Reality → PDT	+0.074	0.479

6   Conclusion

Through a structural equation model, the potential of digitization was examined in the tourism industry. The study of the potential of digitization in tourism reveals six hypotheses that influence the potential of digitization in the tourism industry. Influencing factors such as "sales increase", "sharing economy", "personalized offers", and "social media" do have a positive impact on the potential of digitization in the tourism industry. However, "classical booking" and "customer reviews" do have a significant negative impact on the potential of digitization in the tourism industry.

There are many potential sales figures to consider for the tourism sector and it seems obvious that there will be a concentration of providers in the tourism industry through digitization. As a consequence, the competition between classical bookings in travel agencies and individual offers through digitization might increase in future. Especially, individual offerings like the possibility to promote journeys in social media and to connect with customers are main drivers that do have a positive effect on the potentials of digitization. In this context, customer reviews improve the process of generating offers, increase the transparency on the provider side and facilitate the acquisition of new customers.

The two non-significant hypotheses regarding the factors "process costs" and "virtual reality" had to be rejected within this study. Also, we had to reject the highly significant factor "customer reviews" due to a negative impact on the potentials of digitization in the tourism industry. However, these hypotheses are worth considering, since the survey included only tourism experts from Germany and was limited in terms of location, sample size as well as time period.

This study can serve as a basis for further studies in order to amplify the applications and potentials of digitization technologies (Big Data, Cloud and Mobile Computing etc.) in the tourism industry. There are also additional aspects that can be found in future research considering the number of countries, the duration of the survey and the sample size, for instance. In addition, a qualitative research approach might provide more detailed insights. New developments in the field of digitization offer the possibility to improve business in the tourism industry. Within the increasingly competitive environment of the tourism sector, the use of new techniques can increase profitability [52]. In the foreseeable future, digitization will be an even more important tool for the success of companies in the tourism industry. Obviously, it is expected that digitization will become more important for business as it has not yet reached its full potential in the tourism sector.

References

1. Oskan, L., Boawijk, A.: Airbnb the future of networked hospitality business. J. Tourism Futures **2**(1), 22–42 (2015)
2. Bauer, L., Boksberger, P., Herget, J.: The virtual dimension in tourism: criteria catalogue for the assessment of eTourism applications. In: O'Connor, P., Höpken, W., Gretzel, U. (eds.) Information and Communication technologies in tourism: proceedings of the international conference in Innsbruck 2008, pp. 522–532. Springer, Wien (2008)
3. Rezdy (2017). https://www.rezdy.com/resource/travel-statistics-for-tour-operators/
4. Schubert, P., Williams, S.P.: Realising benefits from current ERP and CRM systems implementations: an empirical study. In: Proceedings of the 23rd Bled eConference (eTrust: Implications for the Individual, Enterprises and Society), pp. 476–480 (2010)
5. Phillips, P., Louvieris, P.: Performance measurement systems in tourism, hospitality, and leisure small medium-sized enterprises: a balanced scorecard perspective. J. Travel Res. **44**(2), 201–211 (2005)
6. Businessdictionary (2017). http://www.businessdictionary.com/definition/digitization.html
7. Härting, R., Schmidt, R., Möhring, M.: Nutzenpotenziale von Industrie 4.0 und Digitalisierung. In: Härting, R. (ed.) Industrie 4.0 und Digitalisierung – Innovative Geschäftsmodelle wagen! Tagungsband, 8. Transfertag, Aalen 2016, pp. 19–32. BOD, Norderstedt (2016)
8. Taylor, F.W.: The Principles of Scientific Management. Harper and Brothers, New York (1911)
9. Schmidt, R., Möhring, M., Härting, R.-C., Reichstein, C., Neumaier, P., Jozinović, P.: Industry 4.0 - potentials for creating smart products: empirical research results. In: Abramowicz, W. (ed.) BIS 2015. LNBIP, vol. 208, pp. 16–27. Springer, Cham (2015). doi:10.1007/978-3-319-19027-3_2
10. Breuer, P., Forina, L., Moulton, J. (2017). http://cmsoforum.mckin-sey.com/article/beyond-the-hype-capturing-value-from-big-data-and-advanced-analytics
11. Schmidt, R., Möhring, M., Maier, S., Pietsch, J., Härting, R.: Big data as strategic enabler – insights from central european enterprises. In: Abramowicz, W., Kokkinaki, A. (eds.) BIS 2014. LNBIP, vol. 176, pp. 50–60. Springer, Cham (2014)

12. Telfer, D.J., Sharpley, R.: Tourism and Development in the Developing World. Routledge, London (2016)
13. Ralph, D., Searby, S.: Location and Personalisation Delivering Online and Mobility Services. Institution of Engineering and Technology, London (2004)
14. Ighalo, M.: The changing concept of advertising and promotional technology. J. Commun. **5** (1), 63–67 (2014)
15. Souto, J.E.: Business model innovation and business concept innovation as the context of incremental innovation and radical innovation. Tour. Manage. **51**, 142–155 (2015)
16. Cooper, D.R., Schindler, P.S., Sun, J.: Business Research Methods. McGraw-Hill, New York (2006)
17. Gretzel, U., Yuan, Y.-L., Fesenmaier, D.R.: Preparing for the new economy: advertising strategies and change in destination marketing organizations. J. Travel Res. **39**, 146–156 (2000)
18. Dellarocas, C.: The digitization of word of mouth: promise and challenges of online feedback mechanisms. Manage. Sci. **49**(10), 1407–1424 (2003)
19. Tiefenbeck, V., Goette, L., Degen, K.: Overcoming Salience Bias: How Real-Time Feedback Fosters Resource Conservation Management Science (Articles in Advance), pp. 1–19 (2016)
20. Mistliis, N., Buhalis, D., Gretzel, U.: Future eDestination marketing: perspective of an australian tourism stakeholder network. J. Travel Res. **53**(6), 778–785 (2014)
21. Azevedo, E.M., Weyl, E.G.: Matching markets in the digital age. Science **352**(6289), 1056–1057 (2016)
22. PWC: The Sharing Economy. PricewaterhouseCoopers LL (2015)
23. Liu, Z. http://www.hotel-online.com/Trends/ChiangMaiJun00/InternetConstraints.html
24. Buhalis, D., O´Connor, P.: Information communication technology revolutionizing tourism. Tour. Recreation Res. **30**(2), 7–16 (2005)
25. Sundararajan, A.: Peer-to-Peer businesses and the sharing (collaborative) economy: overview, economic effects and regulatory issues. NYU Center for Urban Science and Progress (2014)
26. Harasymowicz, P.M.: Outernet technologies in tourism: a conceptual framework and applications for the travel industry. In: Egger, R., Maurer, C. (eds.) Iscontour 2015 - Tourism Research Perspectives: Proceedings of the International Student Conference in Tourism Research, pp. 97–110. BOD, Norderstedt (2015)
27. Buhalis, D., Amaranggana, A.: Smart tourism destinations. In: Xiang, Z., Tussyadiah, I. (eds.) Information and Communication Technologies in Tourism 2014, pp. 553–564. Springer, Cham (2013)
28. Bennet, S. (2017). http://www.adweek.com/socialtimes/social-media-travel-hospitality/466163
29. Hudson, S., Thal, K.: The impact of social media on the consumer decision process: implications for tourism marketing. J. Travel Tour. Mark. **30**(1–2), 156–160 (2013)
30. Kroll, L. (2017). http://socialmedia-institute.com/uebersicht-aktueller-social-media-nutzerzahlen/
31. Book, L.A., Tanford, S., Chen, Y.S.: Understanding the impact of negative and positive traveler reviews: social influence and price anchoring effects. J. Travel Res. **55**(8), 993–1007 (2016)
32. Singh, S., Diamond, S.: Social Media Marketing for Dummies. Wiley, Hoboken (2012)
33. Hu, M., Liu, B.: Mining and summarizing customer reviews. In: Proceedings of the tenth ACM SIGKDD International Conference on Knowledge Discovery and Data Mining, pp. 168–177. ACM, New York (2004)

34. García-Pablos, A., Cuadros, M., Linaza, M.T.: OpeNER: open tools to perform natural language processing on accommodation reviews. In: Tussyadiah, I., Inversini, A. (eds.) Information and Communication Technologies in Tourism 2015, pp. 125–137. Springer, Cham (2015)
35. Spiller, S.A., Belogolova, L.: On consumer beliefs about quality and taste. J. Consum. Res. (2016) (in press)
36. Xiang, Z., Gretzel, U.: Role of social media in online travel information search. Tour. Manage. **31**(2), 179–188 (2010)
37. Giridharan, N. (2017). http://www.sciencemag.org/news/2013/07/virtual-reality-allows-adults-see-world-through-childs-eyes
38. Huang, Y.C., Backman, K.F., Backman, S.J., Chang, L.L.: Exploring the implications of virtual reality technology in tourism marketing: an integrated research framework. Int. J. Tour. Res. **18**(2), 116–128 (2016)
39. Hennig-Thurau, T., Walsh, G., Schrader, U.: VHB-JOURQUAL: Ein Ranking von betriebswirtschaftlich-relevanten Zeitschriften auf der Grundlage von Expertenurteilen. Zeitschrift für betriebswirtschaftliche Forschung **56**(9), 520–545 (2004)
40. Kothari, C.R.: Research Methodology: Methods and Techniques. New Age International, New Delhi (2004)
41. Hewson, C.: Internet research methods: a practical guide for the social and behavioural sciences. Sage, London (2003)
42. Rea, L.M., Parker, R.A.: Designing and Conducting Survey Research: A Comprehensive Guide. Wiley, San Francisco (2014)
43. Projectteam, T.L. (2017). https://www.limesurvey.org/de/
44. Wong, K.K.K.: Partial least squares structural equation modeling (PLS-SEM) techniques using SmartPLS. Mark. Bull. **24**(1), 1–32 (2013)
45. Fornell, C.: Larcker, D.F: Evaluating structural equation models with unobservable variables and measurement error. J. Mark. Res. **18**, 39–50 (1981)
46. Markus, K.A.: Principles and practice of structural equation modeling by Rex B. Kline. Struct. Eqn. Model.: Multidisc. J. **19**(3), 509–512 (2012)
47. Chin, W.W.: The partial least squares approach to structural equation modeling. Mod. Methods Bus. Res. **295**(2), 295–336 (1998)
48. Ringle, C., Wende, S., Will, A. (2017). www.smartpls.de
49. Hubona, G.S.: Structural equation modeling (SEM) using SmartPLS software: analyzing path models using partial least squares (PLS) based SEM. In: AMCIS 2009 Proceedings 80, (2009)
50. Ringle, C.M., Sarstedt, M., Straub, D.: A critical look at the use of PLS-SEM in MIS quarterly. MIS Q. **36**(1), iii–xiv (2012)
51. Ho-Dac, N.N., Carson, S.J., Moore, W.L.: The effects of positive and negative online customer reviews: do brand strength and category maturity matter? J. Mark. **77**(6), 37–53 (2013)
52. Werthner, H., Alzua-Sorzabal, A., Dickinger, A., Gretzel, U., Jannach, D., Neidhardt, J., Pröll, B., Ricci, F., Scaglione, M., Stangl, B., Stock, O., Zanker, M.: Future research issues in IT and tourism. Inf. Technol. Tour. **15**, 1–15 (2015)

Process Management

How to Make Process Model Matching Work Better? An Analysis of Current Similarity Measures

Fakhra Jabeen[1], Henrik Leopold[1(✉)], and Hajo A. Reijers[1,2]

[1] VU University Amsterdam,
De Boelelaan 1081, 1081 HV Amsterdam, The Netherlands
{f.jabeen,h.leopold,h.a.reijers}@vu.nl
[2] Eindhoven University of Technology,
PO Box 513, 5600 MB Eindhoven, The Netherlands

Abstract. Process model matching techniques aim at automatically identifying activity correspondences between two process models that represent the same or similar behavior. By doing so, they provide essential input for many advanced process model analysis techniques such as process model search. Despite their importance, the performance of process model matching techniques is not yet convincing and several attempts to improve the performance have not been successful. This raises the question of whether it is really not possible to further improve the performance of process model matching techniques. In this paper, we aim to answer this question by conducting two consecutive analyses. First, we review existing process model matching techniques and give an overview of the specific technologies they use to identify similar activities. Second, we analyze the correspondences of the Process Model Matching Contest 2015 and reflect on the suitability of the identified technologies to identify the missing correspondences. As a result of these analyses, we present a list of three specific recommendations to improve the performance of process model matching techniques in the future.

Keywords: Process model matching · Performance improvement · Weakness analysis · Activity similarity

1 Introduction

Process model matching refers to the automatic identification of corresponding activities between two process models, i.e. activities that represent the same or similar behavior. By automatically producing such activity correspondences, process model matching techniques are a prerequisite for many advanced analysis techniques. Among others, the identification of activity correspondences is required for the harmonization of process model variants [1,2], process model search [3,4], and the detection of process model clones [5,6]. Recognizing the importance of matching for the automated analysis of process models in general,

© Springer International Publishing AG 2017
W. Abramowicz (Ed.): BIS 2017, LNBIP 288, pp. 181–193, 2017.
DOI: 10.1007/978-3-319-59336-4_13

researchers have defined a plethora of process model matching techniques (see e.g. [7–10]).

Despite the considerable attention that has been devoted to the problem of process model matching, the performance of existing matching techniques is not yet convincing. The results from the Process Model Matching Contests in 2013 and 2015 show that, depending on the data set, the best F-measures range between 0.45 and 0.67 [11,12]. While the need for performance improvements is widely recognized, attempts to further improve the results of process model matching, for example through performance prediction, were not very fruitful [13]. This raises the question of how the performance of process model matching techniques can be further improved.

In this paper, we aim to answer this question by (a) systemically analyzing the technological state of the art and (b) by analyzing the missing capabilities of existing process model matching techniques. To this end, we first conduct a structured literature review on process model matching. We provide an overview of existing techniques and the specific technologies they use to identify similar activities. Then, we analyze the characteristics of the correspondences of the Process Model Matching Contest 2015 that the participating matching techniques failed to identify. In this way, we aim to develop an understanding to what extent current matching performance can be explained by a focus on a limited set of technologies and which directions might be promising to improve process model matching performance in the future.

The remainder of the paper is organized as follows. Section 2 introduces the problem of process model matching using a running example. Section 3 discusses the methodological details and the results of our literature review. Section 4 presents the analysis of the correspondences of the Process Model Matching Contest 2015. Section 5 elaborates on opportunities for improving the performance of process model matching and gives three specific recommendations. Section 6 concludes the paper.

2   The Problem of Process Model Matching

Process model matching techniques aim at automatically identifying activity correspondences that represent similar behavior in both models. Figure 1 illustrates the matching problem by showing the recruitment processes from two different companies. The grey shades highlight the correspondences between the two processes. For example, the activity *"Evaluate"* from company B corresponds to the activities *"Check grades"* and *"Examine employment references"* from company A. The correspondences show that the two models differ with respect to the terms they use (e.g. *"eligibility assessment"* versus *"aptitude test"*) as well as their level of detail (e.g. *"Evaluate"* is described in more detail in the model from company A).

Given such differences, the proper recognition of the correspondences between two process models can become a complex and challenging task. The complexity of the matching task is also highlighted by the rather moderate performance of process model matching techniques. A recent comparative evaluation in the context of the Process Model Matching Contest (PMMC) 2015 showed that the F-measures lie between 0.45 and 0.67 for different data sets [12].

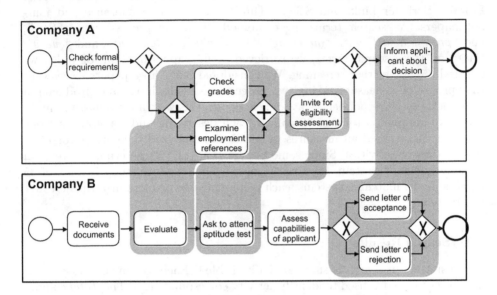

Fig. 1. Two business processes and their correspondences

Following Gal [14], we can subdivide the matching process into first line matching and second line matching. A first line matcher takes the sets of activities A_1 and A_2 from the process models as input and produces a similarity matrix $M(A_1, A_2)$ with $|A_1|$ rows and $|A_2|$ columns. Among others, such a similarity matrix can be obtained by comparing the activity labels. A second line matcher takes one or more similarity matrices produced by first line matchers as input and turns them into a binary similarity matrix $M(A_1, A_2)$ with entries of 0 or 1. The latter indicates a correspondence between two activities.

It is important to note that first line matching plays a particularly important role for the overall matching result. If a first line matcher computes a similarity value of zero for two activities, it is very unlikely that a second line matcher will include this particular activity pair in the final set of correspondences. In the next section, we therefore conduct a systematic literature review and analyze which technologies are employed for first line matching.

3 Review of Existing First Line Matching Measures

To gain insights into the state of the art of process model matching, we conducted a systematic literature review on the measures used in the context of first line matching. In Sect. 3.1, we describe our search strategy. In Sect. 3.2, we elaborate on the results of our review.

3.1   Search Strategy

We conducted a comprehensive *literature review* on process model matching. More specifically, we queried the ACM Digital Library, IEEEXplore Digital Library, Springer Link, and Science Direct for relevant conference and journal papers. As search terms we combined *"business process"*, *"workflow"*, and *"process model"* with *"matching"*, *"similarity"*, *"alignment"*, and *"query"*. Based on these search terms, we retrieved 5,862 papers in total, of which we selected 657 for further screening. We considered only those papers as relevant that proposed a measure taking two process models as input and producing a set of activity correspondences at some stage. In this way, after removing duplicates, we obtained a total of 30 papers, each describing and using at least one measure. We studied all measures in detail and analyzed their usage for identifying similar activities. Since some papers (i.e. [11,12]) described more than a single matching system, the total result of our literature study is a set of 35 process model matching systems, each employing one or more measures for first line matching.

3.2   Search Results

The result of our search is summarized in Table 1. Each row in the table lists a measure type that is used to identify activity correspondences. The *Total* column indicates the total number of matching systems using the respective measure type and the *Reference* column shows the papers discussing these matching systems. Overall, Table 1 shows that we identified a total of 10 measure types, which we categorized into *syntactic* and *semantic* measures.

Table 1. Measures used for first line matching

Measure type	Total	References
Syntactic		
Distance-based	21	5 in [11], 2 in [12], [15–28]
Jaccard/Dice	5	[3, 12, 21, 24, 29]
Cosine similarity	5	3 in [12], [7, 26]
Substring	4	3 in [12], [24]
Jensen-Shannon distance	1	[12, 30]
Semantic		
Synonym-based	16	3 in [12], [3, 11, 15, 16, 19, 21, 24–26, 29, 31, 32]
Lin	12	3 in [12], [8, 10, 11, 21, 22, 22, 27, 33, 34]
Hypernym-based	5	[11], 2 in [10, 12]
Wu & Palmer	3	2 in [11], [12]
Lesk	2	[27, 34]

Syntactic Measures. Syntactic measures relate to simple string comparisons and do not take the meaning or context of words into account. The most prominently employed syntactic measures are *distance-based* measures such as the Levenshtein distance. Given two labels l_1 and l_2, the Levenshtein distance counts the number of edit operations (i.e. insertions, deletions, and substitutions) that are required to transform l_1 into l_2. Another distance-based measure is the Jaro-Winkler distance, which works in a similar way, but produces a value between 0 and 1.

Besides distance-based measures, many matching systems rely on plain word comparisons. Very common measures include the *Jaccard* and the *Dice* coefficient, which both compute the similarity between two activity labels based on the number of shared and non-shared words. An alternative approach based on word comparisons is the *cosine similarity*. To compute the cosine similarity, activity labels are transformed into vectors, typically by weighing words with their frequency of occurrence. The cosine similarity is then given by the cosine of the angle between two activity vectors. An alternative way of taking the word distribution into account is the *Jensen-Shannon* distance, which is a method for measuring the similarity of two probability distributions. However, so far, it has only been employed by the approach from Weidlich et al. [30].

A common pre-processing step is the consideration of *substring* relationships between activities. For instance, Dadashina et al. consider two activities labels l_1 and l_2 to be similar if l_1 is a substring of l_2 (or vice versa) [12]. Such labels are then removed from further similarity considerations and simply receive a similarity score of 1.

Semantic Measures. Semantic measures aim at taking the meaning of words into account. A very common strategy to do so is the identification of *synonyms* using the lexical database WordNet [35]. Typically, matching systems check for synonyms as part of a prepossessing step and then apply other, often also syntactic, similarity measures [12]. The most prominent semantic measure is the *Lin similarity*. The Lin similarity is a method to compute the semantic relatedness of words based on their information content according to the WordNet taxonomy. To use the Lin similarity for measuring the similarity between two activities (which mostly contain more than a single word), approaches typically combine the Lin similarity with the bag-of-words model. The bag-of-words model transforms an activity into a multiset of words, ignoring grammar and word order. The Lin similarity can then be obtained by identifying the word pairs from the two bags with the highest Lin score and by computing their average. Other measures based on the WordNet dictionary include *Wu & Palmer* and *Lesk*. The former computes the similarity between two words by considering the path length between these words in the WordNet taxonomy. The latter compares the WordNet dictionary definitions of the two words. Some approaches also directly check for *hypernym* relationships (a hypernym is a more common word). For instance, Hake et al. [12] consider "*car*" and "*vehicle*" as identical words since "*vehicle*" is a hypernym of "*car*".

Discussion. The findings of our literature highlight two important points. First, our review shows that *syntactic measures play a predominant role*. This means they strongly rely on the use of comparable vocabulary among the considered process models. Even common synonyms, such as *"assess"* and *"evaluate"* or *"conduct"* and *"perform"*, cannot be detected by approaches relying on syntactic measures. Interestingly, 21 out of 35 systems even rely on the most basic syntactic measure: distance-based similarity. The disadvantage of edit-based distance measures is not only their inability to recognize synonymous terms, but also their tendency to consider unrelated words as similar. As an example, consider the unrelated words *"contract"* and *"contact"*. The Levenshtein distance between these words is only 1, indicating a high similarity between the terms. Second, our review shows that the *employed semantic measures are very basic* and exclusively based on the WordNet dictionary. This represents a considerable problem since any WordNet-based measure returns a similarity score of zero if a term is not part of the WordNet dictionary. While the WordNet dictionary is quite extensive, it does not cover complex compound words (e.g. *"problem report"* or *"budget plan"*), which we often find in process models from industry.

Overall, our analysis suggests that current first line matching measures might be not good enough for recognizing the complex notion of similarity between some activities. In the next section, we will empirically investigate whether the choice of syntactic and basic semantic measures can indeed explain the low matching performance.

4   Analysis of the Results of the PMMC 2015

Our literature review in the last section revealed that matching techniques predominantly rely on syntactical and Wordnet-based semantic measures for first line matching. In this section, we investigate to what extend this insight allows us to explain the moderate performance of the matching systems in the Process Model Matching Contest 2015 [12].

To this end, we computed the similarity scores for all 1037 correspondences from the PMMC 2015 gold standard using the most prominently used syntactic and semantic measures, i.e. the Levenshtein distance and the bag-of-words-based Lin similarity. Figure 2 summarizes the results of our computation using box plots. It clusters the results based on the three datasets from the contest (referred to as *"Admission"*, *"Birth"*, and *"Asset"*) and the number of matching systems that identified these correspondences. This means that the first column (0) from Fig. 2(a) shows the distribution of the Levenshtein similarity of the correspondences that have not been identified by any matching system (separately for each dataset from the contest). The second column (1) respectively shows the distribution of the Levenshtein similarity of the correspondences that have been identified by exactly one system etc. Since a total of 12 matching systems participated in the PMMC 2015, each graph has 13 columns.

Analyzing the Levenshtein similarity distributions from Fig. 2(a) shows that there is a clear relationship between the similarity score of the correspondences

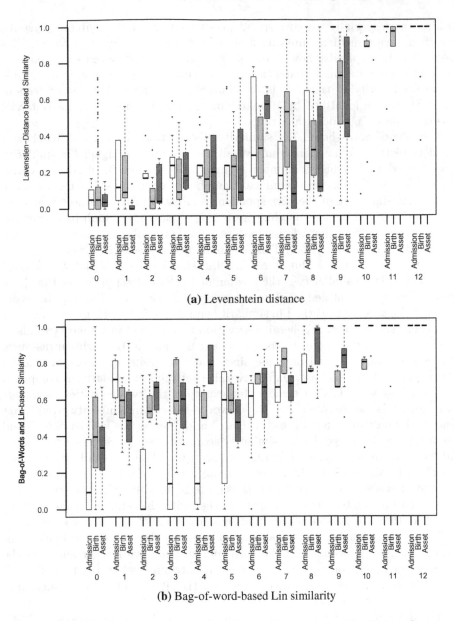

(a) Levenshtein distance

(b) Bag-of-word-based Lin similarity

Fig. 2. Relationship between the number of matching systems identifying the correspondences from the PMMC 2015 gold standard and the similarity score distribution of these correspondences

and the number of matching systems that successfully identified them. The higher the similarity score, the higher the number of matching systems identifying the correspondence. While this may not be completely surprising, it is striking that especially the correspondences that none of the matching systems identified have a median similarity score of below 0.1 among all datasets. This does not only emphasize how strongly existing matching systems rely on syntactic measures, but also that syntactic measures cannot represent a suitable option for identifying them. It is also interesting to note that the correspondences that were identified by 10 matching systems or more are mostly trivial correspondences, i.e. identical strings.

The bag-of-words-based Lin similarity distributions from Fig. 2(b) show a less clear picture. While there is an overall tendency that a higher Lin similarity is associated with a higher number of matching systems identifying a correspondence, we also observe a significant number of deviations. For instance, the median Lin similarity of the correspondences from the Birth dataset that have not been identified by any matching system is already quite high (0.4). For the correspondences from the Admission dataset, we even observe an up and down movement from column 1 to 4. This is, among others, caused by words that are not part of WordNet and, thus, yield in a Lin similarity of 0. A notable analogy to the Levenshtein similarity is the median of 1.0 for correspondences that have been successfully identified by 10 matching systems or more. This can be easily explained by the fact that the Lin similarity equals the Levenshtein similarity for trivial correspondences. A general observation is that the Lin similarity tends to give higher similarity scores to correspondences than the Levenshtein distance. This is quite an expected outcome since the Lin similarity also semantically relates words. The disadvantage, however, is that the Lin similarity is computed on a word by word level. Consider, for example, the two activities "*Evaluate plan*" and "*Assess contract*". A bag-of-words-based Lin similarity would first identify the best word pairs (i.e. "*evaluate*" and "*assess*" as well as "*plan*" and "*contract*") and average the Lin similarity scores of these pairs. Since "*evaluate*" and "*assess*" are synonyms, their Lin similarity is 1. The Lin similarity between "*plan*" and "*contract*" is 0.42, resulting in an average of 0.71. The resulting Lin similarity between these activities is thus quite high, although they are actually not likely to be related. This example together with the numbers from Fig. 2(b) highlights that the mere application of semantic technologies is not sufficient.

To better understand to what extend semantic technology currently contributes to the performance of the matching systems, we further analyzed the correctly identified correspondences by each matching system from the PMMC 2015. Figure 3 illustrates the ration between trivial (i.e. identical strings) and non-trivial correspondences.

The data from Fig. 3 shows that the majority of the correspondences identified by the matching systems from the PMMC 2015 are actual trivial correspondences. Most systems identify between 20% and 50% trivial correspondences. It is interesting to note that the systems with the highest share of non-trivial correspondences (e.g. OPBOT, AML, pPALM, NHCM) also have a particular good performance in the respective datasets. By contrast, the matching systems mainly identifying trivial correspondences (e.g. SMSL, Knoma, NLM, TripleS), also have a relatively bad performance. Looking into the specific techniques the

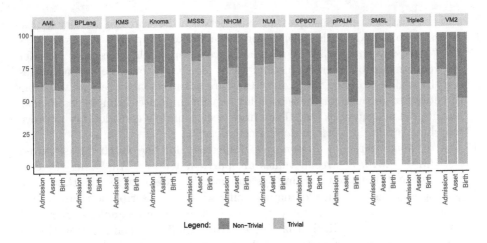

Fig. 3. Ratio between trivial and nontrivial correspondences that the matching systems of the PMMC 2015 correctly identified

more successful systems use, it is apparent that all of them at least partially build on semantic technology.

5 Recommendations for Improving Process Model Matching

Our literature review together with our empirical analysis of the correspondences from the PMMC 2015 revealed that the moderate performance of current matching systems can be well explained by their choices of first line matching measures. The employed measures are based on either syntactic or very basic semantic technology. As a result, a considerable number of complex correspondences cannot be successfully identified by existing matching systems. Based on our analysis, we derive the following recommendations for future work:

1. *Use syntactic technology for preprocessing only*: Syntactic technology is highly useful for recognizing trivial or almost trivial correspondences. We found that the best performing systems mainly use syntactic technology as a preprocessing step: They first match identical and almost identical labels and then apply semantic technology. The large-scale and sole application of distance-based measures, however, did not improve the results. They rather resulted in a high number of false positives, even with high cut-off values for, e.g. the Levenshtein distance.
2. *Apply of semantic technology beyond the word level*: Our analyses illustrated the importance of semantic technology for identifying non-trivial correspondences. However, it is also highlighted that semantic technology on the word level is not sufficient. Comparing activity labels by computing the semantic similarity between individual word pairs does not account for the cohesion

and the complex relationships that exist between the words of activity labels. A first step would be the proper consideration of compound nouns such as *"customer complaint"* [36]. A more advanced step would be to also consider the relationship between nouns and verbs [37]. Only because two activities use the same verb (e.g. *"evaluate"*) they are not necessarily related. Possible directions to account for these complex relationships are technologies such as distributional semantics [38]. They have been found to considerably improve matching results in other matching contexts [39].

3. *Use of domain-specific dictionaries*: Especially for the Birth dataset from the PMMC 2015, the recall values are particularly low. Analyzing the correspondences that the matching systems failed to identify, reveals that these correspondences often use domain-specific words or describe domain-specific procedures. Taking into account how current semantic technology is created and trained, it is not likely that there exists an off-the-shelf solution that is conducive for the identification of these correspondences. Hence, we recommend building on domain-specific dictionaries. They can be used for both inferring relationships between domain-specific words as well as for training statistical approaches, such as the previously mentioned distributional semantics methods. Existing methods for automatically extracting ontologies may represent a promising starting point here [40].

We believe that these three recommendations can appropriately address the weaknesses we identified in our analyses and hope that they provide valuable directions to further improve process model matching.

6   Conclusion

In this paper, we addressed the question of how to improve the performance of process model matching techniques. To this end, we conducted a literature review on existing process model matching systems and the specific technologies they use for identifying activity correspondences. Then, we analyzed the results from the Process Model Matching Contest 2015 in order to learn to what extend the employed technology for the identification of activity correspondences represents a reasonable choice.

Our literature review showed that all existing matching systems mainly rely on syntactic and simple, mostly WordNet-based semantic similarity measures. The analysis of the similarity values these basic measures produce for the correspondences from the Process Model Matching Contest 2015 further illustrated that these measures are not suitable for identifying the correspondences that could not been identified by any of the participating matching techniques. The main reason is that neither the employed syntactic nor the employed semantic similarity measures were able to detect the complex semantic relationships that exist between activity labels.

To provide a basis for improving the performance of process model matching techniques in the future, we derived three specific recommendations. They address the main weaknesses we identified in the context of our analyses. First, we recommend using syntactic technology for preprocessing only. We found that syntactic measures can be useful for filtering highly identical labels but, beyond that, are often responsible for noise. Second, we recommend applying semantic technology beyond the word level. We observed that especially compound nouns and verb-noun combinations require specific attention. Third, we recommend using domain-specific dictionaries. Our analysis revealed that many of the missing correspondences contain words that are unlikely to be covered by general purpose resources such as WordNet.

We hope that the insights and recommendations we provide in this paper can represent valuable directions for future research on process model matching. We plan to build on the insights of this paper by developing a new matching technique that combines distributional similarity technology with a domain ontology.

References

1. La Rosa, M., Dumas, M., Uba, R., Dijkman, R.: Business process model merging: an approach to business process consolidation. ACM Trans. Softw. Eng. Methodol. **22**(2), 11 (2013)
2. Weidlich, M., Mendling, J., Weske, M.: A foundational approach for managing process variability. In: Mouratidis, H., Rolland, C. (eds.) CAiSE 2011. LNCS, vol. 6741, pp. 267–282. Springer, Heidelberg (2011). doi:10.1007/978-3-642-21640-4_21
3. Jin, T., Wang, J., La Rosa, M., Ter Hofstede, A., Wen, L.: Efficient querying of large process model repositories. Comput. Ind. **64**(1), 41–49 (2013)
4. Awad, A., Polyvyanyy, A., Weske, M.: Semantic querying of business process models. In: 12th International IEEE Enterprise Distributed Object Computing Conference EDOC 2008, pp. 85–94. IEEE (2008)
5. Dumas, M., GarcíA-BañUelos, L., La Rosa, M., Uba, R.: Fast detection of exact clones in business process model repositories. Inf. Syst. **38**(4), 619–633 (2013)
6. Uba, R., Dumas, M., García-Bañuelos, L., La Rosa, M.: Clone detection in repositories of business process models. In: Rinderle-Ma, S., Toumani, F., Wolf, K. (eds.) BPM 2011. LNCS, vol. 6896, pp. 248–264. Springer, Heidelberg (2011). doi:10.1007/978-3-642-23059-2_20
7. Weidlich, M., Dijkman, R., Mendling, J.: The ICoP framework: identification of correspondences between process models. In: Pernici, B. (ed.) CAiSE 2010. LNCS, vol. 6051, pp. 483–498. Springer, Heidelberg (2010). doi:10.1007/978-3-642-13094-6_37
8. Leopold, H., Niepert, M., Weidlich, M., Mendling, J., Dijkman, R., Stuckenschmidt, H.: Probabilistic optimization of semantic process model matching. In: Barros, A., Gal, A., Kindler, E. (eds.) BPM 2012. LNCS, vol. 7481, pp. 319–334. Springer, Heidelberg (2012). doi:10.1007/978-3-642-32885-5_25
9. Klinkmüller, C., Weber, I., Mendling, J., Leopold, H., Ludwig, A.: Increasing recall of process model matching by improved activity label matching. In: Daniel, F., Wang, J., Weber, B. (eds.) BPM 2013. LNCS, vol. 8094, pp. 211–218. Springer, Heidelberg (2013). doi:10.1007/978-3-642-40176-3_17
10. Sonntag, A., Hake, P., Fettke, P., Loos, P.: An approach for semantic business process model matching using supervised machine learning. In: Resrach in Progress Papers, p. 47 (2016)

11. Cayoglu, U., et al.: Report: the process model matching contest 2013. In: Lohmann, N., Song, M., Wohed, P. (eds.) Business Process Management Workshops. Lecture Notes in Business Information Processing, vol. 171, pp. 442–463. Springer, Cham (2013)

12. Antunes, G., Bakhshandeh, M., Borbinha, J., Cardoso, J., Dadashnia, S., Di Francescomarino, C., Dragoni, M., Fettke, P., Gal, A., Ghidini, C., et al.: The process model matching contest 2015. In: Proceedings of the 6th International Workshop on Enterprise Modelling and Information Systems Architectures (2015)

13. Weidlich, M., Sagi, T., Leopold, H., Gal, A., Mendling, J.: Predicting the quality of process model matching. In: Daniel, F., Wang, J., Weber, B. (eds.) BPM 2013. LNCS, vol. 8094, pp. 203–210. Springer, Heidelberg (2013). doi:10.1007/978-3-642-40176-3_16

14. Gal, A.: Uncertain schema matching. Synth. Lect. Data Manage. **3**(1), 1–97 (2011)

15. Ehrig, M., Koschmider, A., Oberweis, A.: Measuring similarity between semantic business process models. In: Proceedings of the Fourth Asia-Pacific Conference on Conceptual Modelling, pp. 71–80 (2007)

16. van Dongen, B., Dijkman, R., Mendling, J.: Measuring similarity between business process models. In: Bellahsène, Z., Léonard, M. (eds.) CAiSE 2008. LNCS, vol. 5074, pp. 450–464. Springer, Heidelberg (2008). doi:10.1007/978-3-540-69534-9_34

17. Dijkman, R., Dumas, M., García-Bañuelos, L.: Graph matching algorithms for business process model similarity search. In: Dayal, U., Eder, J., Koehler, J., Reijers, H.A. (eds.) BPM 2009. LNCS, vol. 5701, pp. 48–63. Springer, Heidelberg (2009). doi:10.1007/978-3-642-03848-8_5

18. Dijkman, R., Dumas, M., Garcia-Banuelos, L., Kaarik, R.: Aligning business process models. In: Enterprise Distributed Object Computing Conference, pp. 45–53. IEEE (2009)

19. Dijkman, R., Dumas, M., Van Dongen, B., Krik, R., Mendling, J.: Similarity of business process models: metrics and evaluation. Inf. Syst. **36**(2), 498–516 (2011)

20. Yan, Z., Dijkman, R., Grefen, P.: Fast business process similarity search with feature-based similarity estimation. In: Meersman, R., Dillon, T., Herrero, P. (eds.) OTM 2010. LNCS, vol. 6426, pp. 60–77. Springer, Heidelberg (2010). doi:10.1007/978-3-642-16934-2_8

21. Niemann, M., Siebenhaar, M., Schulte, S., Steinmetz, R.: Comparison and retrieval of process models using related cluster pairs. Comput. Ind. **63**(2), 168–180 (2012)

22. Klinkmüller, C., Leopold, H., Weber, I., Mendling, J., Ludwig, A.: Listen to me: improving process model matching through user feedback. In: Sadiq, S., Soffer, P., Völzer, H. (eds.) BPM 2014. LNCS, vol. 8659, pp. 84–100. Springer, Cham (2014). doi:10.1007/978-3-319-10172-9_6

23. Liu, K., Yan, Z., Wang, Y., Wen, L., Wang, J.: Efficient syntactic process difference detection using flexible feature matching. In: Ouyang, C., Jung, J.-Y. (eds.) AP-BPM 2014. LNBIP, vol. 181, pp. 103–116. Springer, Cham (2014). doi:10.1007/978-3-319-08222-6_8

24. Fengel, J.: Semantic technologies for aligning heterogeneous business process models. Bus. Process Manage. J. **20**(4), 549–570 (2014)

25. Ling, J., Zhang, L., Feng, Q.: Business process model alignment: an approach to support fast discovering complex matches. In: Mertins, K., Bénaben, F., Poler, R., Bourriéres, J.P. (eds.) Enterprise Interoperability VI. Proceedings of the I-ESA Conferences, vol. 7, pp. 41–51. Springer, Cham (2014)

26. Makni, L., Haddar, N.Z., Ben-Abdallah, H.: Business process model matching: an approach based on semantics and structure. In: 12th International Joint Conference on e-Business and Telecommunications (ICETE), vol. 2, pp. 64–71 (2015)

27. Sebu, M.L.: Merging business processes for a common workflow in an organizational collaborative scenario. In: Control and Computing, System Theory, pp. 134–139 (2015)

28. Belhoul, Y., Haddad, M., Duchêne, E., Kheddouci, H.: String comparators based algorithms for process model matchmaking. In: Ninth International Conference on Services Computing, pp. 649–656. IEEE (2012)

29. Humm, B.G., Fengel, J.: Semantics-based business process model similarity. In: Abramowicz, W., Kriksciuniene, D., Sakalauskas, V. (eds.) Bus. Inf. Syst. Lecture Notes in Business Information Processing, vol. 117, pp. 36–47. Springer, Heidelberg (2012)

30. Weidlich, M., Sheetrit, E., Branco, M.C., Gal, A.: Matching business process models using positional passage-based language models. In: Ng, W., Storey, V.C., Trujillo, J.C. (eds.) ER 2013. LNCS, vol. 8217, pp. 130–137. Springer, Heidelberg (2013). doi:10.1007/978-3-642-41924-9_12

31. Corrales, J.C., Grigori, D., Bouzeghoub, M.: BPEL processes matchmaking for service discovery. In: Meersman, R., Tari, Z. (eds.) OTM 2006. LNCS, vol. 4275, pp. 237–254. Springer, Heidelberg (2006). doi:10.1007/11914853_15

32. Koschmider, A., Oberweis, A.: How to detect semantic business process model variants? In: Proceedings of the 2007 ACM Symposium on Applied Computing, pp. 1263–1264 (2007)

33. Pittke, F., Leopold, H., Mendling, J., Tamm, G.: Enabling reuse of process models through the detection of similar process parts. In: La Rosa, M., Soffer, P. (eds.) Business Process Management Workshops. Lecture Notes in Business Information Processing, vol. 132, pp. 586–597. Springer, Heidelberg (2013)

34. Sebu, M.L.: Similarity of Business Process Models in a Modular Design. In: Applied Computational Intelligence and Informatics, pp. 31–36. IEEE (2016)

35. Miller, G.A.: Wordnet: a lexical database for english. Commun. ACM 38(11), 39–41 (1995)

36. Kim, S.N., Baldwin, T.: Automatic interpretation of noun compounds using wordnet similarity. In: Dale, R., Wong, K.-F., Su, J., Kwong, O.Y. (eds.) IJCNLP 2005. LNCS, vol. 3651, pp. 945–956. Springer, Heidelberg (2005). doi:10.1007/11562214_82

37. Kim, S.N., Baldwin, T.: Interpreting semantic relations in noun compounds via verb semantics. In: Proceedings of the COLING/ACL on Main Conference Poster Sessions, pp. 491–498. Association for Computational Linguistics (2006)

38. Bruni, E., Tran, N.K., Baroni, M.: Multimodal distributional semantics. J. Artif. Intell. Res. (JAIR) 49(2014), 1–47 (2014)

39. Leopold, H., Meilicke, C., Fellmann, M., Pittke, F., Stuckenschmidt, H., Mendling, J.: Towards the automated annotation of process models. In: Zdravkovic, J., Kirikova, M., Johannesson, P. (eds.) CAiSE 2015. LNCS, vol. 9097, pp. 401–416. Springer, Cham (2015). doi:10.1007/978-3-319-19069-3_25

40. Kietz, J.U., Volz, R., Maedche, A.: Extracting a domain-specific ontology from a corporate intranet. In: Proceedings of the 2nd Workshop on Learning Language in Logic and the 4th Conference on Computational Natural Language Learning, vol. 7, pp. 167–175. Association for Computational Linguistics (2000)

Semi-automatic Development of Modelling Techniques with Computational Linguistics Methods – A Procedure Model and Its Application

Thorsten Schoormann[1(✉)], Dennis Behrens[1(✉)], Ulrich Heid[2], and Ralf Knackstedt[1]

[1] Department of Information Systems, University of Hildesheim, Universitätsplatz 1,
31141 Hildesheim, Germany
{thorsten.schoormann,dennis.behrens,
ralf.knackstedt}@uni-hildesheim.de

[2] Department of Information Science and Computational Linguistics, University of Hildesheim,
Universitätsplatz 1, 31141 Hildesheim, Germany
ulrich.heid@uni-hildesheim.de

Abstract. In recent years, the number of domain-specific modelling techniques increased. Method engineering already provides text-based and semantic approaches which aim to unify constructs and allocate terminologies. As existing procedures are usually carried out manually, challenges arise such as reproducibility and standardization as well as ensuring quality. Hence, this paper aims to investigate how methods from the Computational Linguistics can be applied to automatically develop domain-specific modelling techniques in order to face these challenges. As a main result, we present a procedure model that was developed and applied in four iterations, recommend tools, methods and resources as well as reflect typical issues.

Keywords: Method engineering · Procedure model · Text-Analytics · Computational linguistics · Conceptual modelling

1 Motivation

In recent years, the number of modelling approaches available has been increased to, for example, support the development of software, the communication between developer and user or the comprehension of a domain (e.g. [1–3]). To contribute to the understanding of a given domain, especially domain-specific modelling approaches have been focused. In contrast to general ones, these provide benefits such as increased effectiveness and efficiency during the modelling process or reusable and easily to manage model instances (e.g. [4, 5]). Due to the great influence of models on enterprises, the assessment of quality became a crucial factor [1, 6].

In the literature, different strategies for developing modelling techniques are discussed such as (a) reuse and adoption of existing ones (e.g. [7]) or (b) text-/semantics-based approaches (e.g. [3, 8]). Regarding (b), modelling techniques should be applicable intuitively – so, they need clear semantics to ensure that relevant notation constructs are considered and they should provide these constructs for the application [4]. In addition,

© Springer International Publishing AG 2017
W. Abramowicz (Ed.): BIS 2017, LNBIP 288, pp. 194–206, 2017.
DOI: 10.1007/978-3-319-59336-4_14

the derivation of modelling techniques should be traceable to reproduce from which source a construct was derived. However, research on the procedure of deriving these constructs from empirical data is rather limited [3].

Existing approaches often apply manually procedures which are confronted with different challenges (traceability, reproducibility, comparability and effectiveness). To semi-automatically identify domain-concepts and their relations and for building representations of knowledge, different methods from Computational Linguistics (CL) can be used. Most of them require large amounts of text (>1.000.000 words) to apply statistical approaches. If, in addition, types of relations and manually annotated material exist, methods of Information Extraction or Information Retrieval are useful (e.g. [38]). However, what to do if these requirements cannot be fulfilled? Particularly in emerging fields such as the sharing economy, only limited amounts of texts are available. Here, methods of symbolic language processing can be applied. *Consequently, this paper aims to investigate what are the potentials of using text-based methods from CL for developing modelling approaches based on a small amount of text data.* Thereby, we do not argue that existing method engineering approaches should be replaced, but rather be extended. A semi-automatic approach can contribute to the standardization, reproducibility and comparability.

Our paper is structured as follows: In Sect. 2, we reviewed literature to identify typical fields of text-based methods in conceptual modelling. In Sect. 3, we outline our methodological approach. In Sect. 4, our procedure model for developing modelling techniques based on text-based analysis is presented. It was (re-)designed in four iterations with demonstrations in modelling Carsharing business models which typically consist of various products and services. In Sect. 5, we evaluate the results by comparing our approach with a manually procedure. In Sect. 6, we derive implications, limitations and research perspectives.

2   Related Work

In order to identify the status quo of text-based methods in Information Systems (IS), we conducted a literature review – following the methodology of [9].

Literature Search.   First, we focused on IS-conferences because they (a) are among to the major ones or (b) can provide articles with text-based analysis, method engineering or natural language processing: ICIS, ECIS, AMCIS, CAiSE, EMMSAD, BPM, ER and BIS. Secondly, we considered IS-journals (AISeL-search). Thirdly, we selected Google Scholar because of the interdisciplinary topic. Due to the decreasing relevance in Google Scholar, only the first 100 articles were considered. In total, 157 articles were found (search items see Fig. 1). We evaluated (a) title, keywords and abstracts, and (b) full texts (Fig. 1; only conferences and journal >0 are presented).

Literature Analysis.   The analysis indicates that text-analytic methods were mostly used to evaluate (14/29), for example, IS-artefacts [10], modelling techniques [5, 11], reference models [12], conceptual models [1, 13], data models [14] and process models [6]. Often, conceptual models were evaluated with the Bunge-Wand-Weber ontology.

Google Scholar	ICIS 1980-2015	ECIS 2000-2015	AMCIS 1995-2016	AISeL Journals	CAiSE 2006-2016	EMMSAD 2006-2010	BIS 2006-2016	Other Sources (e.g., forward/ backward search)
(Text-Based OR Ontology-Based OR Ontology-Driven) AND (Method Engineering OR Method Development)								
N = 1.150 (100!)	N = 8	N = 16	N = 17	N = 4	N = n.a.	N = n.a.	N = 23	
Evaluation I: Title- and Abstract-Based relevance								
N = 15	N = 2	N = 6	N = 11	N = 2	N = 8	N = 2	N = 5	
Evaluation II: Fulltext-based relevance								
N = 5	N = 1	N = 5	N = 7	N = 2	N = 1	N = 1	N = 4	N = 5
After eliminating duplicates: 29 relevant articles								
N = 2	N = 1	N = 1	N = 2	N = 1	N = 1	N = 1		
Inclusion Criteria (Method Engineering/Development and Procedure Model): 9 articles								

Fig. 1. Overview of the literature review (sources, keywords, amount of articles etc.)

Furthermore, semantic and domain-specific modelling (10/29) was addressed which deals with the integration or application of knowledge into IS-artefacts – particularly in business process models (e.g., [2, 6, 8, 15, 16, 36, 37, 40]).

Method engineering (ME) – some articles already deal with using methods of semantics-/text-analysis for method engineering (9/29). However, only few developed a procedure model allowing reproducing and following the derivation. The results of the search can be seen in Table 1 with a short description of the reference found.

Table 1. Literature results related to method engineering approaches.

Ref.	Description	ME	PM
[3]	Derivation of notations from empirical data	X	X
[17]	Ontology-based method engineering; BBW-Ontology	X	(X)
[4]	Ontology-based method engineering	X	X
[18]	Ontology and method engineering	X	–
[5]	Ontologies for evaluation and development of IS (roadmap)	(X)	–
[19]	"Ontologically based development" ([19], p. 624)	(X)	–
[20]	Extension of BBW-Ontology	(X)	–
[21]	Computer-aided method engineering; State of the art	(X)	–
[22]	Framework for construction of method engineering artifacts	X	(X)

Note: (X) = concept was partially/implicitly met.

Procedure models (PM) (4/9) – [3] presented an approach for conceptualizing empirical data with grounded theory which consists of (I) collecting data, (II) identifying concepts, (III) developing concepts, (IV) developing relations and (V) summarizing. [4] provide a procedure for ontology-driven derivation of concepts which consists of (I) selection and development of a domain-ontology (assuming a suitable ontology already exists), (II) extraction and transfer of the ontology into an initial meta model of a technique (transferring relevant concepts), (III) refinement of the meta model (adapting the semantics) and (IV) development of syntax (transferring meta model into concrete syntax and representation). [17] developed an ontology-based method engineering based on the BBW-Ontology (incl. syntax, graphics) and [22] a conceptual framework for constructing, analyzing and comparing ME-artefacts.

Nevertheless, existing methods which deal with text-based analysis mostly require manual procedures. This raises different challenges such as ensuring traceability, reproducibility and quality as well as effectiveness (e.g. cost and time). By executing it manually, the sequence of tools and methods is heterogeneous and with it the process is not clear. Semi-automatic method engineering could contribute to the standardization of the procedure. In addition, current approaches often use semantics analysis to evaluate conceptual models and techniques ex post. However, what about considering these methods already during the construction of modelling techniques?

3   Research Method

Following the design science paradigm we developed, demonstrated and evaluated our artefact – a procedure model – iteratively (Fig. 2) [23]. We applied our approach in the domain of modelling Carsharing. Four iterations were executed (Table 2).

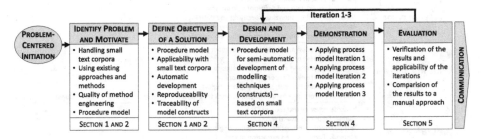

Fig. 2. Research method (adopted from [23])

Table 2. Overview of the iterations executed.

Iteration		Description	Participants	Environment
0	Manual model [24]	Demonstration (Carsharing)	2 IS (4-eye princ.)	Naturalistic
1	Initial model (PMI)	Literature review, workshop	3 IS and 1 CL	Artificial
2	Enhanced model (PMII)	Demonstration (Carsharing)	32 students (CL)	Naturalistic
3	Enhanced model (PMIII)	Demonstration (Carsharing)	28 students (CL)	Naturalistic

Iteration 0 – Current State: Manual Procedure. The starting point of our research is a manual procedure to develop a modelling approach for Carsharing business models [35]. In order to derive modelling constructs, a qualitative method was applied. Based on the literature, websites of Carsharing-providers were analyzed and conceptualized. Afterwards, symbols (representation) were assigned to the relevant concepts. Different challenges came up because each step was executed manually.

Iteration 1 – Initial Procedure Model (PMI). Based on the literature and interdisciplinary researchers, we conducted a workshop to create an initial model which supports semi-automatic method engineering. It consists of (I) building a text corpus, (II) extracting terms, (III) extracting relations via pattern- and syntax-analysis, (IV) building hierarchies and (V) implementing results into a modelling approach.

Iteration 2 – Enhanced Procedure Model (PMII). Based on the first application and evaluation of the results, different improvements could be determined. Hence, we enhanced our model: (I) specifying and defining the domain, (II) extracting and preparing texts, (III) exploring the text corpus, (IV) extracting relevant terms, (V) extracting relationships via syntax-analysis, compounds-analysis and patterns, (VI) consolidating the results and (VII) transferring them into a modelling approach.

Iteration 3 – Enhanced Procedure Model (PMIII). Based on the evaluation of iteration 2, we particularly split the first phase into (I) defining domain and (II) identifying proper text sources as well as provided useful methods and tools. We determined that the texts selected had strong influences on the final results – so, we highlighted the importance by separating these steps in our procedure model (Fig. 3).

Fig. 3. Procedure model (PMIII) of semi-automatic method engineering

4 Design, Development and Demonstration

4.1 Overview

Next, we describe *inputs* (objects/preconditions), *outputs* (results), *tasks* (sub-steps), *methods, tools and resources* (e.g., lexical-semantic nets) as well as *typical/possible sources of error* for each of the eight phases (Fig. 3). We highlighted CL-specific approaches of our model by marking each one with (CL).

For confirming the procedure model (PMIII), we give – bellow – general explanations of the *development* and examples of some positive or negative effects during the *demonstration*. Through continuous theoretical and practical evaluation and reflection, our model could be enhanced iteratively (four iterations). Due to the limited space, only some examples are presented. We assume that the combination of development and demonstration will allow readers to follow more easily.

4.2 Specification of Domain

Development. The first phase aims to specify the purpose of the modelling technique. Moreover, the definition of boundaries to other fields has to be done. Typical tasks deal with researching, analyzing, evaluating and, finally, specifying a domain. Here, literature reviews, internet (re-)search, interviews or Delphi studies can be conducted.

Demonstration. Our demonstration (four iterations) and modelling approach deals with modelling Carsharing business models from a provider perspective.

4.3 Identification and Selection of Text Sources

Development. The identification of suitable text sources is an essential task which affects the entire procedure. Texts can be extracted from sources such as manuals, documentation, interview transcriptions, textbooks or the internet (e.g. [3]). Typical tasks deal with assessing the suitability of sources as well as selecting proper texts. The selected sources are the basis for the identification of relevant domain concepts.

Demonstration. Here, we used websites from Carsharing provider because of the limited availability of textbooks. In doing so, we have to consider that particularly "advertising language" is part of the text corpus – this includes words used to beautify a business model (e.g., "fast cars"). Hence, content often focuses on the value proposition rather than on details such as sharing-conditions. We suggest to define the purpose of a modelling approach early on, to orient the following steps towards it.

4.4 Extraction and Preparing of Texts

Development. This phase aims to generate plain text. We suggest the following tasks:

Extraction of texts (a) – manually or with tools such as crawler (e.g., HTTrack).

Preparing texts (b) – eliminating elements (e.g., HTML-tags) manually or with tools such as HTML2Text or own scripts (e.g., with Perl: $string=~s/<[^>]+>//g;).
Segmenting texts (c) – for further processing it is adequate to segment and annotate texts, for example, headlines, sections and sentences [25]. Due to the fact that in (a) duplicates can be generated, an automatically elimination of them is recommended.

Demonstration. Our demonstration especially indicates that (a) the automatically hyphenation of terms was often faulty because of wrong interpretations of the tools used as well as (b) the elimination of HTML-tags was not consistent.

4.5   Preparation and Building of Corpus

Development. Next, we have to create a corpus which represents and provides all of the material collected. This includes a completely prepared and annotated corpus.

Tokenizing (a) – due to the fact that the next steps and tools require an explicit marking of starting (<s>) and end points (</s>) of sentences, tokenizing is necessary (automatic annotation of starting and end points).
Meta data and linguistic annotation (b) – meta data are different objects which are related to the texts and provide additional information such as specifications of text sources (e.g., Carsharing provider) or publication dates. Typically, *Part-of-Speech-Tagging (POS)* and *Lemmatization* were applied. POS-Tagging assigns labels for grammatical classes of words (e.g., nouns). Usually, tagsets such as STTS[1] (German) are applied. Lemmatization refers to the task of finding the base form of a word. These tasks aim to enable generalized search queries (e.g., consider similar words).
Normalization (c) – normalization aims to specify spelling styles, for example, marking email-addresses or homogenizing measure expressions.

Demonstration. Spelling mistakes of source texts which were not found during the normalization led to wrong results of term extractions because the extractor takes certain words as a special feature (e.g., "Casharing" instead of "Carsharing"). As a result, we created a text corpus with about 45.000 words.

4.6   Extraction of Terms

Development. Based on the corpus, proper terms have to be determined. These "term candidates" (e.g. terms which are marked in the meta data annotating step) need to be confirmed by applying further methods. Two tasks can be distinguished:

Determination of term candidates (a) – term candidates (wordlists) can be extracted by tools which work with statistical or pattern-based approaches.
Assessment of relevance and selection of terms (b) – each term candidate can be assessed (I) manually (e.g., by experts), (II) numerically with quantitative/statistical methods (e.g., calculate domain-specificity [26]) or (III) pattern-based to analyze

---

[1] http://www.ims.uni-stuttgart.de/forschung/ressourcen/lexika/TagSets/stts-table.html .

sequences such as noun + adjective or noun + preposition + noun. Afterwards, terms have to be selected, for example, definition of threshold level (80% yes) or in negotiations.

Demonstration. We decided to assess relevant terms for Carsharing by annotating the relevance in a collaborative way – with groups up to five members. Therefore, different groups assigned "yes" (+), "no" (-) or "unsure" (o) to some candidates. If all team members annotated a yes, we integrated a term into our further procedure (a so called "Gold-standard" was derived). Examples for relevant terms are "rent", "vehicle", "booking" and "next station". Non-relevant terms are, for example, "safe" or "mobile phone" – which certainly are important for access to the cars or for their booking. Often, concrete objects such as names of cars or providers and vehicle classes were evaluated positively by the groups.

4.7 Extraction of Relationships

Development. Relations between concepts are important to determine dependencies. Within a small corpus it is challenging to select appropriate methods which are not based on statistical calculations. We suggest to conduct (a) word-based, (b) intra-sentence-based and (c) inter-sentence-based analysis. These methods require syntactic-analysis, for example, with a dependency-parser such as MATE [27].

Word-based analysis (a) – it is possible to (I) identify groups/similarities by using lexical-semantic nets such as GermaNet[2] which, for example, allows the identification of hypernyms and hyponyms. Besides lexical nets, hypernyms can be derived with (II) the decomposition of compounds, for example, "customer-contract" into two single nouns "customer | contract", where "contract" is the hypernym (e.g. [28]).
Intra-sentence-based analysis (b) – here, additional relations can be extracted. For example, by applying (I) patterns which identify "is-a"-relations (e.g., "VW Polo" and "Opel Corsa" are a "compact car") or (II) complex POS-patterns.
Inter-sentence-based analysis (c) – relations can be in consecutive sentences. For example, the assignment of pronouns to their reference words can support the identification of relations across sentences [29] – here further research is required.

Demonstration. The application indicates that the *analysis of single words*, without relevant contexts, was hard and some of the compounds were assigned wrongly because single elements were not meaningful enough. Regarding the *semantic check* with GermaNet, ambiguous meanings should be considered – for example, the car brand "Fiesta" in German also means "party". The *inter-sentence-based analysis* requires the following structure in particular: "alternative powering for vehicles such as natural gasoline or electro [...]". By applying a *pattern-based analysis* (verb + object and adjective + noun) different relations could be identified, for example, "vehicle: locate | book | rent etc." Adjective + noun-pairs contributed to the identification of subtypes (e.g., "provider: commercial | private"). The POS-tagging had to be rigorous because

---

[2] Ontology of German language, www.sfs.uni-tuebingen.de/GermaNet/ (equivalent: WordNet).

errors affect the application of patterns. Here, especially the following patterns were helpful: "noun-article-adjective-noun"; "adjective-preposition-article-noun"; "noun-preposition-name" (e.g., "booking via Tamyca").

4.8   Consolidation of Results

Development. The consolidation aims at merging and presenting the results of the phases before. It can be supported by methods such as a Zwicky-box which provide relevant dimensions and related characteristics [30], a taxonomy (empirical and inductive), a typology (conceptual and deductive) [39] or an ontology [31].

Demonstration. One of our lessons learned was that specifying the level of abstraction is challenging. How complex should the modelling approach be? This depends on the purpose and has to be defined early on. The text-based analysis only represents selected sources. So, the way of consolidating depends on the modelling approach – here, we used a Zwicky-box. Uncertainties occurred with the definition of characteristics/dimensions and their instances, for example, "is an insurance a service or an own characteristic with related instances?" Overall, this phase is less supported by tools, so most of these methods were applied manually. Tools can be used, for example, to eliminate redundancies or to represent hierarchies of partial results.

4.9   Representation of Results

Development. The last phase deals with the assignment of graphical elements to each construct. *Conceptual aspects* (phase 7, consolidation) and *representational aspects* for a modelling approach are distinguished [32]. Here, designers can create icons and assign them to the building blocks. The result is a prototypical modelling approach for a certain domain. This initial approach can be demonstrated and evaluated (e.g. [33]) to ensure its applicability and to identify enhancements based on user feedback.

Demonstration. In our case, we evaluated, adjusted and extended an existing approach for modelling Carsharing business models – based on a Zwicky box (Fig. 4). We selected this box because it is a proper basis for transferring it to a building-block-based approach [34]. Next, we transferred the results to our approach. By providing predefined blocks, challenges of heterogeneity can be addressed (e.g., of the entire enterprise architecture model, single elements or the abstraction level of usage [35]).

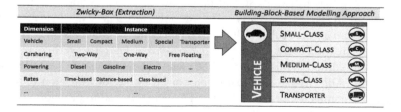

Fig. 4. Example of transferring the results into a building-block-based modelling approach

5   Evaluation

In order to evaluate the results, we compared the findings of iteration 0 (manual [24]) and iteration 3 (semi-automatic; Fig. 3). Based on separately derived Zwicky-boxes of Carsharing business models, we could identify an overlap of 51/74 instances (69%).

Similarities of Results. Especially detailed variations (e.g., car brands) and suitable descriptions of characteristics and their instances were found. In the manual procedure, 31 instances (42%) were found via assessment of relevance in teams, analysis of adjective-noun-pairs (50%) and analysis of compounds (37%). Hence, a lot of instances could be identified with lower effort (e.g., time and costs).

Extensions of Results. By analyzing compounds, different extensions were derived (e.g., car brands). 13 new instances could be identified when analyzing adjective-noun-pairs. In addition, suggestions for integrating model or partner names and information of conditions were provided. The analysis of 589 verb-object-pairs indicates that coherences between objects and possible activities could be identified.

Deviation of Results. Especially the classification of characteristics and instances is different because it depends on selected methods and experts. Approaches which are more standardized are required, to support a consistent (collaborative) assignment.

Overall Assessment. Regarding our modelling technique we (a) could confirm existing – manually identified – concepts as well as (b) suggest useful extensions. The deviations indicate that especially the classification of relevant elements is characterized by individuality. Here, tools of CL can contribute to standardization. Moreover, knowledge from existing lexical nets can be used to derive domain-specific constructs. Furthermore, our group-based assessment highlighted that a high amount of similarities can be derived with low effort, even with non-domain-experts.

Limitations. The selection of tools, resources and methods is based on our own decisions which have limitations – other approaches for analyzing texts can be equally useful. In addition, some steps still require manual decisions – so, the procedure is not fully automatic. Our demonstration and evaluation is initially based on German texts and the transferability to further languages has to be investigated.

6   Conclusion and Outlook

We presented a procedure model which addresses (a) the analysis of a small amount of text data with methods from CL and (b), based on this, a semi-automatically development of modelling techniques. Our approach can contribute to the traceability and reproducibility when deriving domain-specific modelling constructs. Moreover, we (c) evaluated, adjusted and extended an approach for modelling Carsharing business models. Further research should regard further steps, for example:

Effects of Automation. Which effects does the standardization and tool support for deriving concepts on qualitative methods (e.g. Grounded Theory) have? We suggest to automate some steps which could have negative effects on the creativity of researchers. Do established methods need to change? In addition, investigating how processes can be supported by automatic logging of meta data is relevant.

Selection of Suitable Texts. Because the entire development process is affected, it is import to get suitable texts. Therefore, (a) criteria for the assessment of suitability could be determined, (b) approaches for preparing texts towards the development of modelling techniques or (c) guidelines to create proper texts can be designed.

Combination of Tools. We initially applied different types of tools and resources. In further steps, (a) the applicability of tools and methods (e.g. POS-Tagging) can be evaluated as well as (b) suggestions for a combination of tools can be derived.

Comparison with other Procedures. We evaluated our approach by applying it in different iterations as well as by comparing it with a manual approach. Because only limited text sources for modelling Carsharing business models were available, we excluded statistical approaches such as methods from Information Extraction. However, further evaluation could compare with other procedures, for example, from Text/Data Mining [38], Topic Modeling or other processing pipelines.

Demonstration in Further Domains. Besides the Carsharing domain, our approach should be extended to other domains in order to verify its applicability.

Relations of Modelling Building Blocks. Our approach allows creating building blocks. However relations between each block need further investigation.

Acknowledgments. This research is partly funded by the European Regional Development Fund (ERDF) and the State of Lower Saxony (Investitions- und Förderbank Niedersachsen) in the scope of the research project "SmartHybrid – Process Engineering" (ID: ZW 6-85003451). We like to thank them for their support.

References

1. Pfeiffer, D., Niehaves, B.: Evaluation of conceptual models – a structuralist approach. In: Proceedings of the European Conference in Information Systems (ECIS), p. 43 (2005)
2. Evermann, J., Wand, Y.: Ontology based object-oriented domain modelling: fundamental concepts. Requirements Eng. **10**(2), 146–160 (2005)
3. Becker, J., Karow, M., Müller-Wienbergen, F.: Applying lessons learned from counselling: on nurturing relations in design projects. In: Proceedings of the European Conference in Information Systems (ECIS), p. 54 (2009)
4. Guizzardi, G., Ferreira Pires, L., Van Sinderen, M. J.: On the role of domain ontologies in the design of domain-specific visual modeling languages. In: Proceedings of the 17th ACM OOPSLA (2002)

5. Kishore, R., Sharman, R.: Computational ontologies and information systems I: foundations. Commun. Assoc. Inf. Syst. (CAIS) **14**, 158–183 (2004)
6. Fellmann, M., Thomas, O., Busch, B.: A query-driven approach for checking the semantic correctness of ontology-based process representations. In: Abramowicz, W. (ed.) BIS 2011. LNBIP, vol. 87, pp. 62–73. Springer, Heidelberg (2011). doi:10.1007/978-3-642-21863-7_6
7. Brinkkemper, S., Lyytinen, K., Welke, R.J.: Method Engineering Principles of Method Construction and Tool Support. Chapman & Hall, London (1996)
8. Lin, Y.: Semantic annotation for process models: facilitating process knowledge management via semantic interoperability. Ph.D. thesis, University of Trondheim (2008)
9. Vom Brocke, J., Simons, A., Niehaves, B., Riemer, K., Plattfaut, R., Cleven, A.: Reconstructing the giant: on the importance of rigour in documenting the literature search process. In: Proceedings of the European Conference in Information Systems (ECIS), pp. 2206–2217 (2009)
10. Becker, J., Niehaves, B., Pfeiffer, D.: Ontological evaluation of conceptual models. Scand. J. Inf. Syst. **20**(2), 83–110 (2008). Article 4
11. Zhang, H., Kishore, R., Ramesh, R.: Ontological analysis of the MibML grammar using the Bunge-Wand-Weber model. In: Proceedings of the American Conference on Information Systems (AMCIS), p. 553 (2004)
12. Fettke, P., Loos, P.: Ontological evaluation of reference models using the Bunge-Wand-Weber model. In: Proceedings of the American Conference on Information Systems (AMCIS), p. 384 (2003)
13. Becker, J., Pfeiffer, D., Janiesch, C.: Percieved evaluability - development of a theoretical model and a measurement scale. In: Proceedings of the American Conference on Information Systems (AMCIS), p. 153 (2008)
14. Milton, S., Kazmierczak, E., Keen, C.: Data modelling languages: an ontological study. In: Proceedings of the European Conference in Information Systems (ECIS), p. 304 (2001)
15. Hofferer, P.: Achieving business process model interoperability using metamodels and ontologies. In: European Conference in Information Systems (ECIS), pp. 1620–1631 (2007)
16. Becker, J., Pfeiffer, D.: Solving the conflicts of distributed process modelling: towards an integrated approach. In: Proceedings of the European Conference in Information Systems (ECIS), p. 90 (2008)
17. Gehlert, A., Buckmann, U., Esswein, W.: Ontology based method engineering. In: Proceedings of the American Conference on Information Systems (AMCIS), p. 436 (2005)
18. Holten, R., Dreiling, A., Becker, J.: Ontology-Driven Method Engineering for Information Systems Development. In: Business Systems Analysis with Ontologies, pp. 174–217. IDEA Group, Hershey (2005)
19. Rosemann, M., Green, P.: Integrating multi-perspective views into ontological analysis. In: International Conference in Information Systems (ICIS) (2000)
20. Rosemann, M., Wyssusek, B.: Enhancing the expressiveness of the Bunge-Wand-Weber ontology. In: American Conference on Information Systems (AMCIS), p. 438 (2005)
21. Niknafs, A., Ramsin, R.: Computer-aided method engineering: an analysis of existing environments. In: Bellahsène, Z., Léonard, M. (eds.) CAiSE 2008. LNCS, vol. 5074, pp. 525–540. Springer, Heidelberg (2008). doi:10.1007/978-3-540-69534-9_39
22. Leppänen, M.: An ontological framework of method engineering: an overall structure. In: 10th International Workshop on Exploring Modeling Methods in Systems Analysis and Design (EMMSAD 2007), pp. 11–12 (2007)
23. Peffers, K., Tuunanen, T., Rothenberger, M.A., Chatterjee, S.: A design science research methodology for information systems research. J. Manage. Inf. Syst. **24**(3), 45–77 (2007)

24. Schoormann, T., Behrens, D., Knackstedt, R.: Carsharing Geschäftsmodelle – Entwicklung eines bausteinbasierten Modellierungsansatzes. In: Thomas, O., Nüttgens, M., Fellmann, M. (eds.) Smart Service Engineering, pp. 303–325. Springer, Fachmedien (2016)

25. Kliche, F., Blessing, A., Heid, U., Sonntag, J.: The eIdentity text exploration workbench. In: Proceedings of the Ninth International Conference on Language Resources and Evaluation (LREC 2014), pp. 691–697 (2014)

26. Ahmad, K., Rogers, M.: Terminology management: a corpus-based approach. In: Translating and the Computer 14: Quality Standards and the Implementation of Technology in Translation, pp. 33–44 (1992)

27. Björkelund, A., Bohnet, B., Hafdell, L., Nugues, P: A high-performance syntactic and semantic dependency parser. In: Proceedings of the 23rd COLING, pp. 33–36 (2010)

28. Cap, F.: Morphological processing of compounds for statistical machine translation. Ph.D. thesis, University of Stuttgart (2014)

29. Rösiger, I., Schäfer, J., George, T., Tannert, S., Heid, U., Dorna, M.: Extracting terms and their relations from German texts: NLP tools for the preparation of raw material for e-dictionaries. In: eLex Proceedings, Herstmonceux Castle, UK (2015)

30. Zwicky, F.: Discovery, Invention, Research - Through the Morphological Approach. The Macmillan Company, Toronto (1969)

31. Gruber, T.: A translation approach to portable ontologies. Knowl. Acquis. **5**(2), 199–220 (1993)

32. Strahinger, S.: Metamodellierung als Instrument des Methodenvergleichs: Eine Evaluierung am Beispiel objektorientierter Analysenmethoden. Shaker Verlag, Aachen (1996)

33. Schalles, C.: Usability Evaluation of Modeling Languages. Springer Science, Heidelberg (2012)

34. Becker, J., Pfeiffer, D., Räckers, M.: Domain specific process modelling in public administrations – the PICTURE-approach. In: Wimmer, M.A., Scholl, J., Grönlund, Å. (eds.) EGOV 2007. LNCS, vol. 4656, pp. 68–79. Springer, Heidelberg (2007). doi: 10.1007/978-3-540-74444-3_7

35. Bakhshandeh, M., Pesquita, C., Borbinha, J.: An ontological matching approach for enterprise architecture model analysis. In: Abramowicz, W., Alt, R., Franczyk, B. (eds.) BIS 2016. LNBIP, vol. 255, pp. 315–326. Springer, Cham (2016). doi:10.1007/978-3-319-39426-8_25

36. Gailly, F., Poels, G.: Towards ontology-driven information systems: redesign and formalization of the REA ontology. In: Abramowicz, W. (ed.) BIS 2007. LNCS, vol. 4439, pp. 245–259. Springer, Heidelberg (2007). doi:10.1007/978-3-540-72035-5_19

37. Leopold, H., Mendling, J., Reijers, Hajo A.: On the Automatic Labeling of Process Models. In: Mouratidis, H., Rolland, C. (eds.) CAiSE 2011. LNCS, vol. 6741, pp. 512–520. Springer, Heidelberg (2011). doi:10.1007/978-3-642-21640-4_38

38. Rajman, M., Vesely, M.: From text to knowledge: document processing and visualization: a text mining approach. In: Sirmakessis, S. (ed.) Text Mining and Its Applications. STUDFUZZ, vol. 138, pp. 7–24. Springer, Heidelberg (2004)

39. Bailey, K.D.: Typologies and Taxonomies: An Introduction to Classification Techniques. Sage Publications, Thousand Oaks (1994)

40. Pittke, F., Leopold, H., Mendling, J.: Automatic detection and resolution of lexical ambiguity in process models. IEEE Trans. Softw. Eng. **41**(6), 526–544 (2015). doi:10.1109/TSE. 2015.2396895

Obstacle-Aware Resource Allocation in Business Processes

Farah Bellaaj[1(✉)], Mohamed Sellami[2], Sami Bhiri[3,4], and Zakaria Maamar[5]

[1] ReDCAD, FSEGS, University of Sfax, Sfax, Tunisia
bellaajfarah@yahoo.fr
[2] ISEP, Paris, France
[3] National Engineering School of Tunis, OASIS, Univ. Tunis El Manar,
2092 Tunis, Tunisia
[4] University of Monastir, Monastir, Tunisia
[5] Zayed University, Dubai, United Arab Emirates

Abstract. This paper presents an approach for allocating resources to Business Processes (BPs) so that obstacles preventing the execution of these BPs are mitigated. Resources include persons and machines who could be the origins of obstacles due to unexpected call-in-sick and sudden breakdown, for example. The approach uses a log that tracks past obstacles along with their undesirable impacts - time, cost, and quality - on BP execution. Estimating the likelihood occurrence of an obstacle is built-upon this log so that preventive and/or corrective measures are taken. An implementation of the approach is also presented in the paper.

Keywords: Business process · Resource allocation · Obstacle

1 Introduction

Business Processes (BPs) assist enterprises capture their best practices so they can achieve their goals with lower cost, shorter time, and higher quality [1]. A BP is a set of related, structured tasks that produce a specific service and/or product in response to a particular event. A specific occurrence (or execution) of a BP is called BP instance. Typically, a BP's life-cycle consists of creating a process model that is then, translated into a workflow model. Finally, it is assigned to persons and/or machines for execution. At run-time, monitoring oversees the instance's progress so that corrective actions are taken (e.g., substituting a person and fixing a machine) [2].

A good amount of research on allocating resources to BP executors is reported in the literature [3–5]. However, this research has overlooked cases where resources are the source of obstacles that usually delay BP execution and sometimes lead to their failure. Indeed, cases like persons calling-in-sick and machines breaking down are common. Although the BP community treats obstacles as risks [6,7], we differentiate them for the sake of a better understanding of these concepts. An obstacle impedes the execution progress of a BP instance

© Springer International Publishing AG 2017
W. Abramowicz (Ed.): BIS 2017, LNBIP 288, pp. 207–219, 2017.
DOI: 10.1007/978-3-319-59336-4_15

(e.g., degradation of machines and employee high-turnover) while a risk is the impact of obstacles on the execution of a BP instance, usually negative (e.g., BP delays, low quality results, and extra cost).

To prevent obstacle occurrence, we deem necessary understanding their origins so that tackling them happen. In this paper, we first, capitalize on logs to record details on BP past executions like who executed what and what happened [8] and second, define three indicators - time, cost, and quality - to define the impact of obstacles on forthcoming executions making our approach proactive. The rest of the paper is organized as follows: Sect. 2 discusses related work. Section 3 presents a motivating example. Section 4 illustrates the approach's steps. Sections 5 and 6 present a log's structure and time, cost, and quality indicators extracted from this log, respectively. Experiments are reported in Sect. 7 and concluding remarks are drawn in Sect. 8.

2   Related Work

This section discusses risk-aware BP approaches using Suriadi et al.'s classification namely before execution (design-time), during execution (run-time), and after execution [9].

Before-execution approaches. Jakoubi and Tjoa [10] propose the Risk-Oriented Process Evaluation (ROPE) method whose objective is to make BPs tolerant for risks. ROPE combines the advantages of BPM, risk management, and business continuity. In fact, it uses three layers to describe the risks that BPs are exposed to: BPM layer representing BPs, condition-action-resource-environment layer representing BP elements, and threat-impact-process layer representing potential threats, eventual countermeasures, and recovery strategies. ROPE helps identify risks and their possible mitigation at design-time. Fenz et al. [11] propose a design approach for resource-based risk analysis. The approach focuses on the consequence of a risk event and its likelihood, propagation, and overall risk level of a BP. The limitation of this approach is that it focuses only on risks regarding asset availability (e.g., data loss).

During-execution approaches. Kang et al. [12] study real-time risk detection so that abnormal terminations of BP instances are predicted. This consists of two phases: off-line pre-processing and run-time detection. A training set of historical BP instances and risk detection techniques are used in both phases. Kang et al. define the current status of a BP instance as the probability that it will be executed with "no-hassle". Plus, they provide a proactive detection if the probability of occurrence of a given outcome is over a threshold. Conforti et al. [13] detect risks during BP execution by anticipating exceptional conditions in which the process should behave adequately and providing a form of runtime obstacle detection capability based on a probabilistic analysis.

After-execution approaches. Pika et al. [14] predict time-related process risks by identifying (using statistical principles) indicators observable in event logs that highlight the possibility of not meeting deadlines. However, the authors restrict their work to identifying indicators of risks. Suriadi et al. [15] use root-cause analysis based on classification algorithms. After enriching a log with workload, occurrence of delay, and use of resources, decision trees identify the causes of overtime faults. Although both works generate risk predictions, they do not propose possible solutions for risk mitigation.

Compared to the aforementioned works, we consider obstacle management in BPs instead of risk management for many reasons: first, it is more advantageous to prevent risks before they occur by managing their causes (i.e., obstacles). Indeed, it is better to understand the causes of the unavailability of an employee rather than assigning him additional tasks causing delays. Second, risks can be caused by different factors including resource factors. These resources are considered to be *"unequivocally the single most important element that can affect process success"* [16]. So by managing obstacles related to resources, BP managers collect information about resources' behaviors in order to investigate resource related issues and identify best practices or opportunities for performance improvement. Obstacle detection depends heavily on recording events that are observed during BP execution. These events assist BP manager detect execution delays, extra cost, and quality degradation triggered by resources.

3   Motivating Example

We consider a simplified car-manufacturing BP in which human and non-human resources collaborate to get the job done. This BP consists of four sub-processes: stamping, welding, painting, and assembly. In the following we consider the welding sub-process where specialized robots heat stamped parts (task t_3) and join them together (task t_4). At the end, the finished body is sent for painting (task t_5).

In the car manufacturing BP, human (p) and/or non-human (m) resources execute tasks like t_3 done by robots and t_8 (install car parts) done by humans. Other tasks like t_5 is done by both. In such a BP, obstacles could occur:

- Scenario 1: t_3 can be suspended in case m_3 cannot be substituted on a short notice due to an unplanned breakdown causing delay.
- Scenario 2: t_5 can be suspended in case p_3 and p_4 are dissatisfied with the job because of poor working conditions causing high-turnover. Employees turnover is financially costly to the business because it needs resources filling the position by training new employees.
- Scenario 3: in case of the unavailability of p_4, t_5 will be assigned to another person who despite her willingness to accept an additional task might not be able to complete it.

Information collected during such scenarios report about tasks executed, resources involved, and obstacles encountered. Such information is recorded in the logs. Digging into these logs should help prevent obstacles from happening again. We assume that a regular occurrence of the same obstacle due to the same resource, could hint that this resource is likely to be the source of that obstacle again.

4   Approach Overview

This section discusses obstacles along with their categories and impacts, and then provides an overview of the obstacle-aware resource allocation approach.

An obstacle undermines the execution progress of a BP instance (e.g., the use of expensive spare parts and high employee turnover). We distinguish between controllable (the focus of this work) and uncontrollable obstacles. The former are internal factors to an organization and ought to be eliminated. They can be of human (e.g., turnover and strike) and physical (e.g., machine failure) nature. The latter are external factors and are beyond an organization's control. Examples are economical (e.g., market risk and pricing pressure), natural (e.g., flood and earthquake), and political (e.g., change in tax and government policy) [17]. Table 1 provides some controllable obstacles (column 1), their related resource type (column 2) and their impact on BP execution (column 3). For instance, high employee turnover (obstacle related to human resources) could increase the cost of completing a product.

Impacts of obstacles can be grouped into three categories (column 4): time, cost, and quality. Based on this categorization, we define for each resource three obstacle indicators which vary according to its workload. These indicators can be computed from past BP executions traces (Sect. 6). They provide useful insights about the likelihood of a resource to encounter obstacles according to its current workload. The better a resource indicators are the least that resource is obstacle-prone and *vice-versa*.

Table 1. Illustrative obstacles along with their impacts on BP execution

Obstacle	Resource	Impact on BP execution	Category of impact
O_1: Inexperienced employee	Human	Quality of work decreases	Quality
O_2: High employee turnover	Human	Costs increase	Cost
O_3: Shortage of employees	Human	Loss of productivity	Time and cost
O_4: Material costs increase	Human/Non human	Costs increase	Cost
O_5: Broken, destroyed machine	Non human	Costs increase, quality of work decreases	Cost and quality
O_6: Inefficient use of resources	Human/Non human	Loss of productivity	Time

In this paper, we present an approach that assists resource allocation with the aim of avoiding obstacle prone resources (Fig. 1). The approach takes as input (*i*) the event log which contains past execution traces and (*ii*) allocation requirements specifications for each task. Allocation requirements are specified as an objective function implying the three obstacle indicators. By default the three indicators are equally weighted (see Sect. 6.3 for more details). BP designers can however modify their weights to express specific requirements. The obstacle calculator computes obstacle metrics for each resource based on past execution traces. When instantiating a given task, the allocation advisor ranks the candidate resources according to the specified requirements and based on their computed metrics as output.

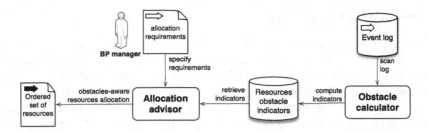

Fig. 1. Approach for obstacle-aware resource allocation

5 Tracking Business Process Past-Executions

In this section, we define the structure of an event log and show how we enrich it with the resources' workload. Such enriched log provides necessary information for estimating obstacle indicator.

5.1 Event-Log Structure

An event log ($\mathcal{E}_{log}$) consists of agreed-upon temporal details during BP execution (Definition 1).

Definition 1. *An event log is an ordered set of business events (evt) recorded during the execution of a BP instance. A sequence of events is a case. evt is a 8-tuple $< Case_{id}, Event_{id}, T_{id}, T_{state}, R_{id}, R_{role}, R_{state}, Timestamp >$ where: $Case_{id}$ is the case (instance) identifier; $Event_{id}$ is the logged event identifier; T_{id} is the task identifier; T_{state} is the execution state of T_{id}; R_{id} is the resource identifier; R_{role} is the role of a resource; R_{state} is the operational state of R_{id}; Timestamp is date and time of day identifying when the event was recorded.*

As per [18], task and resource states are as follows: $T_{state} \in$ {prepared, activated, done, suspended, failed} for task, $R_{state} \in$ {idle, on-leave, busy} for person, and $R_{state} \in$ {idle, used, serviced} for machine.

– A task takes initially on prepared state indicating that it needs to be assigned to resources. At this stage, two transitions are possible: activated state when adequate resources are assigned to the task, or suspended state when no adequate resources are found. When the task takes on activated state it could transition to either done state when no obstacles occur so the task execution succeeds, or suspended state when an obstacle occurs and until it is addressed. When the obstacle is addressed, the task transitions back to activated state. If not, the task transitions back to failed state.
– A person takes initially on idle state indicating that he is ready to perform a task. At this stage, two transitions are possible: busy state when a person is executing a task, or on-leave state when a person is excused from work.
– A machine takes initially on idle state indicating that it is ready to perform a task. At this stage, two transitions are possible: used state when a machine executes a task, or serviced state when it is unavailable due to maintenance.

5.2   Event Log Enrichment

The event log provides information that enable detecting resources which are source of obstacle (i.e., by analysing resources and tasks' states). It is possible that some information may not be explicitly available in the event log (e.g., workload), but can be derived from its existing content (e.g., resources' states). In our work, we enrich the event log's structure with resources' workload computed based on their states. We consider workload as the number of tasks assigned to a resource at a specified time (Timestamp of an entry in the event log). This log enrichment allows considering resources' workload impacts on obstacle occurrence. Even though obstacles are not explicit in event logs, their impact can be implicitly extracted and measured through dedicated indicators (i.e., time, cost, and quality). Table 2 illustrates the enriched log structure of three instances of the car manufacturing BP.

Table 2. Excerpt from an event log

$Case_{id}$	$Event_{id}$	T_{id}	T_{state}	R_{id}	R_{state}	$Timestamp$	$Workload$
$order_{10}$	evt_{b14}	t_3	done	m_3	idle	30-12-2010:22.33	1
$order_{10}$	evt_{b15}	t_4	activated	m_4	used	01-01-2011:12.40	4
$order_{10}$	evt_{b16}	t_4	suspended	m_4	serviced	01-01-2011:15.02	3
$order_{11}$	evt_{b25}	t_2	done	(m_{11}, p_3)	(idle, idle)	29-12-2010:13.48	(1,3)
$order_{11}$	evt_{b26}	t_3	activated	m_5	used	30-12-2010:12.10	2
$order_{11}$	evt_{b27}	t_3	done	m_5	idle	30-12-2010:12.24	1
$order_{12}$	evt_{b38}	t_3	suspended	m_3	serviced	06-01-2011:17.47	1
$order_{12}$	evt_{b39}	t_3	activated	m_3	used	07-01-2011:09.30	2
$order_{12}$	evt_{b41}	t_5	done	(m_{12}, p_4)	(idle, idle)	09-01-2011:13.55	(4,1)

6   Past Obstacle-Aware Business Process Resource Allocation

In this section, we introduce resource obstacle indicators and illustrate how we mine BP event logs to identify/quantify the impact of obstacles on the BP execution. Then, we present how we match between these impacts and BP managers' requirements for a past obstacles-aware resource allocation.

6.1   Resource Obstacle Indicator

A resource obstacle indicator signals obstacle occurrence likelihood per resource. We consider three indicators: *time*, *cost*, and *quality*.

Time: represents the average working time of a resource (r_i) on a given task (t_j) according to a certain workload (w). Considering that r_i with w begins performing t_j at a start time (logged as $evt_{start}.Timestamp$) and achieves it at a finish time (logged as $evt_{end}.Timestamp$), its time (T) is computed as follows:

$$T(r_i, t_j, w) = \sum_{case_k \in F_w} \frac{evt_{end}.Timestamp - evt_{start}.Timestamp}{|case_k|} \tag{1}$$

Such that:

- $F_w \subseteq F$ is the subset of cases with a workload equal to w, where F is the set of logged cases.
- evt_{end} and evt_{start} are two logged events representing the end and the beginning of the task execution, respectively, such that:
  - $evt_{end}.T_{state} = done \wedge evt_{end}.R_{state} = idle$
  - $evt_{start}.T_{state} = activated \wedge (evt_{start}.R_{state} = busy \vee evt_{start}.R_{state} = used)$
- evt_{end} and evt_{start} are related to the same BP case, resource, and task:
  - $evt_{end}.case_{id} = evt_{start}.case_{id}$
  - $evt_{end}.R_{id} = evt_{start}.R_{id} = r_i$
  - $evt_{end}.T_{id} = evt_{start}.T_{id} = t_j$
- $|case_k|$ is the number of cases with a workload equals to w.

Cost: represents the amount of money spent by a resource (r_i) for carrying out a task (t_j), like workers' salaries and machines' maintenance costs, while having a certain workload (w). It is calculated as the product of the hourly cost of the resource and the processing time (in hours) of the task. Other costs corresponding to repair or replacement of failed resources, called corrective costs can be added. Considering a resource r_i with a workload w performing a task t_j, its cost (C) is computed as follows:

$$C(r_i, t_j, w) = (UC_{r_i} \times T(r_i, t_j, w)) + CC_{r_i} \tag{2}$$

Such that:

- UC_{r_i} is r_i's cost per time unit of work;
- $T(r_i, t_j, w)$ is the average working time of r_i performing t_j with a workload w.
- CC_{r_i} is the eventual corrective cost spent for r_i if it was unavailable. It includes maintenance cost for machines and replacement cost for persons. In this work, we assume that human replacement is made within the same company, so no extra costs are added. For the maintenance cost, it is the product of the machine's downtime in hours ($T_{CC_{r_i}}$) and a hourly maintenance cost (UCC_{r_i}) defined by the BP manager:

$$CC_{r_i} = UCC_{r_i} \times T_{CC_{r_i}} \tag{3}$$

Where:

$$T_{CC_{r_i}} = \sum_{case_k \in F_w} \frac{evt_{available}.Timestamp - evt_{unavailable}.Timestamp}{|case_k|} \tag{4}$$

Such that:
- $evt_{available}$ and $evt_{unavailable}$ are two logged events representing the availability of a resource after a corrective maintenance, and the unavailability of a resource under maintenance, respectively, such that:
  - $evt_{available}.R_{state} = idle \lor evt_{available}.R_{state} = used \lor evt_{available}.R_{state} = busy$
  - $evt_{unavailable}.R_{state} = serviced \lor evt_{unavailable}.R_{state} = on-leave$
- $evt_{available}$ and $evt_{unavailable}$ are associated to the same BP instance, resource, and task:
  - $evt_{available}.case_{id} = evt_{unavailable}.case_{id}$
  - $evt_{available}.R_{id} = evt_{unavailable}.R_{id} = r_i$
  - $evt_{available}.T_{id} = evt_{unavailable}.T_{id} = t_j$

Quality: represents the quality of a BP from a resource's perspective (e.g., high successful job and low absenteeism). It is defined as the ratio of complete tasks out of the total tasks allocated to a resource. For a resource (r_i) executing a task (t_j) while having a workload (w), *Quality* (Q) is defined as follows:

$$Q(r_i, t_j, w) = \sum_{case_k \in F_w} \frac{\frac{|evt_{end}|}{|evt_{start}|}}{|case_k|} \tag{5}$$

Such that:

- $|evt_{end}|$ is the number of times t_j was successfully completed by r_i.
- $|evt_{start}|$ is the number of times t_j was allocated to r_i.

6.2   Obstacle-Aware Resource Representation

Based on the defined indicators, we associate each resource with a three-dimension vector $r_i^{(t_j, w)} = [t_j^w, c_j^w, q_j^w]$ reflecting its "history" of obstacles such that $t_j^w \in \mathcal{T}^w$, $c_j^w \in \mathcal{C}^w$, and $q_j^w \in \mathcal{Q}^w$. $\mathcal{T}^w$, $\mathcal{C}^w$, and $\mathcal{Q}^w$ represent r_i's indicators' dimensions for time, cost, and quality, respectively, while having a workload w. These indicators values are measured on different scales and need to be adjusted to a notionally common scale. For this purpose, we use Min-Max normalization to transform each indicator value to fit in a range from 0 to 1. The normalized time indicator (resp., cost and quality) is computed using Formula (6):

$$For\ i = 1 \cdots n\ and\ j = 1 \cdots m, \mathcal{T}^w[i,j] = \frac{T(r_i, t_j, w) - T^w_{min}}{T^w_{max} - T^w_{min}} \tag{6}$$

Such that n is the number of resources, m is the number of tasks, T (resp., C and Q) is the time (resp., cost and quality) function (Formula (1) (resp., (2) and (5)), and T^w_{min}/ T^w_{max} is the minimum/ maximum value for T with workload w. The obstacle calculator uses this formula to compute the obstacle indicators as described in Fig. 1.

6.3   Obstacle-Aware Resource Selection

A resource's vector $r_i^{(t_j,w)}$ allows evaluating a resource's obstacles likelihood while having a workload w. For example, a resource with a time indicator equal to 10 h is more likely to face obstacles delaying its execution time than a resource with time equal to 7 h. However, in some contexts we might need to give more importance to one indicator than another (e.g., time is more important than cost for a given BP). Thus, we propose an objective function (7) estimating and minimizing the obstacle likelihood (O) of a resource with a certain workload for a given task where the BP manager requirements (i.e., time, cost, and quality) are defined using weights. By default the three indicators are equally weighted (i.e., $\alpha = \frac{1}{3}, \beta = \frac{1}{3}, and\ \gamma = \frac{1}{3}$. Since quality is to maximise (contrary to time and cost), its weight needs to be negative. Basically the sum of the absolute values of weights is set to 1.

$$O(r_i, t_j, w) = \alpha \times \mathcal{T}^w[i,j] + \beta \times \mathcal{C}^w[i,j] - \gamma \times \mathcal{Q}^w[i,j] \tag{7}$$

The allocation advisor depicted in our approach (Fig. 1) uses this formula to implement Algorithm 1. The latter allows matching the BP manager requirements with resources obstacle indicators to identify resources considering past obstacles. It takes as input: the index (k) of the task to allocate, the required role assigned to the task, the three indicator's matrices, and the BP manager requirements. It returns an ordered set of resources. From the three matrices, the algorithm considers only resources matching the needed $role$ (line 2). For matched resources, it estimates the obstacle likelihood of each resource for t_k (i.e., $O(r_i, t_k, w)$) based on the BP manager requirements (line 3). After, resources and their obstacle likelihood values are stored as couples $(r_i, O(r_i, t_k, w))$ in $\mathcal{O}$ (line 4). Lastly, Algorithm 1 returns an ordered set of candidate resources with the n-minimum values in $\mathcal{O}$ (line 7).

Algorithm 1. BP manager requirements matching

Input : k - Index of the task to be allocated, $role$ - resource's role , α, β, γ - indicators' weights, $\mathcal{T}, \mathcal{C}, \mathcal{Q}$ - indicators' matrix
Output: R - ordered set of resources
1 **for** $i = 1 \rightarrow n$ **do**
2 $\quad$ **if** $r_i.role = role$ **then**
3 $\quad\quad$ $O(r_i, t_k, w) = \alpha * \mathcal{T}^w[i,k] + \beta * \mathcal{C}^w[i,k] - \gamma * \mathcal{Q}^w[i,k]$
4 $\quad\quad$ $\mathcal{O} \leftarrow \mathcal{O} \cup (r_i, O(r_i, t_k, w))$
5 $\quad$ **end**
6 **end**
7 **return** $getResource(sort(\mathcal{O}))$

6.4   Example

We illustrate now how our approach takes place in practice. We consider the car-manufacturing motivating example and we focus on BP cases where robot m_3 is performing task t_3 with w equal to 2. We suppose that during these executions

m_3 broke down once (i.e., m_3 goes on serviced state and t_3 takes on suspended state) and succeed once while performing t_3. We apply the predefined indicators Formulas ((1), (2), and (5)) on the enriched event log considering only cases where m_3's workload is equal to 2. For the time indicator, we extract from the log both start and end times of the two executions and we compute it as follows:

$$T(m_3, t_3, 2) = \sum_{case_k \in F_2} \frac{evt_{end}.Timestamp - evt_{start}.Timestamp}{|case_k|}$$

$$= \frac{(30/12/2010 : 22.33 - 30/12/2010 : 16.08) + (07/01/2011 : 11.42 - 06/01/2011 : 11.30)}{2}$$

$$= 15.3$$

$T(m_3, t_3, 2)$ is then normalized, supposing that we already calculated the matrix T for all resources and tasks, using Formula (6) as follows:

$$T^2[3,3] = \frac{T(m_3, t_3, 2) - T^2_{min}}{T^2_{max} - T^2_{min}} = \frac{15.3 - 0.1}{20 - 0.1} = 0.76$$

In the same way, we compute indicator's values for each resource in the car-manufacturing company, for the different available tasks, and for different workloads. We compute these indicators based on a sample log[1] associated to our motivating example. The indicator's values are represented in Fig. 2 as three-dimensional matrices where x axis represents the resources that participated in the BP cases, y axis the BP tasks, and z axis the associated workload. We assume that w vary from 1 to 10, for a given resource r_i and a given task t_j and we assume that the event log is complete enough to compute time, cost, and quality for all possible values of w. The so computed time, cost, and quality values, allow extracting the vector representation of a resource. For instance $m_3^{(t_3,2)} = [0.76, 0.91, 1]$ indicates the time, cost, and quality values for m_3 with workload 2 performing t_3.

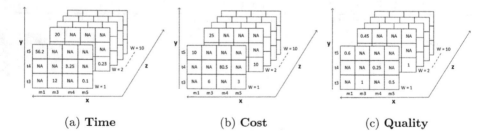

(a) **Time**　　　　　　　　(b) **Cost**　　　　　　　　(c) **Quality**

Fig. 2. Resources' $\mathcal{T}, \mathcal{C}, \mathcal{Q}$ indicators

Supposing that the BP manager would like to allocate t_3 to a robot with more emphasis on quality than time and cost (i.e., $\gamma = 0.6 > \alpha = 0.25 > \beta = 0.15$).

---

[1] The used sample log is available at http://perso.isep.fr/msellami/BIS17/.

Using Algorithm 1, obstacle likelihood is computed for two resources having the adequate role (m_3 and m_5) with a workload equal to 2. m_3 and m_5 are then recommended such that $O(m_5, t_3, 2) = -0.57$ and $O(m_3, t_3, 2) = -0.27$.

7   Approach Development

We operationalized our approach by developing a Java tool allowing to compute resource's indicators (i.e., *time, cost*, and *quality*) using a BP's log. The tool displays computed indicators as charts providing BP managers with an overview of their resources (Fig. 3a). To initialise the application, a BP manager has to provide: a BP log, a list of BP tasks and resources, and resources' unit cost and unit corrective cost. We tested this tool using the sample log of Sect. 6.4. Based on this log, the tool computes indicator's values of all available resources, with their respective tasks, and considering different workloads using Formulas (1), (2), and (5). These values (representing matrices $\mathcal{T}$, $\mathcal{C}$, and $\mathcal{Q}$) are stored in a dedicated JSON file. By clicking on the Decision button, the BP manager is first invited to choose a task to allocate, the resource's role, and to fill his requirements by specifying the *time, cost*, and *quality* weights (Fig. 3b). After validating his choices and following the steps depicted in Algorithm 1, the tool recommends to the BP manager adequate resources considering past obstacles. Finally, a recommendation panel is displayed proposing a set of resources to be allocated according to their obstacle likelihood values.

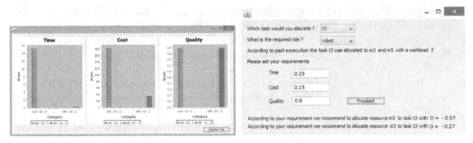

(a) Resources indicators overview   (b) Resource allocation overview

Fig. 3. Screenshots from the developed tool

8   Conclusion and Future Work

This paper proposes an obstacle-aware resource allocation approach that allows BP managers to take obstacle-informed decisions when allocating resources. Using historical data extracted from BP execution logs and enriched with workload information, the proposed approach computes obstacle-likelihood values for BP resources based on three indicators (i.e. *time, cost*, and *quality*). Based on these values, and according to a BP manager's requirements, resources minimizing obstacles occurrence are proposed for allocation. As part of our future work, we aim to minimize the overall obstacle of each process instance (i.e. by considering the impact of a resource's obstacle on the entire process). Another direction for future work is to investigate the incremental updating of the indicator's values each time a new event is recorded.

References

1. van der Aalst, W.M.P.: Business process management: a comprehensive survey. ISRN Softw. Eng. **2013**, 1–37 (2013)
2. Wetzstein, B., Ma, Z., Filipowska, A., Kaczmarek, M., Bhiri, S., Losada, S., Lopez-Cob, J.M., Cicurel, L.: Semantic business process management: a lifecycle based requirements analysis. In: Proceedings of the Workshop on Semantic Business Process and Product Lifecycle Management, pp. 1–11 (2007)
3. Huang, Z., Lu, X., Duan, H.: A task operation model for resource allocation optimization in business process management. IEEE Trans. Syst. Man Cybern. Part A: Syst. Hum. **42**(5), 1256–1270 (2012)
4. Xu, J., Liu, C., Zhao, X., Ding, Z.: Incorporating structural improvement into resource allocation for business process execution planning. Concurr. Comput. Pract. Exp. **25**(3), 427–442 (2013)
5. Cabanillas, C., Knuplesch, D., Resinas, M., Reichert, M., Mendling, J., Ruiz-Cortés, A.: RALph: a graphical notation for resource assignments in business processes. In: Zdravkovic, J., Kirikova, M., Johannesson, P. (eds.) CAiSE 2015. LNCS, vol. 9097, pp. 53–68. Springer, Cham (2015). doi:10.1007/978-3-319-19069-3_4
6. Muehlen, M., Ho, D.T.-Y.: Risk management in the BPM lifecycle. In: Bussler, C.J., Haller, A. (eds.) BPM 2005. LNCS, vol. 3812, pp. 454–466. Springer, Heidelberg (2006). doi:10.1007/11678564_42
7. Dali, A., Lajtha, C.: ISO 31000 risk management "The Gold Standard". EDPACS **45**(5), 1–8 (2012)
8. van der Aalst, W.M.P.: Process Mining: Discovery, Conformance and Enhancement of Business Processes. Springer Publishing Company, Incorporated, Heidelberg (2011)
9. Suriadi, S., Weiß, B., Winkelmann, A., ter Hofstede, A.H., Adams, M., Conforti, R., Fidge, C., La Rosa, M., Ouyang, C., Rosemann, M., et al.: Current research in risk-aware business process management: overview, comparison, and gap analysis. Commun. Assoc. Inf. Syst. **34**(1), 933–984 (2014)
10. Jakoubi, S., Tjoa, S.: A reference model for risk-aware business process management. In: Proceedings of CRiSIS, pp. 82–89 (2009)
11. Stefan, F.: From the resource to the business process risk level. In: Proceedings of SAISMC, pp. 100–109 (2010)
12. Kang, B., Cho, N.W., Kang, S.H.: Real-time risk measurement for business activity monitoring (BAM). Int. J. Innov. Comput. Inf. Control **5**(11), 3647–3657 (2009)
13. Conforti, R., de Leoni, M., Rosa, M.L., van der Aalst, W.M.P., ter Hofstede, A.H.M.: A recommendation system for predicting risks across multiple business process instances. Decis. Support Syst. **69**, 1–19 (2015)
14. Pika, A., van der Aalst, W.M.P., Wynn, M.T., Fidge, C.J., ter Hofstede, A.H.M.: Evaluating and predicting overall process risk using event logs. Inf. Sci. **352**, 98–120 (2016)

15. Suriadi, S., Ouyang, C., van der Aalst, W.M.P., Hofstede, A.H.M.: Root cause analysis with enriched process logs. In: Rosa, M., Soffer, P. (eds.) BPM 2012. LNBIP, vol. 132, pp. 174–186. Springer, Heidelberg (2013). doi:10.1007/978-3-642-36285-9_18
16. Thevendran, V., Mawdesley, M.: Perception of human risk factors in construction projects: an exploratory study. Int. J. Project Manage. **22**(2), 131–137 (2004)
17. Nada R. Sanders, J.D.W.: Risk management. In: Foundations of Sustainable Business: Theory, Function, and Strategy, pp. 157–186, November 2014
18. Maamar, Z., Sellami, M., Faci, N., Lefebvre, S.: Detecting and tackling run-time obstacles in social business processes. In: Proceedings of AINA (2017, to appear)

Ontology-Based Data Access for Extracting Event Logs from Legacy Data: The onprom Tool and Methodology

Diego Calvanese[1], Tahir Emre Kalayci[1(✉)], Marco Montali[1],
and Stefano Tinella[2]

[1] KRDB Research Centre for Knowledge and Data,
Free University of Bozen-Bolzano, Bolzano, Italy
{calvanese,tkalayci,montali}@inf.unibz.it
[2] EBITmax srl, via Macello 63/F, Bolzano, Italy
s.tinella@ebitmax.it

Abstract. Process mining aims at discovering, monitoring, and improving business processes by extracting knowledge from event logs. In this respect, process mining can be applied only if there are proper event logs that are compatible with accepted standards, such as extensible event stream (XES). Unfortunately, in many real world set-ups, such event logs are not explicitly given, but instead are implicitly represented in legacy information systems. In this work, we exploit a framework and associated methodology for the extraction of XES event logs from relational data sources that we have recently introduced. Our approach is based on describing logs by means of suitable annotations of a conceptual model of the available data, and builds on the ontology-based data access (OBDA) paradigm for the actual log extraction. Making use of a real-world case study in the services domain, we compare our novel approach with a more traditional extract-transform-load based one, and are able to illustrate its added value. We also present a set of tools that we have developed and that support the OBDA-based log extraction framework. The tools are integrated as plugins of the ProM process mining suite.

Keywords: Process mining · Ontology-based data access · Event log extraction · Relational database management systems

1 Introduction

Contemporary organizations are increasingly recognizing the importance of analyzing how their business processes are conducted in the real world, towards quality assurance, optimization, and continuous improvement. Process mining [1] is emerging as one of the most promising and effective framework to tackle this need. Process mining stands at the intersection of model-driven engineering and data science: insights are automatically extracted from event data that represent the footprint of process executions inside the company, and used to discover and enrich process models, provide operational support, check compliance,

© Springer International Publishing AG 2017
W. Abramowicz (Ed.): BIS 2017, LNBIP 288, pp. 220–236, 2017.
DOI: 10.1007/978-3-319-59336-4_16

analyze bottlenecks, compare process variants, and suggest improvements [2]. A plethora of process mining techniques and technologies have been developed and successfully employed in several application domains[1].

The applicability of process mining depends on two crucial factors:

- the availability of high-quality event data, that is, logs containing correct and complete data about which cases (process instances) have been executed, which events occurred for each case, and when they did occur;
- the representation of such data in a format that is understandable by process mining algorithms, such as the IEEE XML-based standard eXtensible Event Stream (XES) [3].

In this respect, two main situations typically arise in an industrial setting. In the first situation, the company explicitly adopts a business process or enterprise management system that logs cases, events and corresponding attributes explicitly, facilitating the extraction of an event log and its conversion into XES.

The literature abounds of techniques and tools to handle the log extraction in this setting, such as, e.g., XESame [4] and ProMimport [5]. Additionally, commercial tools like Disco[2], Celonis[3], and Minit[4] support the conversion from CSV or spreadsheet files into XES. Worth mentioning are also [6,7], the first because it tackles the extraction of event logs from redo-logs of relational databases, the second because it is one of the few approaches that leverages the relational technology to access the event log directly, instead of materializing it into XML.

In the second situation, the company adopts a more general management system, configuring it for its own specific needs, and combining it with domain-specific databases and other legacy data sources. In this setting, cases and events may not be explicitly stored in dedicated data structures, but instead implicitly present inside the company information system. In addition, there is typically not a single notion of "case" and related "events", but they change depending on the perspective of interest, and on which aspects of the company one wants to focus on. For example, in an order-to-cash process, one could focus on the flow of orders, to understand why sometimes orders take too much time to be delivered, or on the flow of operations conducted by a warehouse employee, to check whether it complies with internal regulations. Depending on which notion of case is selected, also the relevant events change. E.g., the payment of an order is important when analyzing the flow of orders, but may be irrelevant when focusing on the warehouse.

Unfortunately, the literature lacks techniques, methodologies and tools to support domain experts and process analysts in the extraction of event logs from legacy information systems, and reflecting multiple perspectives. The result is that logs are extracted manually, adopting ad-hoc procedures that are based on extracting a copy of the data and transforming it according to specific requirements. This process, which resembles the extract-transform-load (ETL) approach taken for data warehousing, creates redundancy, and is labor intensive and error prone.

---

[1] http://tinyurl.com/ovedwx4.
[2] https://fluxicon.com/disco/.
[3] http://www.celonis.de/en/.
[4] http://www.minitlabs.com.

In this paper, we tackle this open challenge. Leveraging the technique first presented in [8], we propose an approach based on conceptual modeling to semi-automatize the extraction of event logs from legacy information systems. In our approach, called onprom, humans only focus on the conceptual issues involved in the extraction: *(i)* Which are relevant concepts and relations? *(ii)* How do such concepts/relations map to the underlying information system? *(iii)* Which concepts/relations relate to the notion of case, event, and event attributes?

Once this information is provided, the log extraction process is handled in a fully automatized way, leveraging the paradigm of *ontology-based data access* (OBDA) [9–11]. In OBDA, a high-level representation of the domain of interest, in our case provided in terms of a conceptual schema, is linked to the data sources using a declarative specification, called *mappings*. In this way, information about the event logs can be extracted from the sources by exploiting both the conceptual schema and the mappings.

In the following, we describe a real process mining use case that has been initially handled using an ad-hoc, ETL-like methodology. Employing this use case as a running example, we then introduce our onprom methodology, discussing its different phases and how conceptual models are used both as documentation and computational artifacts. We then show how the preliminary implementation reported in [8] has now been transformed into a complete chain of tools, fully integrated with the well-known ProM process mining framework.

2   Case Study and Motivation

To provide a concrete motivation and explanation for our framework, we introduce the problem of extracting event logs from legacy information systems in a real case study. The case study has been carried out by EBITmax[5], an innovative SME from South Tyrol, Italy. EBITmax provides consultancy services in program management and business process management for a number of small and large enterprises, operating within the territory and abroad. Recently, EBITmax incorporated process mining to complement its standard consultancy services, enriching and comparing models with fine-grained insights automatically extracted from data, and accounting for how business processes are executed in reality. In particular, a pilot project in process mining is currently run by EBITmax on the service provisioning and financial processes of Markas[6], a company with more than 7 000 employees providing a multitude of services for large establishments operating in Italy, Austria, and Romania. Specifically, the pilot consists in the analysis of the *Accounts Payable Process* (APP), used by Markas to handle payments to external suppliers, and their corresponding invoices. To support the internal management of the APP, Markas does not employ a workflow management system, but relies on shared guidelines on how to handle payments, and on an *Enterprise Resource Planning* (ERP) system to track the executed operations. In this setting, Markas management would like to understand whether the

---

[5] http://www.ebitmax.it.

[6] http://www.markas.com/en/home.html.

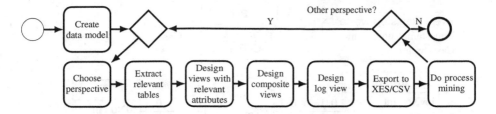

Fig. 1. Traditional methodology for data preparation for process mining

APP is executed as expected and, if not, where do deviations appear, considering all the orders created in 2015.

2.1 Understanding the Problem

The first step followed by EBITmax has been to understand the details of the APP, both conceptually and in terms of IT support. On the one hand, this resulted in the creation of a model for the expected APP that is expressed in the *Business Process Model and Notation*[7] (BPMN) standard. The model has been obtained following a traditional interview-based approach. On the other hand, this resulted in the annotation of the BPMN model, so as to know which tasks are executed manually, and which are performed through the ERP system, and are consequently logged. Among the logged tasks, we mention the following, key ones:

- SubmitOrder: an order is inserted into the ERP and submitted to a supplier.
- GetTD: Markas receives the ordered material; this is traced in the ERP when the transport document (TD) attached to the material is inserted into the system.
- RegisterInvoice: the invoice for the payment is inserted into the ERP, reflecting what is listed in the TD.
- PaySupplier: the payment is confirmed.

Normally, the management expects that *the invoice of an order is not registered (and, consequently, not paid) unless the official TD is obtained and inserted into the ERP*. Within the pilot project, the general question of the alignment between the expected and the actual APP boiled down to check whether the business constraint "no invoice unless TD" is indeed respected by the actual APP.

2.2 Process Mining via Manual Event Log Extraction

To answer the research question through process mining, EBITmax needed to tackle the difficult issue of data preparation, which is considered one of the most challenging, open problems in process mining [1]. In this specific case, the

---

[7] http://www.bpmn.org/.

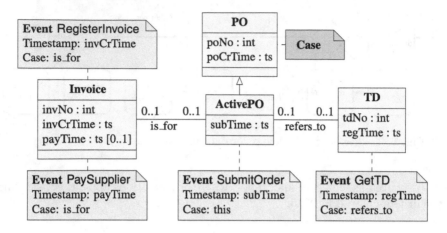

Fig. 2. Excerpt of the data model for APP

concrete problem was to identify the semantics of tables/columns of the ERP, and the whereabouts of relevant data for orders, TDs, invoices, and payments. To this end, EBITmax devised and documented the methodology shown in Fig. 1.

Conceptual data modeling. The first step of the methodology consists in the creation of a conceptual model that accounts for the data maintained in the ERP at a higher level of abstraction, on the one hand making it possible to discuss with managers and domain experts about the semantics of such data, and on the other hand providing the basis to understand where and how they are stored within the ERP. The UML class diagram in Fig. 2 depicts a small excerpt of the resulting model, showing the key concepts of PO (purchase order), TD (transport document), and Invoice. ActivePO represents a PO that has been submitted, consequently triggering the execution of an APP instance.

Choosing perspective. The second step consists in combining the research question with the data model, so as to choose a *perspective* for the analysis, and in particular: (*i*) the "subject" of the analysis, i.e., which notion of case to adopt; (*ii*) which relevant events should be considered in the evolution of cases; and (*iii*) which event attributes should be included.

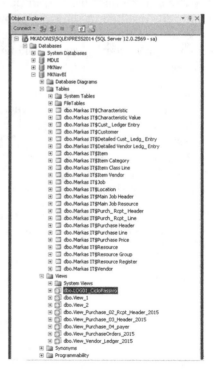

Fig. 3. Manual construction of views

Manual data preparation. Building on this guideline, EBITmax started a fine-grained analysis of the ERP system and its underlying database, towards the extraction of the desired information. The extraction is executed manually in an ETL-like fashion, and is organized in four incremental phases: *(i)* identification of the relevant tables; *(ii)* creation of "filter views" that maintain, and clean up, the relevant information present in such tables (e.g., selection of interesting attributes for purchase orders); *(iii)* merging of filter views into "composite views" that provide a higher level of abstraction, and group together data belonging to the same conceptual classes/relations (e.g., all data referring to a transport document); and *(iv)* creation of a single "log view" that coherently rearranges the information present into composite views in accordance with the chosen perspective for case, events, and attributes. Figure 3 shows the selected tables and views constructed by EBITmax on top of Markas ERP.

2.3   Log Extraction and Process Mining

Finally, EBITmax converted the "log view" into a CSV file, and exploited the Disco process mining toolkit[8] for the analysis. Some interesting deviations departing from the expected APP were consequently detected, and discussed with the Markas management. One of the most interesting, and quite common, deviations was represented by orders that are submitted and paid without registering the transport document at all. Looking at the data, Markas realized that this deviation is due to a recent "drift" in the management of the APP, caused by the introduction of digital invoices in the Italian market. In fact, at some point suppliers equipped with e-invoicing started to digitally send the billing information related to submitted purchase orders. The information contained in the e-invoice mirrors all the TD data needed to execute the payment, and obviously is received by Markas before the ordered material and the corresponding TD. This allows one to speed-up the process, by paying as soon as the e-invoice is received. When the payment is executed, the ERP forbids further changes to the order and its related information, thus making it impossible to register the TD once it is received. Markas appreciated the findings obtained from the pilot, and recently accepted to continue the project at a bigger scale.

Experienced Issues. In spite of the promising results obtained by EBITmax during the pilot project, the company experienced several issues caused by the manual data preparation for process mining. First of all, the manual creation of views requires a detailed knowledge of the ERP tables, and is a demanding and error-prone task. In addition, whenever the perspective of the analysis is changed, it is necessary to go through all the data preparation phases again, since there is no guarantee that the selected tables, and designed views, will also be useful in relation with the new perspective. This contrasts with the process mining best practices in two respects: quality assurance of the input event log, and feasibility of quickly going through several batches of analysis by changing perspective on the company's data.

---

[8] https://fluxicon.com/disco/.

This experience has spawn a collaboration between EBITmax and the Free University of Bozen-Bolzano, so as to face the next phase of the Markas project in a more systematic way, starting from the log extraction approach first introduced in [8]. In particular, the ongoing collaboration is centered around the methodology and tool support described in the remainder of the paper.

3   The onprom Methodology

We now present our methodology for the semi-automated extraction of logs from legacy information systems, starting from the seminal ideas proposed in [8]. The methodology is backed up by the chain of tools described in Sect. 4.

As a starting point, we assume the existence of a legacy information system $\mathcal{I} = \langle \mathcal{R}, \mathcal{D} \rangle$, with schema $\mathcal{R}$ and set $\mathcal{D}$ of facts about the domain of interest. In the typical case where the information system is a relational database, $\mathcal{R}$ accounts for the schema of the tables and their columns, and $\mathcal{D}$ is a set of data structured according to such tables. On top of $\mathcal{I}$, our methodology is centered on the usage of conceptual models in two respects. First, they are used as documentation artifacts that explicitly capture not only knowledge about the domain of interest, but also how legacy information systems relate to that knowledge. This facilitates understanding and interaction among human stakeholders. Second, conceptual models are used as computational artifacts, that is, to automatize the extraction process as much as possible.

The overall methodology is illustrated in Fig. 4. We review the different phases next, leveraging the APP case study illustrated in Sect. 2.

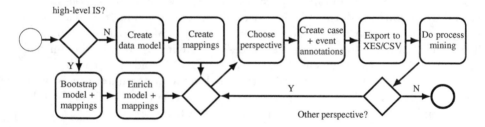

Fig. 4. The onprom methodology

3.1   Conceptual Modeling

The first phase of our methodology consists in the creation of two conceptual models. The first one is the *conceptual data model* $\mathcal{T}$, already discussed in Sect. 2.2. It accounts for the structural knowledge of the domain of interest, i.e., relevant concepts and relations, consequently providing a high-level view of $\mathcal{I}$ that is closer to domain experts. More specifically, we employ UML class diagrams as a concrete language for conceptual data modeling, and we provide their logic-based, formal encoding in terms of the OWL2 QL ontology language. OWL2 QL is one of the profiles[9] [12] of the W3C standard Semantic Web language OWL2, and it has been specifically designed to capture the essential features of conceptual modeling formalisms (see, e.g., [10]). In the following, depending on the context, we refer to $\mathcal{T}$ as a UML class diagram or as the corresponding OWL2 QL ontology.

The second conceptual model, the *mapping specification* $\mathcal{M}$, is a distinctive feature introduced by our approach, borrowed from the area of ontology-based data access (OBDA) [9,10]. $\mathcal{M}$, which explicitly links $\mathcal{I}$ to $\mathcal{T}$, consists of a set of logical implications that map patterns of data over schema $\mathcal{R}$ to high-level facts over $\mathcal{T}$. Patterns over the data $\mathcal{D}$ are expressed as queries over $\mathcal{R}$ (e.g., SQL SELECT statements, when $\mathcal{R}$ is relational), while facts over $\mathcal{T}$ are expressed as logical terms involving objects.

Example 1. Consider again the App case study. In the Markas ERP, Markas-Purchase table contains information about purchase orders, including the primary key No to store the order number, and columns Order_Date and Posting_Date to store the dates at which the order was respectively created and submitted. In addition, by interacting with the domain experts of Markas, EBITmax discovered the following two important facts about such a table:

- Sometimes when the order is created, the Order_Date field is left unspecified[10]; however, it is still possible to reconstruct the orders created in 2015 (i.e., the year targeted by the analysis), as those whose order number starts with 15.
- When the order is created, its posting date is left unspecified, and gets a value when the order is actually submitted.

This knowledge can be made explicit by establishing dedicated mappings. In particular, the following mapping, specified using the mapping syntax of the ontop OBDA framework[11] [11]:

```
order/{oid} poNo {oid} .
 ←  SELECT No AS oid FROM Markas-Purchase WHERE No LIKE '15%'
```

declares that each value *oid* that is stored in the No column of Markas-Purchase and that begins with number 15, corresponds to an object term order/*oid* in the ontology[12], and determines relationship of type poNo between such object and the value *oid*. Since relationship poNo is actually an integer attribute[13] of

---

[9] In W3C terminology, a *profile* is a sublanguage.

[10] It is important to notice that the possible absence of an actual value for Order_Date does not contrast with the class diagram of Fig. 2, which dictates that every purchase order has exactly one creation time. In fact, conceptual models are interpreted under *incomplete information*: the absence of the creation date for an order does not mean that the order has no creation date, but that such an order has a creation date that is not certainly known.

[11] http://ontop.inf.unibz.it/.

[12] In the left-hand side of a mapping, curly brackets are used to denote answer variables of the SQL query in the right-hand side.

[13] In OWL terms, it is a *data property*.

class PO in $\mathcal{T}$, this mapping is implicitly declaring also that order/*oid* belongs to class PO, and that its order number is *oid*. An example of mapping explicitly populating a class is:

```
order/{oid} rdf:type ActivePO.
  ←   SELECT No AS oid FROM Markas-Purchase
      WHERE No LIKE '15%' AND Posting_Date IS NOT NULL
```

This mapping expresses that each order tuple in Markas-Purchase identified by value *oid* is mapped to an object order/*oid* of type ActivePO in $\mathcal{T}$, whenever its posting date has a non-null value. ∎

When $\mathcal{M}$ is fully defined, it can be used for two purposes. On the one hand, it explicitly documents how the structure of the company information system has to be conceptually understood in terms of domain concepts and relations, and thus constitutes an asset for the company that itself might be worth an investment [13]. On the other hand, $\langle \mathcal{I}, \mathcal{T}, \mathcal{M} \rangle$ constitutes what is called an *OBDA system*, which completely decouples end users from the details of the information system: whenever a user poses a conceptual query Q (e.g., expressed using the semantic web query language SPARQL) over $\mathcal{T}$, the OBDA system *(i)* leverages $\mathcal{T}$ and $\mathcal{M}$ to automatically reformulate Q as a corresponding concrete query Q' over $\mathcal{I}$; *(ii)* submits Q' to $\mathcal{I}$; and *(iii)* automatically translates the so-obtained answers into meaningful answers over the vocabulary of $\mathcal{T}$. Notably, this "virtual" approach is conceptually identical to the one in which the mapping $\mathcal{M}$ is used a là ETL to materialize data from $\mathcal{D}$ as facts over $\mathcal{T}$, with the advantage that: *(i)* users do not need to code procedures for data extraction, *(ii)* data are not replicated, and *(iii)* data are retrieved using the standard query engine of the information system.

Bootstrapping. The creation of a suitable data model and mapping specification is a labor-intensive and challenging task. As shown in Fig. 4, if the information system has a "high-level" structure, that is, a structure that is understandable by domain experts, such a phase can be (partially) automatized through bootstrapping techniques [14], which synthesize a conceptual data model that mirrors the structure of the information system, together with suitable mappings. The result of bootstrapping can then be manually improved and enriched towards the creation of the final OBDA system.

3.2   Event Data Annotations

Once the OBDA system is set up in the previous phase, our methodology allows one to abstract away the information system. In this way, the process mining expert is not required to manually construct views for the extraction of an event log, as done in Fig. 1. Instead, she focuses on $\mathcal{T}$ only, and uses it as the basis for discussion with the company stakeholders, in particular to decide which perspective for process mining to consider. Concretely, choosing a perspective amounts to *annotate* $\mathcal{T}$ with a set $\mathcal{L}$ of event data annotations, where each annotation is used either to: *(i)* indicate which class in $\mathcal{T}$ (possibly with additional restrictions) represents a *case*, *(ii)* which events are present in $\mathcal{T}$ and to which classes they refer, *(iii)* which attributes are attached to events, and where they are located in $\mathcal{T}$. We consider each type of annotation next.

Case annotation. The case annotation specifies which class constitutes the reference point for the analysis. Each object instantiating the case class represents an instance of the process according to the chosen perspective, and provides the basis for correlating events: the set of all events referring to the same case object form a *trace* for such an object. In Fig. 2, the case annotation is shown as a red UML note, and indicates that each purchase order is a case. Additional restrictions on which instances to consider can be applied. E.g., one could specify to consider only orders referring to suppliers from a given geographical area, or orders involving at least a given amount of money.

Event annotations. An event annotation specifies that the annotated class provides information about the occurrence(s) of a type of event that is relevant for the chosen perspective. To "discover" which classes in $\mathcal{T}$ may be subject to an event annotation, our methodology combines two constraints:

- Each event class has to be directly or indirectly linked to the case class. Technically, there must be a UML association, or a chain of concatenated associations (possibly involving IS-A generalizations), that lead to navigate from the event class to the case class.
- Each event class has to be directly or indirectly linked to a timestamp attribute, providing the information on "when" instances of such an event occurred. Technically, there must be a *functional* attribute of the event class, or a chain of one or more *functional* associations (possibly involving IS-A generalizations), that lead to navigate from the event class to its timestamp attribute.

Two observations are in place regarding the aforementioned navigations and the multiplicities attached to the involved associations. For event-case navigations, we allow arbitrary multiplicities since, in general, an event may belong to multiple cases. Consider, e.g., the economic transaction of a purchase between two persons, in a situation where the person is marked as case class. In this setting, the transaction may be considered as an event belonging to the trace of the buyer *and* to that of the seller.

Event-timestamp navigations are of a different nature: since each event must be unambiguously associated to a single timestamp, the navigation can only traverse *functional* associations, that is, many-to-one or one-to-one associations, and lead to a functional attribute.

In both situations, we allow for optionality, i.e., for navigations traversing associations whose minimum multiplicity is 0. This is needed to reflect that a trace may be incomplete (thus missing events), and that multiple traces of the same case class may indeed contain different events. The following example clarifies this aspect.

Example 2. Consider again the class diagram in Fig. 2. The `Invoice` class may be annotated with two event types, one for the invoice registration, and one for its payment. The first event exists for every invoice, as it is associated to the mandatory `invCrTime` timestamp attribute. The second event, instead, does not necessarily exist for every invoice: there may be invoices that are not yet paid, and invoices that may never be paid. This is reflected by the fact that the `payTime` attribute of `Invoice` is optional.   ■

In general, there exist several possible event-case and event-timestamp navigational paths. To disambiguate which ones are actually used to define an event class, attribute annotations are used.

Attribute annotations. Attribute annotations decorate event annotations with information about their features. Each attribute annotation consists of an attribute and of the specification of a navigational path to (functionally) reach its value. Mandatory attributes are: *(i) case* (how to reach the case class from the event class); *(ii) timestamp* (how to reach the timestamp attribute for the event class); and *(iii) activity* (a constant string or a navigational path specifying which is the activity to which the event refers). Optional attributes are related to *resources*, i.e., how to reach the identifier and/or the role of the resource responsible for the event.

Example 3. Consider again the APP case study. Differently from the manual approach illustrated in Sect. 2.2, in our methodology the four event types elicited by EBITmax in Sect. 2.1 may be elicited as shown in the orange UML notes of Fig. 2:

- Each instance of `ActivePO` (directly corresponding to a case) may determine a **SubmitOrder** event that occurs at the submission time (attribute `subTime`) for that order.
- Each instance of `TD` may determine a **GetTD** event for the order obtained by navigating the `refers_to` association, and that occurs at the registration time (attribute `regTime`) for that document.
- Each instance of `Invoice` may determine two events for the order obtained by navigating the `is_for` association: a **RegisterInvoice** event occurring at the creation time (attribute `invCrTime`) for that invoice, and a **PaySupplier** event occurring at the payment time (attribute `payTime`) for that invoice.   ■

3.3   Automated Event Log Extraction

Thanks to the technique introduced in [8], once the conceptual data model is suitably annotated, it is possible to automatically extract from the legacy information system an event log that reorganizes the data contained there according to the specified annotations. Intuitively, this is done by combining the mapping specification with the annotations, and by computing the answers to a series of queries that ask for all the cases present in the information system and, for each case, all the events referring to that case, together with the corresponding attribute values. The so-obtained data structure can then be represented using the IEEE XES standard for event logs.

Differently from the manual extraction methodology of Sect. 2.2, this approach does not only help the process mining expert in working at the conceptual level without coding ad-hoc views, but also facilitates multi-perspective process mining, that is, the extraction of several event logs reflecting different perspectives over the same information system. Each perspective, in fact, requires to change the annotations, while the conceptual data model T and the mapping specification $\mathcal{M}$ from the information system to the data model remain unchanged.

4   Implementing the Framework: The onprom Tool

To support the various phases of the OBDA-based event log extraction framework illustrated in Sect. 3, we have developed a tool suite named onprom, consisting of various plug-ins of the ProM extensible process mining framework[14] [4]. Specifically, onprom consists of the following components: *(i)* a *UML Editor*, used to design the domain ontology (cf. Sect. 3.1); *(ii)* an *Annotation Editor*, allowing domain expert to specify the event data annotations (cf. Sect. 3.2); *(iii)* a *Log Extractor*, used to extract from the underlying database the XES event log, based on the annotated domain ontology and the mapping specification (cf. Sect. 3.3). The different tools are implemented as separate projects in Java. When used as ProM plug-ins, they exchange data relying on the mechanisms built in ProM. However, both the UML Editor and the Annotation Editor can be used also as stand-alone tools that operate using files for input and output. We now describe the tools in more detail, relying on the APP case study for examples.

4.1   UML Editor

The log extraction framework based on OBDA makes use of a domain ontology expressed in the OWL2 QL profile [12] of OWL2, which is the profile supported by the OBDA system ontop. Ontologies expressed in OWL2 QL admit a natural graphical representation in the form of UML class diagrams [10]. Hence, to provide domain experts support for the design of such ontologies, we have developed a graphical editor for UML class diagrams. Actually, since the editor can import standard OWL2 QL ontologies, it can also be used to modify and enhance an already existing or independently developed OWL2 QL ontology.

To maintain the UML Editor lightweight, and to guarantee at the same time that the designed UML class diagrams can indeed be expressed in OWL2 QL, we have made some natural simplifying assumptions on the form of the UML class diagrams supported by the tool:

– we do not support *completeness* of UML generalization hierarchies, since the presence of such construct would fundamentally undermine the virtual OBDA approach based on query reformulation [10];

---

[14] http://www.promtools.org/.

- in line with Semantic Web languages, we support explicitly only associations of arity 2, and do not support association classes currently;
- multiplicities in associations (resp., of attributes) are restricted to be either 0 or 1. Hence, we can express functionality and mandatory participation;
- we do not support ISA between associations;
- we ignore all those features of UML class diagrams that are more relevant for the software engineering perspective, and less for the conceptual perspective of UML, such as stereotypes, method specifications, and aggregations.

The developed UML class diagram can be saved in a proprietary JSON format for further processing and as input for the Annotation Editor. It can also be exported as a standard OWL2 QL ontology, hence ready to be processed by ontop. We observe that the graphical layout information, which is not part of the OWL2 QL language, is maintained in the form of OWL2 annotations, thus resulting in an ontology fully compliant with the W3C standard.

A screenshot of the UML Editor with the domain ontology of the App use case is shown in Fig. 5.

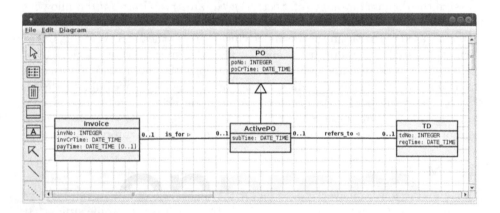

Fig. 5. The UML Editor showing the domain ontology of the App use case

4.2   Annotation Editor

To provide process mining experts with the possibility of specifying in a simple, intuitive way the over the domain-specific ontology $\mathcal{T}$, we have developed an Annotation Editor that supports the different forms of annotation.

A screenshot of the Annotation Editor, with the domain ontology of the App use case annotated with the case and various event elements, is shown in Fig. 6. Specifically, the domain expert has annotated PO as the case of the log, while four different events are defined:

1. GetTD accesses the case using the refers_to association and navigating the *IS-A* relationship; it has regTime in the TD class as timestamp (see Fig. 7a).

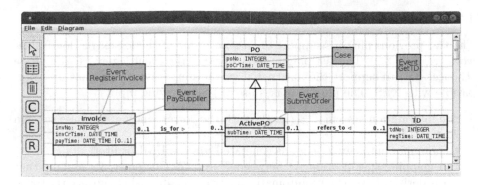

Fig. 6. The Annotation Editor showing annotations for the APP use case

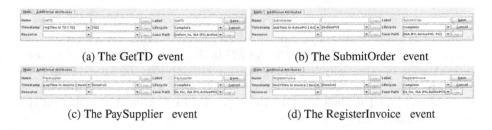

(a) The GetTD event

(b) The SubmitOrder event

(c) The PaySupplier event

(d) The RegisterInvoice event

Fig. 7. The properties of event annotations defined for the APP use case

2. SubmitOrder accesses the case directly through an *ISA* relationship; it has subTime in the ActivePO class as timestamp (see Fig. 7b)
3. PaySupplier accesses the case using the is_for association and *IS-A*; it has payTime in the Invoice class as timestamp (see Fig. 7c).
4. RegisterInvoice also accesses the case using the is_for association and *IS-A*, but it has invCrTime in the Invoice class as timestamp (see Fig. 7d).

We observe that all events have a *complete* life-cycle.

To simplify the annotation task, the editor supports some advanced operations:

– Properties and paths can be chosen using navigational selection over the diagram via mouse-click operations.
– The editor takes into account multiplicities on associations and attributes; when the user is selecting properties of the case and of events (in particular the timestamp), the editor enables only navigation paths that are functional, thus guaranteeing that the properties to include in the extracted log are uniquely determined.

The annotated domain ontology can be exported using a proprietary JSON format, that can then be imported by the log extraction plug-in.

4.3   Log Extraction Plug-in

The two previous plug-ins support the design phase of the log extraction framework (cf. Sect. 3.2). The last plug-in is deployed in the event log extraction phase (cf. Sect. 3.3) to support the automated extraction of event logs that are compatible with XES. The plug-in makes use of the following inputs:

- the information system $\mathcal{I} = \langle \mathcal{R}, \mathcal{D} \rangle$, with the corresponding database schema $\mathcal{R}$;
- the domain ontology $\mathcal{T}$, e.g., as generated via the UML Editor;
- the mapping specification $\mathcal{M}$ between $\mathcal{T}$ and $\mathcal{R}$. Currently, we assume that $\mathcal{M}$ is derived (either semi-automatically or manually) using third-party tools, such as the ontop mapping editor;
- annotations $\mathcal{L}$, which are created using the Annotation Editor.

The tool exploits the *query rewriting* functionalities provided by ontop to generate from the above inputs a *new mapping specification* $\mathcal{M}_{log}$, which establishes a direct correspondence between the data $\mathcal{D}$ in $\mathcal{I}$ and the elements of XES, i.e., trace, event, ..., essentially bypassing the domain ontology $\mathcal{T}$. An ontology capturing the main elements of XES, together with $\mathcal{M}_{log}$ and $\mathcal{I}$, constitutes a new OBDA system that is then used for the event log extraction, by relying on the *data access* functionalities of ontop.

5   Conclusion

In this work, we have presented the onprom framework for the extraction of event logs from legacy information systems, using a real case study to illustrate the limitations of manual extraction, and to show the features of our approach. The framework comes with a methodology centered around conceptual models to capture domain knowledge, to link such knowledge to the underlying data, and to annotate such knowledge with event-related information, reflecting the chosen perspective for process mining. In addition, the framework comes with a toolchain to handle such conceptual models, and automatically extract a XES event log in accordance with the chosen perspective, leveraging ontology-based data access (OBDA) techniques. The toolchain exploits the OBDA features offered by the ontop system, and is fully integrated with the ProM process mining framework. It can be downloaded from http://onprom.inf.unibz.it.

We are currently investigating, together with the EBITmax company, the concrete application of onprom to the Markas case study reported here, and plan to use the results of this case study as a way to further validate the methodology, and to conduct an extensive experimental evaluation, extending the preliminary results obtained in [8].

In addition, we are actively working on extending the annotation editor, on the one hand to provide more guidance to the user in discovering meaningful event classes, and on the other hand to support more sophisticated navigational queries. Finally, we observe that, currently, we do not offer an editor for the specification of mappings that is fully integrated with our toolchain, but we rely instead to what is natively offered by the ontop OBDA framework. A natural next step is to manage the specification of mappings within our toolchain, leveraging recent approaches on the graphical specification of mappings, developed within the recently concluded Optique EU Project[15].

Acknowledgement. This research has been partially supported by the Euregio IPN12 *"KAOS: Knowledge-Aware Operational Support"* project, which is funded by the "European Region Tyrol-South Tyrol-Trentino" (EGTC) under the first call for basic research projects and by the UNIBZ internal project *"OnProm"*. We thank Ario Santoso for the development of the log extraction plug-in of onprom, and Wil van der Aalst for the interesting discussions and insights on the problem of extracting event logs from legacy information systems.

References

1. van der Aalst, W.M.P., et al.: Process mining manifesto. In: Daniel, F., Barkaoui, K., Dustdar, S. (eds.) BPM 2011. LNBIP, vol. 99, pp. 169–194. Springer, Heidelberg (2012). doi:10.1007/978-3-642-28108-2_19
2. van der Aalst, W.M.P.: Process Mining - Data Science in Action, 2nd edn. Springer, Heidelberg (2016)
3. IEEE Computational Intelligence Society: IEEE Standard for eXtensible Event Stream (XES) for Achieving Interoperability in Event Logs and Event Streams. IEEE Std 1849-2016, i–50 (2016)
4. Verbeek, H.M.W., Buijs, J.C.A.M., Dongen, B.F., van der Aalst, W.M.P.: XES, XESame, and ProM 6. In: Soffer, P., Proper, E. (eds.) CAiSE Forum 2010. LNBIP, vol. 72, pp. 60–75. Springer, Heidelberg (2011). doi:10.1007/978-3-642-17722-4_5
5. Günther, C.W., van der Aalst, W.M.P.: A generic import framework for process event logs. In: Eder, J., Dustdar, S. (eds.) BPM 2006. LNCS, vol. 4103, pp. 81–92. Springer, Heidelberg (2006). doi:10.1007/11837862_10
6. van der Aalst, W.M.P.: Extracting event data from databases to unleash process mining. In: vom Brocke, J., Schmiedel, T. (eds.) BPM - Driving Innovation in a Digital World. Management for Professionals, pp. 105–128. Springer, Cham (2015). doi:10.1007/978-3-319-14430-6_8
7. Syamsiyah, A., van Dongen, B.F., van der Aalst, W.M.P.: DB-XES: enabling process discovery in the large. In: Proceedings of the 6th International Symposium on Data-Driven Process Discovery and Analysis (SIMPDA). CEUR, vol. 1757, pp. 63–77. ceur-ws.org (2016)
8. Calvanese, D., Montali, M., Syamsiyah, A., van der Aalst, W.M.P.: Ontology-driven extraction of event logs from relational databases. In: Reichert, M., Reijers, H.A. (eds.) BPM 2015. LNBIP, vol. 256, pp. 140–153. Springer, Cham (2016). doi:10.1007/978-3-319-42887-1_12
9. Poggi, A., Lembo, D., Calvanese, D., Giacomo, G., Lenzerini, M., Rosati, R.: Linking data to ontologies. In: Spaccapietra, S. (ed.) Journal on Data Semantics X. LNCS, vol. 4900, pp. 133–173. Springer, Heidelberg (2008). doi:10.1007/978-3-540-77688-8_5

---

[15] http://optique-project.eu.

10. Calvanese, D., Giacomo, G., Lembo, D., Lenzerini, M., Poggi, A., Rodriguez-Muro, M., Rosati, R.: Ontologies and databases: the *DL-Lite* approach. In: Tessaris, S., Franconi, E., Eiter, T., Gutierrez, C., Handschuh, S., Rousset, M.-C., Schmidt, R.A. (eds.) Reasoning Web 2009. LNCS, vol. 5689, pp. 255–356. Springer, Heidelberg (2009). doi:10.1007/978-3-642-03754-2_7

11. Calvanese, D., Cogrel, B., Komla-Ebri, S., Kontchakov, R., Lanti, D., Rezk, M., Rodriguez-Muro, M., Xiao, G.: Ontop: answering SPARQL queries over relational databases. Semant. Web J. **8**(3), 471–487 (2017). doi:10.3233/SW-160217

12. Motik, B., Cuenca Grau, B., Horrocks, I., Wu, Z., Fokoue, A., Lutz, C.: OWL 2 Web Ontology Language profiles, 2nd edn. W3C Recommendation, W3C, December 2012. http://www.w3.org/TR/owl2-profiles/

13. Antonioli, N., Castanò, F., Coletta, S., Grossi, S., Lembo, D., Lenzerini, M., Poggi, A., Virardi, E., Castracane, P.: Ontology-based data management for the Italian public debt. In: Proceedings of FOIS. Frontiers in Artificial Intelligence and Applications, vol. 267, pp. 372–385. IOS Press (2014)

14. Jiménez-Ruiz, E., Kharlamov, E., Zheleznyakov, D., Horrocks, I., Pinkel, C., Skjæveland, M.G., Thorstensen, E., Mora, J.: BootOX: bootstrapping OWL 2 ontologies and R2RML mappings from relational databases. In: Proceedings of ISWC Posters & Demonstrations Track. CEUR, vol. 1486. ceur-ws.org (2015)

Subgroup Discovery in Process Mining

Mohammadreza Fani Sani[1]([✉]), Wil van der Aalst[1], Alfredo Bolt[1],
and Javier García-Algarra[2]

[1] Eindhoven University of Technology, Eindhoven, The Netherlands
{M.Fani.Sani,w.m.p.v.d.aalst,a.bolt}@tue.nl
[2] Telefonica, Madrid, Spain
fco.javier.garciaalgarra@telefonica.com

Abstract. Process mining enables multiple types of process analysis based on event data. In many scenarios, there are interesting subsets of cases that have deviations or that are delayed. Identifying such subsets and comparing process mining results is a key step in any process mining project.

We aim to find the statistically most interesting patterns of a subset of cases. These subsets can be created by process mining algorithms features (e.g., conformance checking diagnostics) and serve as input for other process mining techniques. We apply subgroup discovery in the process mining domain to generate actionable insights like patterns in deviating cases. Our approach is supported by the ProM framework. For evaluation, an experiment has been conducted using event data from a large Spanish telecommunications company. The results indicate that using subgroup discovery, we could extract interesting insights that could only be found by spitting the event data in the right manner.

Keywords: Process mining · Subgroup discovery · Pattern mining · Performance management · Quality of metrics

1 Introduction

Our society, organizations and IT systems depend on processes. Products and services can only be delivered efficiently and effectively when processes are running as planned. Process mining aims to discover, monitor, and enhance processes by extracting knowledge from event data that can be extracted from almost all modern [1].

Process Mining is able to bridge the gap between Business Process Modeling (BPM) and data driven methods like data mining and machine learning [2]. Process mining is able to analyze the actual processes without relying on simplistic models. There are basically two main types of data-driven analysis [3]:

- **Predictive analysis:** involving techniques that extract knowledge and rules to predict or classify samples, such as classification, regression and time series algorithms.

© Springer International Publishing AG 2017
W. Abramowicz (Ed.): BIS 2017, LNBIP 288, pp. 237–252, 2017.
DOI: 10.1007/978-3-319-59336-4_17

- **Descriptive analysis:** involving techniques that discover interesting knowledge about samples and their attributes to explain the data (e.g. association rules).

In other words, descriptive analysis techniques extract patterns from the data with respect to properties and their values. For example, a manager wants to know in which situations customers have complaints. Descriptive analysis will not be able to predict the complaints; however, it will provide insights about various factors that may cause the complaints [5].

The lion's share of process mining research has been devoted to descriptive forms of analysis. Next to process discovery techniques, there have been approaches to group traces. The approach presented in [4] clusters traces thereby characterizing each cluster. However, in this method class of samples cannot be used. The approach presented in [5] extracts interesting patterns based on a class attribute. In many applications, stakeholders prefer to analyze and know more about a subset of cases rather than all the cases. Examples of interesting subsets (or target group) include:

- Deviating cases from the reference model
- Cases with high or low performance
- Cases with high profits for the company
- Unfinished or canceled cases
- Cases from a particular period
- Cases that pertain to users complaints
- Events related to particular products or services

Given such subsets of cases, it is of the utmost importance to see what kind of attributes they share. For example, discovering that deviating cases are caused by particular resources or limited to specific groups of customers. According to our knowledge, there is no research has been done to extract such information from event data. The main contribution of this paper is that we apply subgroup discovery techniques in the context of process mining domain, to discover the statistically most interesting patterns in a subset of cases called the target groups. The attributes and also the target group can be created based on features extracted using process mining algorithms (e.g., conformance checking or performance analysis). Moreover, our approach also produces insightful collections event logs that can be used as input for a range of existing process mining techniques (e.g., process discovery). In short, this approach will help process analyst to find what are distinctive attributes in a subgroup of cases. Such information assists further investigations like root cause analysis.

To evaluate the possibility of using this method in the reality, we provide a case study where we applied our proposed method on the ticket handling process of Telefonica. Figure 1, shows the process model of this process. The ticket handling process in Telefonica consists of the following main steps. First, a ticket is created through the 'New' activity and then it should be activated by conducting the 'Active' activity. After this activation, a ticket should be handled appropriately and consequently closed through the 'Solved' and 'Closed'

Table 1. Small fragment of the dataset provided by Telefonica, related to the ticket handling process.

CaseID	EventID	Operation	Resource	Group	Severity	Type	Creator	Date-Time
A1001	1	New	Sara	G17	Major	Claim	G1	20150711-10:12
A1001	2	Active	Jon	G17	Major	Claim	G1	20150711-10:19
A1001	3	Solved	Alex	G10	Major	Claim	G1	20150711-16:01
A1001	4	Closed	Alex	G10	Major	Claim	G1	20150711-16:21
A1002	5	New	Sara	G17	Minor	Order	G1	20150713-08:32
A1002	6	Active	Tim	G17	Minor	Order	G1	20150713-08:51
A1002	7	Canceled	Leo	G19	Minor	Order	G1	20150713-14:04
A1003	8	New	Sara	G17	Slight	Claim	G2	20150711-11:20
A1003	9	Active	Tim	G17	Slight	Claim	G2	20150711-11:27
A1003	10	Active	Tim	G17	Slight	Claim	G2	20150711-11:28
A1003	11	Canceled	Alex	G10	Slight	Claim	G2	20150712-09:51

activities. It is possible to interrupt the handling of a ticket by the 'Delayed' activity. Also, a ticket could be restored to the customer via the 'Restored' activity. There is also another possibility, namely: the cancellation of tickets by the 'Canceled' activity. This can happen at any point in their lifetime. We consider 'Canceled' and 'Closed' as the possible final activities of a ticket.

Every process may be executed for multiple cases (also called process instances). Each case is composed of a set of events that are stored in the event log. The standard format for storing an event log which is supported by the majority of process mining tools is XES [6]. In Table 1, a simple example of the event log for the Fig. 1 is shown that contains 3 cases. Cases A1001 and A1003 have 4 events and A1002 has 3 events. By using the CaseID field we know which events are related to particular cases. Note that, case A1002 is not completely "fitting" in the process model (there is one so-called "move in log" showing an event that happened in reality but could not happen according to the model). Furthermore, both events and cases may have attributes that can be used. For

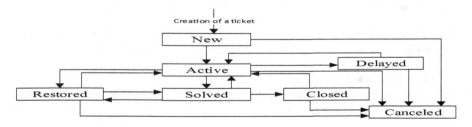

Fig. 1. A normative process model that describes the ticket handling process. This model was designed by Telefonica (of course such models can also be discovered based on the event log).

example, in Table 1, Resource and Group are event attributes. These attributes indicate that who is handled each event and do it in which organizational part of the company. Also, Severity that is a case property, shows the importance of different tickets (cases) in the event log.

The remainder of this paper is organized as follows. In Sect. 2, subgroup discovery is explained. In Sect. 3, we describe how we map and use subgroup discovery in the process mining domain. Section 4 describes the implementation of our approach. Next, Sect. 5 illustrates the usefulness of our approach through the application of our techniques to a real life dataset obtained from Telefonica. Finally, Sect. 6 concludes the paper.

2   Subgroup Discovery

Subgroup discovery was originally proposed by [7,8] and it is based on the idea of *local exceptionality detection* [9]. In contrast with most classification or prediction algorithms, subgroup discovery does not try to find rules that are used to decide or predict things for new instances of the problem. Also, unlike clustering methods, in this technique, we assume that we have a population of samples that have already a class label (e.g., deviating or not). As mentioned before, the aim of subgroup discovery algorithm is to discover patterns for particular class labels (target groups) [8]. In other words, we try to find the common characteristics in a subset of cases that are fewer happened in the other cases. For example, discovering cases that are delayed caused by particular resources or limited to specific type of tickets. Subgroup discovery is used in various domains including the filed of Bioinformatic, e-learning and medical domain [11]. Also in [12] this technique is extended to used multi class data.

We define a subgroup as $(ValueSet \rightarrow Target)$ where $ValueSet$ is an ordered list of independent attributes having specific values. In addition, $Target$ is the desired class of samples that we are interested in analyzing them like deviated cases. For example, S_1, S_2 and S_3 are three examples of possible subgroups:

$$S_1 : Type = \text{``Claim''} \wedge Severity = \text{``Minor''} \rightarrow Target = Deviating$$
$$S_2 : Creator = \text{``G2''} \rightarrow Target = Deviating$$
$$S_3 : Severity = \text{``Major''} \rightarrow Target = \overline{Deviating}$$

Using subgroup discovery we want to discover interesting subgroups. According to [8], a subgroup is interesting if it satisfies the following conditions:

– it is of considerable size and
– it has the most unusual statistical distribution characterization (distribution of different classes in the subgroup compared to their distribution in whole samples)

In Fig. 2, this concept is illustrated. Consider that the class feature is depicted by a red dash or a blue plus. In this figure, three subgroups are shown. Subgroup (a) is not an interesting one because there are too few samples included in it. In other

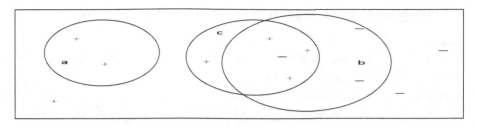

Fig. 2. Three different subgroups. Subgroup (a) is very specific, subgroup (b) has a class distribution similar to the whole and thus not "unusual" enough. The subgroup (c) is an interesting subgroup because it has sufficient samples with a class distribution sufficiently different from the rest. (Color figure online)

words, this subgroup is too specific. In contrast, subgroup (b) has more samples, but the distribution of samples in it is not unusual, because it is the same as the whole population. Finally, the subgroup (c) has a substantial number of samples and has an atypical distribution at the same time, therefore it is considered as an interesting subgroup. It should be noted, it is not required that all samples included in a subgroup have the same class (see for example subgroup (c)). Also, one sample could be placed in more than one subgroup simultaneously (or not placed in any subgroup), i.e., subgroups are not a partitioning of the whole set.

In this approach, we consider only the standard definition of interestingness (based on size and statistical difference), however other definitions could be applied that incorporate domain or business knowledge.

Many measures have been proposed in the literature to quantify the quality of a subgroup and its interestingness. Table 2 summarizes several of the proposed metrics mentioned in papers like [3]. Many of these measures have also been applied in the association rules mining field. To illustrate them in a better way, we use the contingency table presented in Table 3. This table is a useful way to examine relations between categorical variables [10]. A sample matches a particular *ValueSet* if its attributes have values in the ranges defined by *ValueSet*. Similarly, a sample matches a particular *Target* if its class attributes has a value defined by *Target*. In this table, the number of samples that match the *ValueSet* and the number of samples that match the defined *Target* are indicated by n_V and n_T respectively. Also, the number of samples that match both the selected *ValueSet* and *Target* is indicated by n_{VT}. In addition, n_S is the total number of samples. Note that $n_{\overline{T}}$ and $n_{\overline{V}}$ are define the number of samples do not match the *Target* and the selected *ValueSet* respectively. Therefore, $n_S = n_T + n_{\overline{T}} = n_V + n_{\overline{V}}$.

A higher value of the *coverage* metric means that the subgroup has more samples. *coverage* = 1 indicates that the corresponding subgroup includes all the samples. Therefore, an interesting subgroup should have a *coverage* that is high enough. A value of 1 (or 0) in the *support* metric indicates all the samples (or none of them) are match both the *ValueSet* and *Target* class. If a subgroup has a value of 1 in its *confidence* metric, it indicates that if a sample match

Table 2. List of various measures used in subgroup discovery domain.

Measure	Formula	Range
Coverage	$Cov(Subgroup) = \frac{n_{ValueSet}}{n_{Samples}} = \frac{n_V}{n_S}$	$[0, 1]$
Support	$Supp(Subgroup) = \frac{n_{ValueSet \wedge Target}}{n_{Samples}} = \frac{n_{VT}}{n_S}$	$[0, 1]$
Confidence	$Conf(Subgroup) = \frac{n_{ValueSet \wedge Target}}{n_{ValueSet}} = \frac{n_{VT}}{n_V}$	$[0, 1]$
Lift	$Lift(Subgroup) = \frac{Supp(Subgroup)}{Supp(Valueset) \times Supp(Target)} = \frac{n_{VT} \times n_S}{n_V \times n_T}$	$(0, \infty)$
Added value	$Added\,Value(Subgroup) = \frac{n_{VT}}{n_V} - \frac{n_T}{n_S}$	$(-1, 1)$
Precision [7]	$Q_g(Subgroup) = \frac{TP}{FP+g} = \frac{n_{VT}}{n_{V\overline{T}}+g}$	$(0, \infty)$
Unusualness [13]	$WRAcc(Subgroup) = \frac{n_V}{n_S} \times \left(\frac{n_{VT}}{n_V} - \frac{n_T}{n_S} \right)$	$[-0.25, 0.25]$
PS [14]	$PS(Target \rightarrow ValueSet) = \frac{n_{VT}}{n_S} - \frac{n_V \times n_S}{n_S{}^2}$	$[-0.25, 0.25]$

Table 3. Contingency table shows counts of all possible conjunctions of *ValueSet* and *Target* group.

	Target	$\overline{Target}$	
$ValueSet$	n_{VT}	$n_{V\overline{T}}$	n_V
$\overline{ValueSet}$	$n_{\overline{V}T}$	$n_{\overline{VT}}$	$n_{\overline{V}}$
	n_T	$n_{\overline{T}}$	n_S

the selected *ValueSet* it should match the *Target* too. The *lift* metric computes how dependent (or independent) are the *Valueset* and *Target*. If *Lift* equals 1 then they are independent. However, a value higher than 1 suggests a positive correlation and a value lower than 1 indicates a negative correlation. If the *added value* metric has a value of 0, it suggests that the distribution of the classes are similar in both subgroup and total samples and consequently, the *ValueSet* has no influence on the *Target* distribution. In addition, a higher positive (or lower negative) value for this measure, suggests higher positive (or negative) effect on the distribution of the target feature.

Precision measures the quality of a subgroup by computing ratio of different classes when samples match the selected *ValueSet*. In its formula, g is the generalization parameter which is usually in the range $[0.5, 100]$. The *unusualness* value of a subgroup is computed based on both the *coverage* and *added value* of it. It could be proven that $Unusualness(subgroup) = WRAcc(ValueSet \rightarrow Target)$ is equal to $PS(Target \rightarrow ValueSet)$ which is widely used in field of association rules mining. Both of them equal to $\frac{n_V \times n_{VT}}{n_S \times n_V} - \frac{n_V \times n_S}{n_S{}^2}$ (one of difference between association rule and subgroup discovery is in association rule we extract $Target \rightarrow ValueSet$ pattern, but here we are interested in $ValueSet \rightarrow Target$ patterns.) These measures account for *coverage* (size of a subgroup) and *added value* (unusual statistical distribution of subgroup) at the same time ($Conf(Target \rightarrow ValueSet) \times Supp(Target)$).

In this paper, we mainly use the *unusualness* measure and it's range is in $[-0.25, 0.25]$. *Unusualness* equals 0, suggests that a subgroup would not be interesting; however, a higher positive value indicates that the *ValueSet* has higher effect on the *Target* compare to the whole samples. Also, lower negative value for this measure, shows that the samples match the selected *ValueSet* have lower fewer in the *Target* class compare to other class. In many applications, discovering subgroups with negative *unusualness* would be also valuable. Thus, we use the absolute value of *unusualness* ($|WRAcc(subgroup)|$) or *RuleInterestVariant* [15].

3   Applying Subgroup Discovery in Process Mining

In this section, we formally define how to apply subgroup discovery in the field of process mining. The architecture of proposed method is illustrated in Fig. 3. The starting point of our method is an event log. An event log may contain many cases and each case has a set of associated events. Most of the process mining techniques consider events as the starting point for process analysis. In this research, we focus on cases rather than events.

Therefore, in the next step we extract properties for all cases. There are three types of properties in process mining: properties that are related to (a) cases, (b)events, and (c) processes mining properties. In general, a case property is the same for all the events of a specific case. However, for event attributes, the values could be different (or simply missing) for individual events within a case. For example, in Table 1, CaseID, Severity, Type, and Creator are case properties and EventID, Operation, Resource, and Group are event properties. Properties of events can also be mapped to cases properties indirectly. In Fig. 4, an example of such mapping is shown. All possible values of each event property are mapped to a case property. If in any event of a case this value occurred, then the corresponding property of case equals 1, otherwise, it will be 0 (here we use existence function, but other functions like frequency could be used as well). To explain more, a resource of event 6 is "Tim" and because this event belongs to case "A1002", the value of "R:Tim" for this case equals 1.

The third type of properties, the so-called process mining properties, are obtained by performing some kind of computation over the events within a case. Examples include performance metrics (sojourn time, waiting time, etc.) or conformance checking metrics (fitness, precision, counts of move on logs and model, etc.). To extract some of the mentioned features we can optionally provide a process model (that could be given as a reference model or discovered by some process discovery algorithm). Some examples of process mining properties for the event log of Table 1 are given in Table 4. Note that process mining techniques

Table 4. Some process mining properties for the event log of Table 1. To compute alignment costs we use standard cost.

CaseID	Event count	VariantID	Case duration	Fitted model	Alignment cost	Completeness
A1001	4	X1	369 min	Yes	0	Complete
A1002	3	X2	332 min	Yes	0	Complete
A1003	4	X3	22.5 h	No	1	Complete

like conformance checking can be used as input for subgroup discovery. However, the very same techniques can be applied to the discovered subgroups in a later phase. This shows the close interaction between process mining techniques and subgroup discovery.

According to Fig. 3, the output of *Property Extractor* component will be a matrix where each row corresponds to a case and each of its columns refers to a property.

Definition 1 (Universes). $U_S = \mathcal{P}(U_V)$ *is the universe of value collections.* $U_H = \mathcal{P}(U_S)$ *is the universe of sets of value collections (set of sets). Note that* $v \in U_V$ *is a single value (e.g.* $v = Claim$*),* $V \in U_S$ *is a value collection (e.g.,* $V = \{Claim, Order, Query\}$*).*

Definition 2 (Case Base). *A case base* $CB = (C, P, \pi)$ *defines a set of cases* C*, a set of properties* P*, and a function* $\pi \in (P \rightarrow (C \rightarrow U_V))$*. For any properties* $p \in P$*,* $\pi(p)$ *(denoted* π_p*) is a partial function mapping cases onto values. If* $\pi_p(c) = v$*, then case* $c \in C$ *has a property* $p \in P$ *and the value of this property is* $v \in U_V$*.*

Therefore, P includes case, events and process properties and each property $p \in P$ corresponds to column in the extracted matrix shown in Fig. 3. According

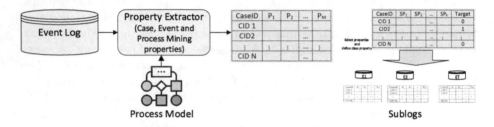

Fig. 3. The architecture of proposed method.

CaseID	EventID	Operation	Resource	Group
A1001	1	New	Sara	G17
A1001	2	Active	Jon	G17
A1001	3	Solved	Alex	G10
A1001	4	Closed	Alex	G10
A1002	5	New	Sara	G17
A1002	6	Active	Tim	G17
A1002	7	Canceled	Leo	G19

CaseID	...	New	Active	Solved	Closed	Canceled	R: Sara	R: Jon	R: Alex	R: Tim	R: Leo	G: G17	G: G10	G: G19
A1001		1	1	1	1	0	1	1	1	0	0	1	1	0
A1002		1	1	0	0	1	1	0	0	1	1	1	0	1

Fig. 4. Mapping the properties of events to case (trace) properties. Values of event attributes are transformed to a case property. For each case, it is computed whether the property is present. The values that are indicated with red color, explain that we use the existence of an attribute value. (Color figure online)

to the method's architecture (Fig. 3), for each case in the *case base* (that is created by the property extractor), a class attribute and intended properties are selected. The class attribute is a binary property that helps us to divide cases into two subsets (two classes). The first subset contains cases that we are particularly interested in analyze them (e.g., cases with delay or deviation). The rest of cases are placed in the other subset. By defining a class attribute, we specify which of the cases are interesting for analysis. In addition, not all of the properties in the *case base* may be noticeable and we should set aside them from properties that will be analyzed them in this subset. We name this subset *target group* and define it formally as follows:

Definition 3 (Target Group). $TG(CB, Att, \pi_{class})$ *is a target group of* $CB = (C, P, \pi)$ *where* π_{class} *is a membership function mapping cases to their relative classes. If a case* $c \in C$ *belongs to our intended subset, then* $\pi_{class}(c) = 1$ *otherwise* $\pi_{class}(c) = 0$. *Attributes* $Att \subseteq P$ *is a subset of the case base properties that we are interested to analyze their effect on the intended subset of cases.*

Therefore, we could say that TG specifies a subset of properties in the *case base* that we want to analyze them and class of each case. Using this definition, we take a case base as an input and returning a subset of it's properties and the class value of each case.

At last, by applying subgroup discovery on the *target group* we will discover many subgroups. Here we formally define a subgroup as the following definition.

Definition 4 (Subgroup). $S(TG, att, vs)$ *is a subgroup of attribute* $att \in Att$ *when* $\pi_{att} = vs$ *on the target group* TG. *Each subgroup is a subset of cases in the* TG *that in these cases, the value of attribute att equals to vs.*

As an example, *att* can be *Type* and *vs* equals *Claim*. The resulted *subgroup* is the subset of cases in the *TG* and the value of "Type" property for these case is *claim*.

Considering several properties in the *target group*, we will have many subgroups. However, the discovered subgroups are different based on their size, interestingness, distribution, and effects of them on the *target group*. We use *unusualness* measure to compute the interestingness of discovered subgroups on the *target group*. We name this measure *Impact Effect* and denote it by $IE(subgroup)$. The higher value of IE suggests higher positive Influence of the subgroup. As mentioned before, we aim to discover subgroups with higher $|IE|$ values. Using this definition we can compute the interestingness (or *unusualness*) of each subgroup on the *target group*.

Until now, we just considered one attribute in the *ValueSet* of a *subgroup*. However, it is possible to have a subgroup with multiple attributes. The complexity of a subgroup could be defined by the number of attributes in its *ValueSet* [3]. For example, $S : Type = "Calm" \wedge Severity = "Major" \rightarrow Target = Deviating$ is a subgroup with multiple attributes and its complexity equals 2. Note that in combination of properties, each property should not appear more than one time in a subgroup.

However, computing all possible subgroups would be very time-consuming. There are many methods proposed to overcome this issue [21]. Here, we use minimum coverage of subgroups. So, subgroups with $Cov(subgroup)$ lower than the minimum threshold are not considered. Note that if the coverage value of ($ValueSet_1 \rightarrow Target$) is Cov_1, the *coverage* value of ($ValueSet_1 \wedge ValueSet_2 \rightarrow Target$), by definition, is less than or equal to Cov_1. Thus, if a subgroup does not contain sufficient samples to have the minimum coverage, no other subgroups included in this subgroup have a higher coverage and there is no need to consider them.

In the final step of our approach we apply process mining techniques to the subgroups created. For each subgroup, we could extract a sublog (i.e., a subset of the main event log). A wide range of process mining algorithms ranging from dotted chart [16] and process comparator [17] to the inductive miner [18] and various conformance checkers [19] could be applied on these sublogs for further analysis.

4   Implementation

To make it possible to apply subgroup discovery approach in the process mining context, a Subgroup Discovery plugin has been developed in ProM framework. ProM is an open source tool that allows to use and implement lots of different techniques in the field of process mining [20]. This tool can be freely downloaded from www.promtools.org.

The *Subgroup Discovery* plugin takes two event logs as input, one contains all the case samples (*Case Base*) and the other one is related to the subset of cases that we want to characterize (i.e., target group). Therefore, the second event log should be a subset of the first event log. Furthermore, regarding the output range of *unusualness* metric (and also *PS* metric) is in $[-0.25, 0.25]$, we use range bar chart (it is also called *Tornado* chart) to visualize the impact of each subgroup on the target group. A screenshot of an example output result of subgroup discovery obtained using our tool is shown in Fig. 5.

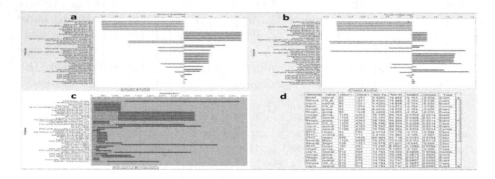

Fig. 5. An example of the output of the *Subgroup Discovery* plugin. (Color figure online)

Our plugin provides four types of results. First and foremost, we provide the impact effect analysis that is shown in Fig. 5(a). Each subgroup is shown in one row and its effect on the target group is indicated by a bar. The blue bars indicate a positive influence and red ones depict a negative impacts on the target group. The result presented in Fig. 5(b) illustrates Added Value that suggests how the percentage of classes are changed in a subgroup compared to whole samples. The chart presented in Fig. 5(c) shows how many samples in each subgroup are placed in the target class or another class (red bars correspond to target class). At last, in Fig. 5(d) the plugin shows a table with measured values for coverage, support, and confidence for each subgroup.

5   Evaluation

To evaluate the usefulness of applying subgroup discovery in the field of process mining we applied our approach and implementation to a dataset of Telefonica. As mentioned before, this data relates to the ticket handling process of three services provided by Telefonica and its corresponding process model is shown in Fig. 1. Also, a few statistics for this dataset are shown in Table 5. Guided by our assumption about complete cases, we just consider cases that contain "Canceled" or "Closed" activities. All other cases are removed from dataset.

The business questions that will be answered in the remainder of this section are:

1. Which attribute values often appeared in cases that have a long duration (cases with delays)?
2. Is there any difference in the property values of different services? If yes, what is the difference and which attribute values have more impact on such differences?

To answer each question, we should first define the intended cases that make our target group. Our target group for Question 1 is defined by cases that take more than 80 days to finish. Also, for answering Question 2, we consider the Jasper service (i.e., one of the three provided services) as our target group. Some statistics for these target groups are shown in Table 6.

The results of applying our new ProM plugin on these target groups are shown in Figs. 6 and 7 respectively. In the remainder of this section, we explain some of our findings for each question.

Question 1: Figure 6 indicates 37 subgroups for the class of slow cases. It shows that in the class of slow cases there is an under representation of $Service =$ "CS_M2M" and an under representation of modification by "$Operator\_071$" (in fact there is no case with this service in the slow case class). In contrast, in this class $Responsible\_group =$ "$Group\_016$", $Modified\_by =$ "$Operator\_172$", and $Service =$ "$CS - SM2MS$" are more represented and therefore, they have a higher positive effect. Therefore, if stakeholders want to collate with slowing cases they should pay more attention to these properties. For example, they

Table 5. Statistical information of Telefonica dataset

Service name	Case#	Events#	Activities#	Median case duration
All	7,426	146,597	7	12.6 day
SM2MS	5,269	110,536	7	7 day
GSIM	794	12,538	7	37.3 day
Jasper	1,363	23,523	7	22.7 day

Table 6. Statistical information of target groups. For each question, we have two classes.

	Case#	Events#	Activities#	Class%
Slow cases (Q1)	1,433	33,543	7	22.78%
Jasper cases (Q2)	1,251	22,022	7	19.89%
All (filtered)	6,290	125,728	7	

should think about the relation of "$Group\_016$" or "$Operator\_172$" with the slowness of cases. Also, in this class, the case with $Conformance =$ "$Fitted$" are more presented (for conformance checking we use "Replay a Log on Petri Net for Conformance Analysis" plugin.)

Question 2: In Fig. 7, again 37 subgroups for Jasper service class are illustrated (in our experiments accidentally the number of discovered subgroups be similar). This chart indicates that $Modified\_by =$ "$Operator\_071$" and $Assigned\_group =$ "$Group\_007$" have higher influence on this service. Also, the unfitted cases or cases with $Conformance =$ "$unfitted$" are more presented in this target class. Although, some of these subgroups may be obvious (like "$Service = CS\_SM2SM$" has negative impact and $Service = $ "CS_M2M" has positive impact, because we just consider "CS_M2M" service in this target group), the extracted rules indicate that this approach could extract interesting and correct patterns in subgroups that could not be uncovered by looking at the whole log.

We also present these subgroups to Telefonica experts who have business knowledge. They confirmed that all of the discovered subgroups are correct, but not all of them were considered as surprising. They also recommended us to define other target groups and reapply our approach using these new target groups.

Even though other techniques like correlation, association rule mining and decision tree have similarities with subgroup discovery algorithm, they could not find these discovered subgroups. For example, when we apply correlations we typically do not consider the sizes of subgroups. Also, when applying decision trees, we aim to discover rules for predicting future samples not describing current ones. In association rule mining variations that consider a class feature, the focus is on *coverage*, *support* and *confidence* of a class and an item set and

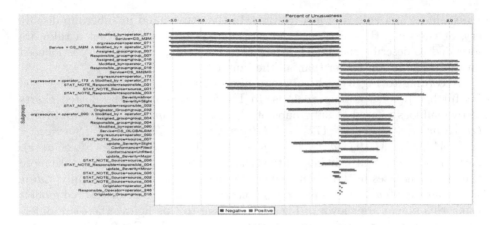

Fig. 6. Interesting patterns discovered by subgroup discovery technique for cases with long duration. The red bars indicate negative effect and coloration and blue bars suggest positive influence and correlation. (Color figure online)

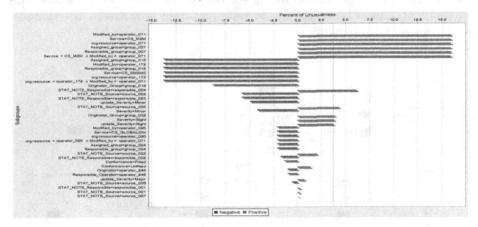

Fig. 7. Interesting patterns for Jasper service cases. Longer bars show higher influence for corresponding subgroup.

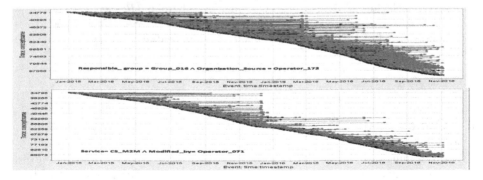

Fig. 8. The Dotted charts of two sublogs extracted from two discovered subgroups.

unusualness of a rule is of less relevant. In the other hand in subgroup discovery, at the same time *coverage* and *unusualness* of a pattern (not a rule) are intended to describe the model of samples. Also, subgroup discovery is focused on the target group rather than all the samples.

According to Fig. 3, each subgroup also is an event log that could be used for further process mining analysis. In Fig. 8, we compare the dotted charts of two sublogs related to subgroups of (*Responsible_group = "Group_016" ∧ Operator = "Operator_172" → Target = SlowCases*) and (*Service = "CS_M2M" ∧ Modified_by = "Operator_071" → Target = SlowCases*). According to Fig. 8, it is indicated that the cases of the first subgroup take more time whereas cases in the second subgroup take less time. We also apply *"Mine Petri Net with Inductive Miner"* plugin on these sublogs. The discovered models using this plugin are shown in Fig. 9. According to this figure, there are difference in their process. For example, in the process model in Fig. 9(a) it is possible for a "Delayed" ticket to be "Active" again, but it is impossible in the in the process model in Fig. 9(b). These kinds of analysis could give valuable information to stakeholders for understanding the reasons for difference in behavior in subgroups of cases. It is also possible to apply any other process mining technique on the discovered subgroups.

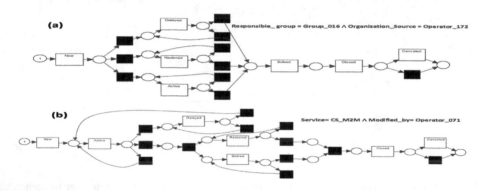

Fig. 9. The process models of two sublogs discovered by *"Mine Petri Net with Inductive Miner"* plugin: (a) is the process model of subgroup (*Responsible_group = "Group_016" ∧ Operator = "Operator_172" → Target = SlowCases*) and (b) is the process model of (*Service = "CS_M2M" ∧ Modified_by = "Operator_071" → Target = SlowCases*)

6   Conclusion

Process mining can be used to extract knowledge from event logs. However, event logs may contain information on cases with very different characteristics.

Analyzing these different group of cases together may conceal important phenomena. Delays and deviations may be linked to very particular subgroups that are not known beforehand.

To address this problem, we applied subgroup discovery technique to find the statistically interesting patterns in subsets of cases belonging to a predefined target class (e.g., cases that are delayed). In this regard, properties of the event log are extracted with their corresponding values. These properties could be related to the case, its events or computed by other process mining techniques. Afterwards, interesting subgroups of the target group can be extracted by applying well-known measures like *Added Value* and *WRAcc*. Interesting subgroups that contribute to the target group positively or negatively may be discovered. Importantly, any process mining algorithms can be applied to the discovered subgroups to extract surprising insights and behaviors.

To evaluate the proposed approach we developed a plugin in a *ProM* platform and applied it in a case study conducted together with Telefonica. Two target groups are defined for this purpose, one for slow cases and another for cases related to a specific service. This case study indicates that the proposed approach could is able to discover interesting patterns. However, not all of them were surprising for business experts.

In the current implementation we do not consider attributes with continues values. In *ProM* and other data mining tools there are techniques to make these attributes discrete. Not doing this up-front, but trying to integrate this in the approach itself may be very time consuming, especially for time and date attributes. Here, we also define target groups manually, however defining a suitable target group would be a challenging task for continues attributes.

References

1. van der Aalst, W.M.P.: Process Mining: Data Science in Action. Springer, Heidelberg (2016)
2. Van der Aalst, W.M.P.: Using process mining to bridge the gap between BI and BPM. IEEE Comput. **44**(12), 77–80 (2011)
3. Herrera, F., Carmona, C.J., González, P., Del Jesus, M.J.: An overview on subgroup discovery: foundations and applications. Knowl. Inf. Syst. **29**(3), 495–525 (2011)
4. Bose, R.P.J.C., van der Aalst, W.M.P.: Context aware trace clustering: towards improving process mining results. In: Proceedings of the 2009 SIAM International Conference on Data Mining, pp. 401–412. SIAM (2009)
5. Novak, P.K., Lavrač, N., Webb, G.I.: Supervised descriptive rule discovery: a unifying survey of contrast set, emerging pattern and subgroup mining. J. Mach. Learn. Res. **10**, 377–403 (2009)
6. Verbeek, H.M.W., Buijs, J.C.A.M., Dongen, B.F., Aalst, W.M.P.: XES, XESame, and ProM 6. In: Soffer, P., Proper, E. (eds.) CAiSE Forum 2010. LNBIP, vol. 72, pp. 60–75. Springer, Heidelberg (2011). doi:10.1007/978-3-642-17722-4_5
7. Klösgen, W.: Explora: a multipattern and multistrategy discovery assistant. In: Advances in Knowledge Discovery and Data Mining, pp. 249–271. American Association for Artificial Intelligence (1996)

8. Wrobel, S.: An algorithm for multi-relational discovery of subgroups. In: Komorowski, J., Zytkow, J. (eds.) PKDD 1997. LNCS, vol. 1263, pp. 78–87. Springer, Heidelberg (1997). doi:10.1007/3-540-63223-9_108
9. Atzmueller, M.: Subgroup discovery. Wiley Interdisc. Rev. Data Min. Knowl. Discov. **5**(1), 35–49 (2015)
10. Kateri, M.: Contingency Table Analysis. Springer, Heidelberg (2014)
11. Herrera, F., et al.: An overview on subgroup discovery: foundations and applications. Knowl. Inf. Syst. **29**(3), 495–525 (2011)
12. Duivesteijn, W., et al.: Subgroup discovery meets Bayesian networks-an exceptional model mining approach. 2010 IEEE 10th International Conference on Data Mining (ICDM). IEEE (2010)
13. Atzmueller, M., Baumeister, J., Puppe, F.: Introspective subgroup analysis for interactive knowledge refinement. In: FLAIRS Conference, pp. 402–407 (2006)
14. Piatetsky-Shapiro, G.: Discovery, analysis, and presentation of strong rules. Knowl. Disc. Databases, 229–238 (1991). https://www.bibsonomy.org/bibtex/26fa5f6987b667b728c7e94f7c68b52d7/enitsirhc
15. Huynh, X.-H.: Interestingness Measures for Association Rules in a KDD Process: Postprocessing of Rules with ARQAT Tool. Université de Nantes (2006)
16. Song, M., van der Aalst, W.M.P.: Supporting process mining by showing events at a glance. In: Proceedings of the 17th Annual Workshop on Information Technologies and Systems (WITS), pp. 139–145 (2007)
17. Bolt, A., Leoni, M., Aalst, W.M.P.: A visual approach to spot statistically-significant differences in event logs based on process metrics. In: Nurcan, S., Soffer, P., Bajec, M., Eder, J. (eds.) CAiSE 2016. LNCS, vol. 9694, pp. 151–166. Springer, Cham (2016). doi:10.1007/978-3-319-39696-5_10
18. Leemans, S.J.J., Fahland, D., van der Aalst, W.M.P.: Process and deviation exploration with inductive visual miner. In: BPM (Demos), p. 46 (2014)
19. Van der Aalst, W., Adriansyah, A., van Dongen, B.: Replaying history on process models for conformance checking and performance analysis. Wiley Interdisc. Rev. Data Min. Knowl. Disc. **2**(2), 182–192 (2012)
20. Verbeek, H.M.W., Buijs, J., Van Dongen, B.F., van der Aalst, W.M.P.: Prom 6: the process mining toolkit. Proc. BPM Demonstration Track **615**, 34–39 (2010)
21. Agrawal, R., Srikant, R.: Fast algorithms for mining association rules. In: Proceedings of 20th International Conference Very Large Databases, VLDB, vol. 1215, pp. 487–499 (1994)

Business Process Comparison: A Methodology and Case Study

Alifah Syamsiyah[1][✉], Alfredo Bolt[1], Long Cheng[1], Bart F.A. Hompes[1],
R.P. Jagadeesh Chandra Bose[2], Boudewijn F. van Dongen[1],
and Wil M.P. van der Aalst[1]

[1] Eindhoven University of Technology, Eindhoven, The Netherlands
{a.syamsiyah,a.bolt,l.cheng,b.f.a.hompes,b.f.v.dongen,
w.m.p.v.d.aalst}@tue.nl
[2] Xerox, Bangalore, India
jagadeesh.prabhakara@conduent.com

Abstract. Business processes often exhibit a high degree of variability. Process variants may manifest due to the differences in the nature of clients, heterogeneity in the type of cases, etc. Through the use of process mining techniques, one can benefit from historical event data to extract non-trivial knowledge for improving business process performance. Although some research has been performed on supporting *process comparison* within the process mining context, applying process comparison in practice is far from trivial. Considering all comparable attributes, for example, leads to an exponential number of possible comparisons. In this paper we introduce a novel methodology for applying process comparison in practice. We successfully applied the methodology in a case study within Xerox Services, where a forms handling process was analyzed and actionable insights were obtained by comparing different process variants using event data.

Keywords: Process comparison · Process mining · Business analytics

1 Introduction

Modern information systems and devices collect and store large amounts of event data. For instance, ERP systems record business transaction events, and high-tech systems such as X-ray machines record an abundance of events [10]. Such historical *event data* can be used to extract non-trivial knowledge and interesting insights that can be used for further analysis. Increasingly process mining techniques are used to analyze such data [20]. Process mining covers three types of analysis [19]: *process discovery* automatically extracts a process model from an event log; *conformance checking* measures how well the behavior recorded in an event log fits a given process model and vice versa; and *process enhancement* is concerned with extending or improving an existing a-priori process model using event data.

© Springer International Publishing AG 2017
W. Abramowicz (Ed.): BIS 2017, LNBIP 288, pp. 253–267, 2017.
DOI: 10.1007/978-3-319-59336-4_18

In process mining, process characteristics such as *waiting times, through-put times,* and *utilization rates* are typically of interest, and can be obtained from real-life event data. Many processes exhibit a high degree of variability. There may be major differences between processes and their variants, due to an abundance of factors such as temporal changes, the geographical location of the process' execution, the involved resources and the overall context in which a process is executed [1,6]. In such scenarios, our research question is: how can we conduct a comparative analysis of different processes and their variants in real businesses? Based on the results of the analysis, we should be able to find the differences between multiple processes and also find root causes for ineffi-ciencies such as delays and long waiting times for the interpretation of process behaviors. Moreover, domain experts should also able to identify the precise events that correspond to unusual behavior, and consequently devise concrete measures to improve their business processes [17].

In this paper, we present a methodology for business process comparison. We do so by presenting an overall methodology and an instantiation thereof in the context of a large service delivery organization: Xerox Services. This organiza-tion caters similar processes across several clients, hence process variants may manifest due to the differences in the nature of clients, heterogeneity in the type of cases, etc. Moreover, the organization's operational Key Performance Indi-cators (KPIs) across these variants may widely vary. We show that, using our method, we gain insights into the differences between variants and we leverage these insights on non-performing variants by means of process comparison.

The highlighted contributions of this paper are as follows:

- Present a methodology for process comparison which focuses on the analysis of multiple processes. This methodology considers multiple perspectives, such as control flow, organizational, data, performance, etc.
- Validate the methodology in a case study using real-life data.

The remainder of this paper is organized as follows. In Sect. 2, we discuss related work in process comparison and process mining methodologies. Then, we explain the proposed process comparison methodology in Sect. 3 and apply the methodology in a case study in Sect. 4. Section 5 concludes the paper.

2   Related Work

In recent years, the value of process mining techniques has been demonstrated in *case studies* across different domains such as healthcare [11,21,25], industry [12,13,15], insurance [17], and finance [7,8]. However, few *methodologies* have been proposed to carry out process mining projects in a structured manner. In [2], the *Process Diagnostics Method* (PDM) is proposed to quickly obtain a broad overview of the process at hand, without the need for any domain knowledge. As such, it can be used to steer a process mining project by providing initial insights and analysis opportunities. For example, the method has been adopted for the analysis of healthcare processes in [14]. In [19], the *L* life-cycle model* is proposed as an approach for mining processes. L* covers many techniques, and describes the life-cycles of a typical process mining project aiming to improve processes. Since PDM focuses on providing a broad overview using a limited set of process mining techniques and because L* is aimed at the analysis of structured processes, the authors of [23] proposed *PM²: a Process Mining Project Methodology*. PM² is designed to support projects aiming to improve process performance or compliance, and focuses on iterative analysis. Its applicability was shown by a case study conducted on data provided by IBM. Like L*, PM² covers a wide array of process mining techniques. Contrary to L*, however, PM² is suitable for the analysis of both structured and unstructured processes.

A common pitfall of the discussed process mining methodologies is that the focus is on the analysis of a single process, as mentioned in [23]. As such, *process comparison* remains an interesting but insufficiently researched topic. The comparison of processes based on event logs has been the focus of several papers [1,4,5,18]. However, most process comparison approaches take into consideration only the control-flow aspect (i.e., presence, routing and frequency of activities), while ignoring other dimensions.

Given the increased interest in process comparison from perspectives other than just control flow, and the lack of methodological support for applying process comparison in a process mining project in practice, we propose The Process Comparison Methodology (PCM). In this work, different from existing process mining methodologies, we introduce a novel methodology by considering multiple aspects, such as the organizational aspect (i.e. the involved resources, roles, and groups), the data aspect (attribute values), the performance aspect, etc. We validate our methodology in a case study using real-life data provided by Xerox Services. To the best of our knowledge, this is the first work that methodologically considers business process comparison from multiple perspectives.

3 The Process Comparison Methodology (PCM)

When comparing multiple processes, it is common that those processes have mutual attributes for categorization. When process comparison methods are applied to highlight the differences between similar categorized processes, the results are more detailed and representative than when comparing dissimilar or unrelated processes. Based on these underlying assumptions, in this section, we introduce a methodology for process comparison which considers many perspectives. Our methodology comprises of five main phases as depicted in Fig. 1.

First, the data pre-processing phase transforms raw data to standardized event log formats such that existing process mining techniques can be applied. Next to the event logs, the so-called α-attributes are selected. These attributes are case-level attributes that identify the variants of interest. Next, in the scoping analysis phase, the interesting cases to be used for answering the analysis questions are identified. In the third phase, comparable sub-logs are generated by aggregating similar cases. Fine-grained analysis of the generated sub-logs takes place in the in-depth comparison phase. Finally, the discovered results are delivered to the process owners.

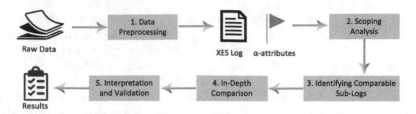

Fig. 1. Process Comparison Methodology (PCM)

Phase 1. In the data pre-processing phase, raw data is translated to standardized event log formats such that existing process mining techniques can be applied. We have two main objectives for the data pre-processing phase: (1) refine event data collected by information systems; and (2) create an event log and identify a set of case attributes α to be used in the comparison process.

Typically, raw event data are collected by different information systems at different levels of granularity. To conduct a meaningful analysis, we combine all collected event data and merge them into a single collection of events. Here, standard data cleaning techniques can be used if the raw data contains noise. From this event collection, an event log is devised. To get an event log from an event collection, a notion of cases is introduced. A case refers to a time-ordered sequence of events relating to the some underlying concept, for example a purchase order or a single run of a machine, (i.e. events need to be correlated to form traces of events). We follow the XES standard [24] as the format for the generated event log, to make existing process mining techniques (implemented in tools such as ProM[1]) accessible in the following phases of our methodology. Finally, next to the case notion, attributes of interest are selected as the so-called α-attributes. In the further comparison, we consider the α-attributes to denote the process variant.

Phase 2. Once the event log and the α-attributes are defined, we scope our analysis. The goal of the scoping phase is to limit the number of comparisons to be executed later. Typically, scoping is done based on the α-attributes, for example by selecting the most frequent values of these attributes. However, in general, the scoping decision must follow the business questions and the goal of doing process comparison. As a result of scoping, a collection of sub-logs is generated, again in the XES format.

Phase 3. The next phase in the analysis is the identification of comparable sub-logs. Each of the sub-logs obtained during scoping refers to a variant of the process under investigation. However, these variants are not always directly comparable. They may, for example, consist of disjoint sets of activities. Therefore, in this phase, we select comparable variants (i.e. variants that have enough commonalities).

The identification of comparable sub-logs can be done in several ways. For example, we can use domain knowledge to manually select sub-logs to be compared. Alternatively, if domain knowledge is not available, *clustering* techniques can be used to group sub-logs based on a quantifiable *similarity* notion [16].

---

[1] See http://processmining.org and http://promtools.org.

Phase 4. After sets of comparable sub-logs are identified, we treat each set as the starting point for the in-depth comparison phase. In this process, the sub-logs in each set will be pairwise compared and the output of this phase will be a collection of observed and interesting differences between the input sub-logs.

For the in-depth comparison, the pairwise analysis of the sub-logs should often not be limited to control flow only. Instead, other aspects of processes, such as performance characteristics, resource utilization and compliance aspects should be considered. Most importantly, the *influence* of these aspects on each other should be investigated. For example, cases in which different resources were involved could have significantly different durations, which might be an actionable insight.

It should be noted that only the *relevant* and *impactful* differences are of interest to the process owner. For example, a difference in case duration of several seconds may be irrelevant in processes where the average case duration is in the order of days, while in processes that generally last minutes this difference can be significant.

Phase 5. After completing the in-depth comparison for each cluster and having identified relevant and impactful differences, the relevant results will be reported to the process owner. We identify two activities for this phase:

1. *Presentation and Interpretation.* After the process mining analysis has been performed, we obtain facts about the process. Most of the time, these facts are raw and disconnected with each other. Therefore, to provide meaningful information at the business level, an additional presentation and interpretation step is needed. The significance of the results depends on how well the analysis and interpretation step is executed.
2. *Validation.* The results from the in-depth comparison have to be validated with the process owner and participants in the process.

In the remainder of this paper, we show how this high-level methodology can be executed in a concrete case study within Xerox. We use publicly available tools and techniques on proprietary data and we closely involved the Xerox stakeholders in the analysis.

4   Xerox Case Study

This section discusses the application of PCM on a case study conducted within Xerox Services. The study involved a real-life data set with millions of events. First, we explain the data set in terms of its structure and its origin, and give a description of the process contained in it. Then, we present the application of our proposed methodology in detail. As demonstrated in Fig. 2, the instantiation of each phase in our application corresponds to the five phases in Fig. 1, respectively. However, for the case study, the phase *Identifying Comparable Sub-Logs* is refined into three smaller phases here: *Discovery, Cross Comparison*, and *Clustering*. This refinement choice is one of many ways to identify comparable sub-logs.

In our implementation, we used both ProM (for steps 1 and 4 in Fig. 2) and RapidProM (for steps 2, 3a, 3b, and 3c). The used RapidProM workflow is depicted in Fig. 3 and available at https://goo.gl/BCq1uO.

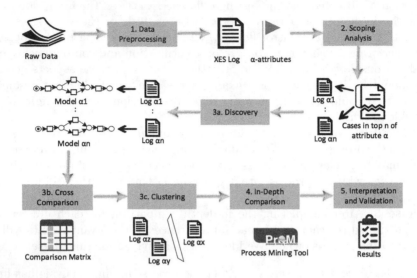

Fig. 2. Process Comparison Methodology applied to a Xerox dataset

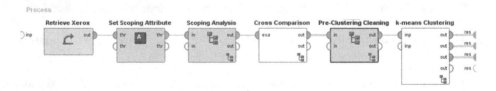

Fig. 3. The RapidProM workflow used for scoping analysis and identifying comparable sub-logs

4.1 Data Set

We analyzed event logs pertaining to the transaction processing business unit within Xerox Services. More specifically, we analyzed the process pertaining to the data entry back-office operations of insurance claim forms. Forms submitted to the insurance providers need to be digitized before the claims can be processed. *Business Process Outsourcing* (BPO) organizations assist the insurance providers in this process. Forms received by the BPO organization are classified and sorted depending on the type of form (e.g. HCFA, UB04, Dental, etc.). More fine-grained classifications further refining each type are possible (e.g. HCFA standard, HCFA careplus, etc.), thereby defining a taxonomy. Different classes in the taxonomy are divided into so-called batches, where each batch caters to one type of insurance claim form (e.g. HCFA standard).

A transaction refers to the data entry operations of one instance of an insurance claim form. The organization handles data entry operations of millions of such instances of insurance claim forms. In this paper, we only consider the transactions of one month pertaining to one client, but the approach can be applied to even larger data sets. Furthermore, different attributes concerning the execution of events such as involved resourced and properties of the forms are recorded as well. The complete dataset used here contains information on hundred transactions comprising 20 million events divided across 94 batches. The organization is interested in analyzing the processes followed across different batches and wants to obtain insights on their executions. In this paper, we focus on the analysis of three batches, where two are similar but not identical and the third batch is different from the other two.

4.2   Data Preprocessing

We transformed the raw event data obtained as CSV file to a standard XES log with the *Convert CSV to XES* plugin in ProM. To make this transformation meaningful and successful, we have done the following three main pre-processing steps. (1) We enriched the set of attributes based on anticipated questions. Since we are interested in analyzing different batches (see Subsect. 4.1), we set the attribute BATCHNAME as the α attribute to be used in comparison process. (2) We refined data into event level. Each activity in the input log includes two timestamps, indicating its start and end point, therefore we divide each activity into two events based on that. (3) We removed uninteresting/uncompleted cases from the log. Based on statistics on the start and end activities for all cases, we removed those case that have a start or end activity that does not appear frequently enough. Through this process, we removed 318,002 cases, and the output XES log contains 936,720 cases and 31,660,750 events.

4.3   Scoping Analysis

We implemented our scoping analysis using RapidProM (as depicted in Fig. 4) to select the interesting batches (batches that are infrequent will not be considered in our analysis). For the generated XES log in the preprocessing phase, we first aggregated all the events based on their BATCHNAME values. Then, we filtered out the popular batches based on their occurrence frequency. There are 94 different batches in our log. We selected the 10 most frequent ones. Their corresponding batch identifiers are 1, 4, 2, 11, 7, 18, 3, 58, 23, and 30 respectively, each having between 424,560 and 8,684,476 cases. We divided the XES log into 10 sub-logs according to the chosen batch names, and conducted our process analysis using these sub-logs.

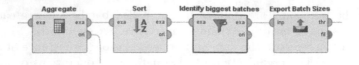

Fig. 4. Scoping analysis to select the most frequent batches

4.4   Identifying Comparable Sub-Logs

Given a collection of sub-logs from the previous phase, the next step is to identify subsets such that sub-logs within each subset share similar behavior (i.e. they are comparable to each other). In the next phase, the sub-logs within one such subset will be compared to obtain more refined comparison results. In this section, we explain the different steps of the techniques we used to identify comparable sub-logs. In Fig. 2, these steps refers to phases 3a, 3b, and 3c.

Discovery. Based on our goals, we compared the extracted batches based on the analysis of their high-level process models. These models can be retrieved by current process mining techniques. Various process discovery algorithms such as the Alpha Algorithm [20], ILP Miner [22] and Inductive Miner [9] have been proposed in the past years. Considering the amount of events in our logs as well as the quality of discovered processes (e.g., soundness and fitness), we have chosen the Inductive Miner. Besides the fact that the Inductive Miner is the state-of-the-art process discovery algorithm, other techniques are inclined to produce models that are unable to replay the log well, create erroneous models, or have excessive run times for event logs of this size. The output of this phase is a collection of process models per sub-log.

Cross Comparison. In [3], Buijs coined the term *cross comparison* and presents the so-called comparison table, which has been evaluated in a case study using event data from five municipalities. A comparison table consists of three types of metrics, namely *process model metrics, event log metrics,* and *comparison metrics. Process model metrics* are metrics calculated using only the process model, such as total number of nodes in the process model, cyclicity, or concurrency in the process model. *Event log metrics* are metrics calculated based on event log, such as the total number of traces and events, average trace duration, etc. *Comparison metrics* are used to compare modeled and observed behavior and include metrics such as fitness, precision, generality, and simplicity [19].

In this phase, we apply the fitness comparison metric to the sub-logs and their corresponding discovered models. We choose fitness rather than the other metrics due to the need of Xerox Services to have process models which allow for most of the observed behavior. Table 1 shows an excerpt of the cross comparison using fitness metric between logs and models in different batches. Each row represents a log of a particular batch n (Log_n), and each column represents a

Table 1. Example cross-comparison table showing the cross comparison between logs and models in different batches.

	$Model_1$	$Model_2$	$Model_3$	
Log_1	0.64	0.37	0.25	...
Log_2	0.26	0.68	0.6	...
Log_3	0.25	0.69	0.61	...
...	...	...	...	...

discovered model from a particular batch m ($Model_m$). Each cell contains the fitness value after replaying a log into a process model.

Clustering. Based on the cross-conformance checking results, we grouped the sub-logs (i.e. batches) into clusters using k-means clustering. We chose this clustering algorithm because of the information from domain expert that a batch belongs to a single cluster and thus cluster overlap is not possible. Concretely, we used the rows of the cross-conformance matrix (Table 1) as observations to perform a *k-means* clustering. In our experiments, we used k = 3 clusters, and we identify the clusters as follows: (1) cluster 0: batches 3, 4, 18, 23, (2) cluster 1: batches 1, 2, 7, 11, 52, (3) cluster 2: batch 30.

The resulting clusters contain groups of similar (i.e. comparable) batches. Comparative analysis can be performed within any of the clusters. Note that cluster 2 contains only one batch (batch 30). This can be caused by the fact that batch 30 is very different from all other batches.

4.5 In-Depth Comparison

Once clusters of comparable batches have been identified, we can proceed to compare the batches in each cluster. To illustrate this, we apply two process comparison techniques to the batches contained in cluster 0. The first technique (as introduced in [1]) detects statistically significant differences between two sub-logs in terms of control-flow and overall performance. The results of this technique identify those parts of the process where differences occur. However, they do not explain why such differences manifest. The second technique (as introduced in [6]) tackles this issue by analyzing, for each sub-log, the effect that different contexts (e.g. involved resources, data attributes, control-flow, etc.) have on process performance, and whether that effect is statistically significant. Using these two techniques on the batches contained in cluster 0, we could obtained valuable insights.

We first applied the process comparison technique [1] to the four sub-logs of cluster 0 to get a ranked list of pairs of sub-logs. After sorting the list based on the percentage of control-flow differences between the pairs of sub-logs, we found that: batches 3 vs. 18 (38.04% control-flow difference), batches 4 vs. 23 (42.03%), batches 3 vs. 23 (72.16%), batches 18 vs. 23 (73.40%), batches 3 vs. 4 (78.43%), and batches 4 vs. 18 (78.64%). This means that batches 4 and 18 are the most dissimilar pair within cluster 0, and batches 3 and 18 are the most similar. In order to illustrate the in-depth comparison phase, in the remainder of this section we will analyze the differences between batches 4 and 18 and between batches 3 and 18.

In Fig. 5, we provide an example of the control-flow differences found between batches 4 and 18. The dark-blue colored states are executed only in batch 18, and never in batch 4. These states are related to *optical character recognition* (OCR) in forms. Moreover, an example of the performance differences found between batches 4 and 18 is also shown in Fig. 6. We can see that the duration of the activity *Entry* is statistically significantly higher in batch 18 than in batch 4. This activity refers to manual entry of form content.

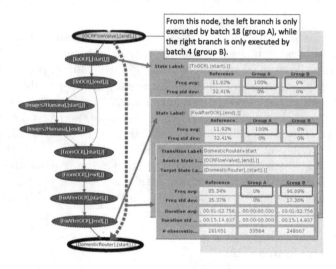

Fig. 5. Control-flow differences between batch 18 (group A) and batch 4 (group B). The activities *ToOCR, Images2Humana, FromOCR, FixAfterOCR* are executed only in batch 18.

	Reference	Group A	Group B
Freq avg:	86.48%	94.75%	85.36%
Freq std dev:	34.2%	22.31%	35.35%
Duration avg:	00:10:25.153	00:44:05.089	00:05:21.610
Duration std ...	00:28:49.769	00:26:57.270	00:25:30.823

Fig. 6. Performance differences between batch 18 (group A) and batch 4 (group B). The average duration of the *Entry* activity is 44 mins for batch 18 and 5 mins for batch 4.

To see whether there are any significant differences manifested in the behavior of the similar batches 3 and 18, we also conducted another comparison using the same technique. From Fig. 7, we can see a significant difference in the frequency of execution of the process fragment, corresponding to the transformation of data from XML to the X12 format, and the transmission and acknowledgment of that data. This fragment is almost always executed in batch 18, it is executed only in approximately 93% of the cases in batch 3. Similarly, the *Cleanup* activity is executed in only 5% of the cases in batch 18 against 12% in batch 3. From a performance point of view, we see that there is a significant difference in the average duration of cases until the execution of the *Cleanup* activity (22 days vs. 10 days). Note that besides this difference, for both batches, the standard deviation of the duration until *Cleanup* is very high relative to the average duration.

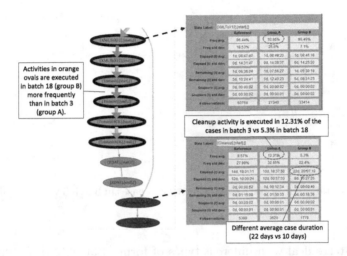

Fig. 7. Example of differences found between batch 3 (group A) and batch 18 (group B).

We analyzed the observed differences between the three batches in more detail using the context-aware performance analysis technique from [6]. This analysis revealed that, for batch 18, significant differences exist between the resources that execute the *Entry* activity (in terms of the waiting time for the activity). This observation is shown in Fig. 8. The waiting times range from several hours to multiple days, and hence might be worth looking into. As explained, the standard deviation for the case duration until the *Cleanup* activity between batches 18 and 3 is quite high relative to the average duration. This observation was analyzed in more detail as well. We found that the duration until *Cleanup* showed big differences between the days in which the cleanup activity happened. In some dates, the duration until *Cleanup* took several days while in other dates, it took multiple weeks. This is illustrated in Fig. 9.

4.6   Delivering Results

Our results discussed above have been presented to and confirmed by a domain expert. (1) The control-flow differences in Fig. 5 are attributed to the fact that

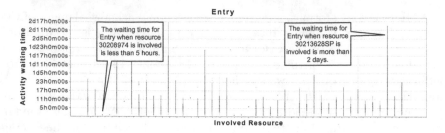

Fig. 8. The resources involved in the *Entry* activity in batch 18 lead to different waiting times.

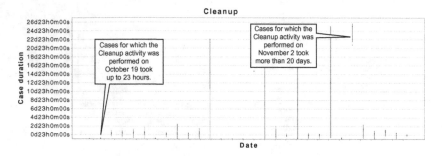

Fig. 9. The duration until the *Cleanup* activity in cases in batch 3 varies highly between days.

the two batches deal with different types of forms. Batch 18 deals with UB-04 forms, a claim form used by hospitals, nursing facilities, in-patient, and other facility providers. These forms are filled by healthcare providers and they can contain handwriting (e.g. disease codes, diagnosis, etc.), so OCR is needed. In contrast, batch 4 deals with claim correspondence forms (i.e. reply forms from the provider). These forms are typically digital. Hence there is no need for OCR. (2) The performance difference in Fig. 6 is attributed to the fact that the forms related to batch 4 (i.e. correspondence forms) are usually smaller than the forms related to batch 18 (i.e. UB-04 forms), and have little content to be entered manually. Hence, the average duration of *Entry* activity in batch 4 is lower. Although these differences between batch 18 and 4 are insightful, they are not very surprising. Similarly, the differences in duration in the manual entry of smaller vs. larger forms in terms of page and image count is to be expected as well. However, the differences in waiting times for different resources are surprising and need to be investigated in order to avoid delays.

The differences between batches 3 and 18 have also provided interesting actionable insights. Both batches 3 and 18 correspond to a similar type of form (UB-04) and are expected to have very similar behavior. The remarkable differences in the frequencies in the process fragment are statistically significant and moreover unexpected by the domain expert, and hence need further investigation. The observed differences in duration until the *Cleanup* activity can be explained by the fact that, in the analyzed process, a lot of (sub) batch processing is involved, and as such, cases sometimes need to wait for other cases in order to be processed.

5   Conclusion

In this paper we have introduced a novel methodology for process comparison within the process mining context, which aims at efficiently examining the differences between multiple business processes and process variants. The proposed methodology, called The Process Comparison Methodology (PCM), considers multiple perspectives during comparison, such as control flow, organizational, data, and performance.

PCM consists of five main phases. First, the *data pre-processing* phase transforms raw data to standardized event log formats. Secondly, the *scoping analysis* phase creates sub-logs based on some case attributes values. Next, the interesting sub-logs to be used for answering analysis questions are identified. Then, in the *identifying comparable sub-logs* phase, similar sub-logs are aggregated to generate comparable sub-logs. In the *in-depth comparison* phase, fine-grained analysis is conducted within comparable sub-logs. Finally, the results are interpreted and validated in the *interpretation and validation* phase and the discovered insights and actions are delivered to the process owners.

The practical relevance of PCM is shown in a case study using real-life data provided by Xerox Services. The process pertains to the data entry back-office operations of insurance claim forms. The organization is interested in analyzing the processes followed across different batches. As there are 94 batches it was unfeasible to compare each pair in detail. Through the application of our methodology, however, very meaningful results were obtained, confirmed by a domain expert, and transformed into actionable insights such as studying the root causes and contextual circumstances for the aberrant instances.

In the future, we would like to investigate more techniques related to the comparison of business processes in order to further refine our methodology. Moreover, we would also like to study more relevant business questions through our collaboration with Xerox Services.

References

1. Bolt, A., de Leoni, M., van der Aalst, W.M.P.: A visual approach to spot statistically-significant differences in event logs based on process metrics. In: Nurcan, S., Soffer, P., Bajec, M., Eder, J. (eds.) CAiSE 2016. LNCS, vol. 9694, pp. 151–166. Springer, Cham (2016). doi:10.1007/978-3-319-39696-5_10
2. Bozkaya, M., Gabriels, J., van der Werf, J.M.E.M.: Process diagnostics: a method based on process mining. In: Kusiak, A., Lee, S. (eds.) eKNOW 2009, pp. 22–27. IEEE Computer Society (2009)
3. Buijs, J.C.A.M.: Flexible evolutionary algorithms for mining structured process models. Ph.D. thesis, TU Eindhoven, p. 179 (2014)

4. Buijs, J.C.A.M., Reijers, H.A.: Comparing business process variants using models and event logs. In: Bider, I., Gaaloul, K., Krogstie, J., Nurcan, S., Proper, H.A., Schmidt, R., Soffer, P. (eds.) BPMDS/EMMSAD -2014. LNBIP, vol. 175, pp. 154–168. Springer, Heidelberg (2014). doi:10.1007/978-3-662-43745-2_11
5. Cordes, C., Vogelgesang, T., Appelrath, H.-J.: A generic approach for calculating and visualizing differences between process models in multidimensional process mining. In: Fournier, F., Mendling, J. (eds.) BPM 2014. LNBIP, vol. 202, pp. 383–394. Springer, Cham (2015). doi:10.1007/978-3-319-15895-2_32
6. Hompes, B.F.A., Buijs, J.C.A.M., van der Aalst, W.M.P.: A generic framework for context-aware process performance analysis. In: Debruyne, C., et al. (eds.) On the Move to Meaningful Internet Systems. LNCS, vol. 10033, pp. 300–317. Springer International Publishing, Cham (2016)
7. Jans, M., Alles, M., Vasarhelyi, M.A.: Process mining of event logs in internal auditing: a case study. In: ISAIS (2012)
8. Jans, M.J., Alles, M., Vasarhelyi, M.A.: Process mining of event logs in auditing: opportunities and challenges (2010). SSRN 2488737
9. Leemans, S.J.J., Fahland, D., van der Aalst, W.M.P.: Discovering block-structured process models from event logs containing infrequent behaviour. In: Lohmann, N., Song, M., Wohed, P. (eds.) BPM 2013. LNBIP, vol. 171, pp. 66–78. Springer, Cham (2014). doi:10.1007/978-3-319-06257-0_6
10. Leemans, S.J.J., Fahland, D., van der Aalst, W.M.P.: Scalable process discovery with guarantees. In: BPMDS/EMMSAD, pp. 85–101 (2015)
11. Mans, R.S., Schonenberg, H., Song, M., van der Aalst, W.M.P., Bakker, P.J.M.: Application of process mining in healthcare - a case study in a Dutch hospital. In: BIOSTEC, pp. 425–438 (2008)
12. Paszkiewicz, Z.: Process mining techniques in conformance testing of inventory processes: an industrial application. In: Abramowicz, W. (ed.) BIS 2013. LNBIP, vol. 160, pp. 302–313. Springer, Heidelberg (2013). doi:10.1007/978-3-642-41687-3_28
13. Puchovsky, M., Di Ciccio, C., Mendling, J.: A case study on the business benefits of automated process discovery. In: SIMPDA, pp. 35–49 (2016)
14. Rebuge, A., Ferreira, D.R.: Business process analysis in healthcare environments: a methodology based on process mining. Inf. Syst. 37(2), 99–116 (2012)
15. Rozinat, A., de Jong, I.S.M., Günther, C.W., van der Aalst, W.M.P.: Process mining applied to the test process of wafer scanners in ASML. IEEE Trans. Syst. Man Cybern. Part C 39(4), 474–479 (2009)
16. Song, M., Günther, C.W., Aalst, W.M.P.: Trace clustering in process mining. In: Ardagna, D., Mecella, M., Yang, J. (eds.) BPM 2008. LNBIP, vol. 17, pp. 109–120. Springer, Heidelberg (2009). doi:10.1007/978-3-642-00328-8_11
17. Suriadi, S., Wynn, M.T., Ouyang, C., Hofstede, A.H.M., Dijk, N.J.: Understanding process behaviours in a large insurance company in Australia: a case study. In: Salinesi, C., Norrie, M.C., Pastor, Ó. (eds.) CAiSE 2013. LNCS, vol. 7908, pp. 449–464. Springer, Heidelberg (2013). doi:10.1007/978-3-642-38709-8_29
18. van Beest, N.R.T.P., Dumas, M., García-Bañuelos, L., La Rosa, M.: Log delta analysis: interpretable differencing of business process event logs. In: BPM, pp. 386–405 (2015)
19. van der Aalst, W.M.P.: Process Mining - Data Science in Action. Springer, Heidelberg (2016)
20. van der Aalst, W.M.P., Weijters, A.J.M.M., Maruster, L.: Workflow mining: discovering process models from event logs. IEEE 16(9), 1128–1142 (2004)

21. van der Spoel, S., van Keulen, M., Amrit, C.: Process prediction in noisy data sets: a case study in a Dutch hospital. In: Cudre-Mauroux, P., Ceravolo, P., Gašević, D. (eds.) SIMPDA 2012. LNBIP, vol. 162, pp. 60–83. Springer, Heidelberg (2013). doi:10.1007/978-3-642-40919-6_4

22. van der Werf, J.M.E.M., van Dongen, B.F., Hurkens, C.A.J., Serebrenik, A.: Process discovery using integer linear programming. Fundam. Inform. **94**(3–4), 387–412 (2009)

23. Eck, M.L., Lu, X., Leemans, S.J.J., van der Aalst, W.M.P.: PM^2: a process mining project methodology. In: Zdravkovic, J., Kirikova, M., Johannesson, P. (eds.) CAiSE 2015. LNCS, vol. 9097, pp. 297–313. Springer, Cham (2015). doi:10.1007/978-3-319-19069-3_19

24. Verbeek, H.M.W., Buijs, J.C.A.M., van Dongen, B.F., van der Aalst, W.M.P.: XES, XESame, and ProM 6. In: Soffer, P., Proper, E. (eds.) CAiSE Forum 2010. LNBIP, vol. 72, pp. 60–75. Springer, Heidelberg (2011). doi:10.1007/978-3-642-17722-4_5

25. Zhou, Z., Wang, Y., Li, L.: Process mining based modeling and analysis of workflows in clinical care - a case study in a Chicago outpatient clinic. In: ICNSC, pp. 590–595. IEEE (2014)

Smart Infrastructures

A Smart Energy Platform for the Internet of Things – Motivation, Challenges, and Solution Proposal

Robert Wehlitz[1(✉)], Dan Häberlein[1], Theo Zschörnig[1], and Bogdan Franczyk[2,3]

[1] Institute for Applied Informatics (InfAI), Leipzig University,
Hainstr. 11, 04109 Leipzig, Germany
`{wehlitz,haeberlein,zschoernig}@infai.org`
[2] Information Systems Institute, Leipzig University, Grimmaische Str. 12,
04109 Leipzig, Germany
`franczyk@wifa.uni-leipzig.de`
[3] Business Informatics Institute, Wrocław University of Economics,
ul. Komandorska 118-120, 53-345 Wrocław, Poland

Abstract. The increasing number of smart appliances equipped with sensors, actuators and tags connected to the Internet and their ability to communicate with one another forms the basis for an Internet of Things (IoT). The data recorded by smart appliances can be utilized to detect energy savings potential and to increase energy efficiency in households. Despite these possibilities, value-added services which successfully incentivize customers to save energy are missing or are still immature. In this paper, we propose a solution for a smart energy platform that aims at reducing this gap by providing the infrastructure and tools which enable the integration of smart appliances and the development of value-added services for an automated energy management in households. In this context, we give an overview of the main challenges for IoT platforms in general and show that existing solutions currently do not fully meet them.

Keywords: Internet of Things · Smart energy · Value-added services

1 Introduction

In a nutshell, the Internet of Things (IoT) consists of heterogeneous smart appliances equipped with sensors, actuators and tags that are capable to communicate with one another over the Internet [1, 2]. The total number of smart appliances is estimated to increase to around 24 billion in 2020 [3]. This will lead to massive amounts of recorded data, offering the possibility for companies to develop data-driven services which create added value to their customers, e.g. in terms of smart home applications [4].

Future trends and recent developments in the energy domain, such as smart metering, e-mobility, or smart grid, can be summarized under the heading of smart energy [5]. Against this background, we investigate possible contributions of the IoT regarding the achievement of smart energy objectives in terms of increasing consumption transparency, saving energy, and energy efficiency. Therefore, we conduct research on a smart energy platform which, as a first step, provides the infrastructure and tools for integrating

W. Abramowicz (Ed.): BIS 2017, LNBIP 288, pp. 271–282, 2017.
DOI: 10.1007/978-3-319-59336-4_19

smart appliances and developing value-added services which allow for an automated energy management in households. The major contributions of this paper are answers to the following questions:

- RQ1: What are the main challenges for IoT platforms?
- RQ2: How could a smart energy platform for the IoT be designed in order to tackle these challenges?
- RQ3: To what extent are these challenges met by existing IoT platforms?

The paper is structured as follows: At first, we describe the motivation of our research work and possible applications of IoT services in the energy domain from a customer perspective (Sect. 2). We then highlight the main challenges for IoT platforms that we have identified by conducting a comprehensive literature analysis to answer RQ1 (Sect. 3). In Sect. 4, we introduce our proposal of a smart energy platform which is primarily derived from the results of the literature analysis (RQ2). Furthermore, we compare existing IoT platforms against the identified challenges in order to answer RQ3 and underpin the need for new research contributions (Sect. 5). The paper concludes with a short summary and outlook (Sect. 6).

2   Motivation

In terms of energy savings, numerous studies show that the overall energy consumption decreases if consumers frequently receive feedback about their energetic behavior, e.g. through smart metering applications [6, 7]. However, the currently available selection of value-added services, e.g. the visualization of energy consumption through web portals, has not lead to a massive consumer demand for smart metering systems. This fact is illustrated by the missing success of market-driven smart meter rollouts in European countries [8].

Currently, smart metering technology is primary used for energy billing which finally does not cover the expenses for deploying the infrastructure necessary for this purpose [9, 10]. Hence, we suggest that traditional energy suppliers adapt their business models in such a way that they offer complementary data-driven services that attract customers in order to achieve a return on investment. Referring to [10, 11], real added value only arises if services automate the energy management without excessively encroaching on the daily life of customers. Since household appliances become more intelligent because of additional sensing, actuation, and communication capabilities, we are confident that they can provide the basis for such innovative services and that a development similar to mobile service ecosystems is possible.

The fields of application of IoT services in households can be divided into the categories monitoring, control, optimization, and autonomy. Monitoring services collect data from sensors and other sources in order to get information about the current state of smart appliances or their environment, e.g. the acquisition, transmission and visualization of smart meter data. Control services allow smart appliances to be remotely controlled, e.g. turning lights off by using a mobile application. Optimization services use algorithms to analyze monitoring data and controlling smart appliances in order to improve processes, such as the adaptation of home lighting according to individual customer habits. The most sophisticated services are arranged as automatic services where smart appliances act autonomously and are able to cooperate with other appliances without human interference, e.g. automatic demand-side management systems [12].

In this context, we aim to utilize the potentials of the IoT by investigating a smart energy platform which enables service providers to integrate, process, analyze, and link energy data from distributed heterogeneous data sources, especially smart appliances. The results could be used, for example, to give service consumers detailed feedback about their energetic behavior, to allow for an individual energy consulting, or to monitor and control different smart appliances automatically according to predefined rules (e.g. due to tariff changes) in order to save energy and lower costs.

3   Challenges

Since the smart energy platform basically represents an IoT platform for the energy domain, we investigated the main challenges for IoT platforms in general by conducting a comprehensive literature analyses following [13]. In order to reduce the risk of missing relevant aspects of other application domains, we broadened the search scope and did not only focus on smart energy related platform challenges. The process of the literature search is depicted in Fig. 1. Due to limited space, in the following, we are concentrating on the key aspects of our approach.

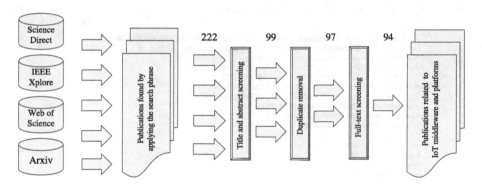

Fig. 1. Literature search process following [13].

In September 2016, we queried the electronic databases *Science Direct*, *IEEE Xplore*, *Web of Science*, and *Arxiv*, which possess numerous publications on computer science and business information systems issues, in order to find papers of conference proceedings and journal articles relevant to our search scope [14]. We tested and applied the search term *"Internet of Things AND platform AND integration"* to all fields for each database without any time restriction. We yielded an unfiltered result set of 222 publications in total. In the next step, we conducted a title and abstract screening to select only relevant publications for further analysis. In order to meet our inclusion criteria, the full text of found papers and articles had to be accessible. Furthermore, we only included contributions on IoT platform comparisons, architecture proposals, or state of the art publications. In contrast, contributions dealing with IoT operation systems, virtual machines, or concrete communication technologies and protocols were excluded. Thus, the number of relevant publications could be further reduced to 99. After checking duplicates, 97 contributions were left over. During full-text screening, we observed that the concepts of IoT platform and IoT middleware are often used interchangeably in the literature. Referring to [15], IoT middleware provides an interlayer between different software applications, communication protocols, and hardware in order to overcome heterogeneity. IoT platforms extend IoT middleware by infrastructural aspects and generally provide additional capabilities, e.g. monitoring, user management, or data analysis [16]. Therefore, we consider IoT middleware as an important part of IoT platforms. However, since some authors refrain from making a distinction (cf. [17, 18]), publications on both concepts were included in our study.

After screening the full texts, 94 relevant publications remained as input for a quantitative analysis. We analyzed and synthesized the contents of all 94 publications and, as shown in Fig. 2, identified 20 different challenges for IoT platforms by detecting, encoding, and grouping every single mention. For each challenge, we determined the absolute frequency of mentions to use as an indicator for its importance. In addition, all identified challenges were classified into a functional or non-functional category [19]. In Sects. 3.1 and 3.2, we shortly describe the most frequently mentioned challenges (9 or above; average is at 8.8), which are *device integration*, *security and privacy*, *data management*, *service abstraction*, *marketplace*, *device and service discovery*, *scalability*, *quality of service*, and *context-awareness*.

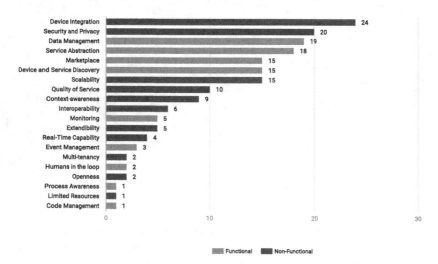

Fig. 2. Absolute frequency of mentioned challenges. (Color figure online)

3.1 Functional Dimension of IoT Platform Challenges

The large amounts of data recorded and transmitted by smart appliances need to be processed in order to obtain new useful insights [20]. Therefore, an adequate *data management* with comprehensive analytical capabilities is required from IoT platforms. Especially methods and tools for data filtering, aggregation, data mining, and advanced machine learning algorithms can accelerate and automate information extraction from data in order to provide meaningful services. Referring to [15], possibilities for data storage and data visualization should at least be provided by IoT platforms.

Smart appliances are characterized by different properties, capabilities, and purposes [21]. They usually provide various interfaces for accessing their functionalities through services. IoT platforms should support the design principle of *service abstraction* in order to enable IoT applications to use these services. This allows for the reuse and composition of services in order to create more sophisticated services [22].

In order to apply new innovative business models in the IoT, a whole ecosystem, commonly described as a *marketplace*, is needed for matching the supply and demand of different stakeholders [23]. The concept is comparable to the application store ecosystem of mobile platforms, such as Google Android or Apple iOS [23]. The presence of a marketplace is rated to be "crucial" for innovations within the IoT domain [16].

The IoT consists of a wide range of distributed smart appliances [24]. Therefore, mechanisms for *device and service discovery* are needed to integrate them with IoT platforms [25]. These mechanisms support the finding, configuration and connection of smart appliances to IoT platforms in order to enable data exchange between them [17]. This is made more difficult by the fact that network topologies containing smart appliances are "both unknown and dynamic" [24].

3.2 Non-functional Dimension of IoT Platform Challenges

Different hardware, operating systems, service interfaces, message protocols, and data formats in the IoT are examples for its fars-reaching heterogeneity [21]. Hence, an IoT platform has to provide the means to allow for simple *device integration* in order to overcome this heterogeneity.

Another challenge to be addressed concerns *privacy and security*. Smart appliances record massive amounts of data which might be related to individuals. This sensitive data as well as control capabilities need to be protected against unauthorized access and misuse. Also, storage and processing applications of platforms need to be aware of user-specific security settings [20]. In this context, data owners should be able to define fine grained security restrictions [16]. This implies that a user can decide which data is accessible by platform applications or other users on schema or even dataset level [23].

IoT platforms can be implemented at the device, fog and cloud level, having to handle device numbers and data amounts from small to very large [26]. Especially cloud-based platforms should be able to handle large amounts of data and device numbers, thus requiring providers to tackle the *scalability* challenge [27]. Platforms need to be scalable in terms of request handling, data ingestion rate, data storage and analytics capabilities [22].

Quality of service is a challenge which originates from the fact that smart appliances are associated with real-world objects. Hence, it might occur that they suffer from connection issues and malfunction [17]. In case of any unexpected behavior, this must be detected and other connected devices may need to be informed since the initiation of compensation measures could be necessary [28].

Context-awareness describes a property of "adaptive systems" [15] which makes them able to find the most appropriate service to do a task or derive the best decision for a specific problem. IoT platforms can enable appliances to become context-aware, and through their interaction with their surroundings create an environment, which contains inherent implicit knowledge. The inference of this knowledge is possible by integration and processing of instance and metadata from them as well as user and infrastructure related data.

4 Solution Proposal

After we highlighted the main challenges for IoT platforms in Sect. 3, we now introduce our proposal of a smart energy platform that aims at providing the infrastructure and tools necessary in order to overcome these issues.

The platform architecture shown in Fig. 3 was primarily derived from our literature analysis. It follows the service-oriented architecture and microservice paradigm, and consists of the four central system areas *interfaces*, *data*, *services*, and *processes*, which comprise various subsystems and technologies. These areas (colored in Fig. 2) are already partly implemented as software prototypes for evaluation purposes.

An IoT middleware, which supports different communication protocols, e.g. HTTP, CoAP and MQTT, and therefore enables *device integration*, is part of the *interfaces* area. It is based on Node-RED and allows for the modeling of reusable data flows between smart appliances and the platform by using a graphical flow designer. The data can be either pushed by smart appliances or pulled by the middleware in certain time intervals, in response to events, or on demand. Incoming and outgoing data streams are handled by Apache Kafka as a message broker so they can be consumed later in case of the unavailability of platform components or connected smart appliances. This contributes to a better *quality of service*. Platform users are able to integrate smart appliances by themselves based on *device and service discovery* mechanisms. This can be carried out manually through a user interface or semi-automatically, e.g. by using the capabilities of CoAP, ZigBee, or Z-Wave. An IoT device repository stores the service descriptions and semantic data of registered smart appliances in a triplestore and shares them with other platform components if needed. A service description contains the subject of a service as well as its technical details, such as how the service can be invoked. Semantic data, for example, can be helpful in terms of finding services that are suitable to be used together and aware of a specific context (*context-awareness*), e.g. location.

The data, frequently received from a variety of smart appliances, requires a high-performance infrastructure to process, store and analyze it, even in real-time [25]. In this context, the *data* area manages the ingestion and processing of large amounts of different data streams and provides capabilities for automated data processing and

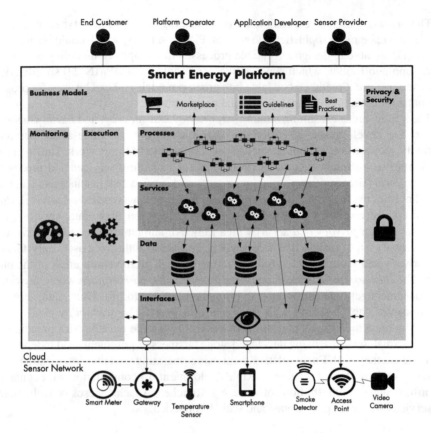

Fig. 3. System areas of the smart energy platform.

analysis (*data management*). It is implemented using the Kappa Architecture [29] concept, which, in a nutshell, is based around the assumption that all data in an environment can be ingested as a stream. Therefore, it is possible to only use stream processing technologies thus eliminating the need to develop and maintain processing applications for batch jobs as done in a Lambda Architecture [30]. This enables fast and flexible data ingestion and processing in an environment with rapidly changing requirements and concept drift. Data persistence is achieved using the Data Lake concept. This allows to store large amounts of unstructured and structured data without knowing the context or later use of it while still maintaining data exploration flexibility [31, 32].

The *services* area abstracts different service interfaces of smart appliances and platform components (*service abstraction*). In this context, we distinguish between services for raw data, transformation, analytics, actuation and external content. Raw data services return unprocessed data as it is recorded by smart appliances. Incoming data streams can be preprocessed by using transformation services, e.g. for unit conversion. Analytical services extract useful information from raw or preprocessed data while actuation services trigger actions, e.g. turning lights off. External services are Web services of third-party providers, e.g. weather data or geolocation services.

The *processes* area is responsible for orchestrating these different types of services in order to create more sophisticated services. Platform users, which could be even end customers, are able to design executable processes in a graphical modeling tool based on the bpmn.io toolkit, which supports the commonly used BPMN 2.0 specification. Thus, services are directly embeddable into process models and can be used for process automatization without the need for special programming.

The system areas *execution, monitoring, privacy and security*, and *business models* support the development, provisioning, and consumption of value-added services. Currently, they are not yet implemented and will be subject of future work. The *execution* area comprises different runtime environments for executing services and processes. The *monitoring* area supervises running process and service task instances and, among others, notifies users or other devices in case of any problems (*quality of service*). Platform users are enabled to grant different access rights on their data, services, and processes by the *privacy and security* area [33]. An authentication and authorization management system ensures that access will be granted to authorized users only. Generally, privacy and security are considered throughout all architectural areas of the platform. The *business models* area comprises an electronic *marketplace* where providers and consumers get together in order to bring the platform to life. Thus, data, services, and processes can be shared with other platform users. The *scalability* challenge is tackled by implementing all platform services following the microservice paradigm. It promotes small services, being developed along business requirements and independent from one another [34]. Encapsulating them using operating-system virtualization technologies, such as Docker, enables flexible deployment. In combination with container and infrastructure orchestration software, e.g. Rancher, Apache Mesos or Kubernetes, the services are easily provisioned and scaled as needed.

5 Related Work

In this section, we present existing solutions which are similar to our proposal and give an overview of their shortcomings. Recent studies have already compared IoT platforms regarding some of the most frequently mentioned functional and non-functional challenges we covered in Sect. 3 [15, 16]. We merged the results of these and extended the findings, in terms of platforms and met challenges, with research of our own. It was conducted by screening documentation, source code and other artifacts, such as project websites and grey literature. The combined results can be seen in Table 1.

Different colors represent to what extend the capabilities of a platform satisfy *data management* (**F1**), *service abstraction* (**F2**), *marketplace* (**F3**), *device and service discovery* (**F4**), *device integration* (**N1**), *privacy and security* (**N2**), *scalability* (**N3**), *quality of service* (**N4**), and *context-awareness* (**N5**). Green was used if a challenge is fully met, red if not. Yellow was used for partially met challenges and white denotes that we did not have access to any further information.

In this context, we observed that almost all platforms use *service abstraction* as a concept for providing a unified way to access different kinds of services. The *data management* as well as the *device and service discovery* challenges are at best partly

Table 1. IoT platform comparison against the identified main challenges, which is partly based on the results of [15, 16].

Platforms / Challenges	F1	F2	F3	F4	N1	N2	N3	N4	N5
FIWARE									
ThingWorx									
H.A.T.									
Ubiware									
TinyMAPS									
Maté									
TinyLIME									
Thing Broker									
OpenIoT									
Node-RED									
Nimbits									
Extended Maté									
ThingSpeak									
MiLAN									
Impala									
GREEN									
Echelon (IzoT)									
Adap. middlew.									
OpenHAB									
OSGP									
Xively									
SwissQM									
SENSEI									
PRISMA									
MidFusion									
MASPOT									
EMMA									
DAViM									
AutoSec									
AFME									
TinyCubes									

Platforms / Challenges	F1	F2	F3	F4	N1	N2	N3	N4	N5
SOCRADES									
RUNES									
MagnetOS									
CarrIoT									
VM*									
UbiROAD									
MAPS									
KASOM									
Hydra									
DVM									
Agilla									
A3-TAG									
TeenyLIME									
Sensewrap									
Mires									
MOSDEN									
Hermes									
TS-Mid									
Smart messages									
Ericsson IoT									
CHOReOS									
Sensorware									
Melete									
ActorNet									
MUSIC									
LinkSmart									
LIME									
Kahvihub									
SensorBus									
IrisNet									
Sparkfun									

met. It is noteworthy, that just about every considered platform lacks an IoT marketplace. This fact prevents service providers from applying innovative business models and customers from using services that add real value to them, which might have a negative impact on IoT market stimulation.

Considering non-functional challenges, most of the platforms allow to create context-aware applications using sensor data to respond to changing environmental conditions. However, the integration of a wide range of heterogeneous smart appliances is only supported partially. The same applies for *scalability* which is a major issue if the amount of connected smart appliances is increasing in the next years as it is estimated by market analysts (c.f. Sect. 1). Furthermore, *privacy and security* as well as *quality of service* still can be regarded as important challenges of the future.

Overall, the platforms FIWARE and ThingWorx stand out, as they satisfy most of the investigated challenges. However, ThingWorx represents a proprietary system which focuses on application scenarios within the industry sector. It provides platform tools for technicians and business analysts for automating and optimizing manufacturing processes. Furthermore, both FIWARE and ThingWorx are huge IT-based ecosystems, which exhibit high complexity and a steep learning curve. Hence, we conclude that they are not suitable to be used by end customers for implementing an automated energy management in households as intended by our solution.

6 Conclusion and Outlook

In this paper, we described our motivation to conduct research on possible applications of the IoT in terms of increasing energy consumption transparency, energy efficiency, and saving energy. We identified and described the main challenges for IoT platforms and presented a solution proposal for a smart energy platform which aims at providing the infrastructure and tools for developing value-added services that allow for an automated energy management in households. We compared IoT platforms against the identified challenges and showed that no examined platform fully meets them.

However, we only investigated challenges which were discussed in publications that we found during our literature search. As a consequence, we cannot eliminate the possibility that we missed relevant literature. The same applies for the number of IoT platforms which was involved in our analysis. Further, challenges and their importance to platform providers might be subject of change as the IoT gradually gains broader acceptance of end customers. For instance, the unconsidered challenge "Openness" (cf. Fig. 2) seems to get more important as the evolution of the IoT gets faster and single companies are hardly able to respond to constant change. Therefore, open, community-driven platforms seem promising to cope with that [16]. Thus, even though existing platforms may have already adapted in part to the challenges we described, the changing environment of the IoT requires new solutions and further research. Also, the development and application of new business models and incentives for end customers to use an IoT platform are issues which need to be addressed. This requires research not only from the perspective of platform providers but also looking at the customer as well as the business side. More specific, businesses which develop and provide services on an IoT platform will face organizational challenges from changed application development and maintenance lifecycles.

In order to start evaluating our own research we have implemented key components of the *interfaces*, *data*, *services*, and *processes* areas and were able to provide working prototypes for these parts. However, comprehensive tests and experiments to further validate the usefulness and effectiveness of our approach still need to be conducted. Furthermore, additional work on the design and development of further platform components, such as the marketplace for applying *business models*, *privacy and security* mechanisms, *execution* environments, and *monitoring* systems, need to be done in order to create a fully integrated and highly automated platform which is able to tackle the challenges presented in this paper as well as the ones that will get more important in the near future.

Acknowledgments. The work presented in this paper is partly funded by the European Regional Development Fund (ERDF) and the Free State of Saxony (Sächsische Aufbaubank – SAB).

References

1. Han, S.N., Khan, I., Lee, G.M., Crespi, N., Glitho, R.H.: Service composition for IP smart object using realtime web protocols: concept and research challenges. Comput. Stand. Interfaces. **43**, 79–90 (2016)
2. Lynggaard, P.: Artificial Intelligence and Internet of Things in a "Smart Home" Context: A Distributed System Architecture: A Thesis Submitted in Partial Fulfilment of the Requirements for the Degree of Doctor of Philosophy (2013)
3. Greenough, J.: How the "Internet of Things" will impact consumers, businesses, and governments in 2016 and beyond. http://www.businessinsider.com/how-the-internet-of-things-market-will-grow-2014-10?IR=T
4. Zanella, A., Bui, N., Castellani, A., Vangelista, L., Zorzi, M.: Internet of Things for smart cities. IEEE Internet Things J. **1**, 22–32 (2014)
5. Aichele, C., Doleski, O.: Smart Meter Rollout: Praxisleitfaden zur Ausbringung intelligenter Zähler. Springer (2012)
6. Hosek, J., Masek, P., Kovac, D., Ries, M., Kropfl, F.: Universal smart energy communication platform. In: 2014 International Conference on Intelligent Green Building and Smart Grid (IGBSG) (2014)
7. Faruqui, A., Sergici, S., Sharif, A.: The impact of informational feedback on energy consumption—a survey of the experimental evidence. Energy **35**, 1598–1608 (2010)
8. Ernst & Young: Cost-benefit analysis for the comprehensive use of smart metering (2013). http://www.ey.com/Publication/vwLUAssets/EY-Kosten-Nutzen-Analyse_Roll-out_Smart_Meter-Eng/$FILE/EY-BMWI-Endbericht-KNA-Smart-Metering-2013-ENG.pdf
9. Aichele, C., Doleski, O.D.: Smart Market: Vom Smart Grid zum intelligenten Energiemarkt. Springer (2014)
10. Grandel, M.: Das "Smart Metering Dilemma" – Strategische Überlegungen zum flächendeckenden Einsatz von Smart Metering. In: Smart Energy, pp. 221–231 (2011)
11. Balta-Ozkan, N., Boteler, B., Amerighi, O.: European smart home market development: public views on technical and economic aspects across the United Kingdom, Germany and Italy. Energy Res. Soc. Sci. **3**, 65–77 (2014)
12. Porter, M.E., Heppelmann, J.E.: How smart, connected products are transforming competition. Harv. Bus. Rev. **92**, 64–88 (2014)
13. Vom Brocke, J., Simons, A., Niehaves, B., Riemer, K., Plattfaut, R., Cleven, A., Others: Reconstructing the giant: on the importance of rigour in documenting the literature search process. In: ECIS, pp. 2206–2217 (2009)
14. Knackstedt, R., Winkelmann, A.: Online-Literaturdatenbanken im Bereich der Wirtschaftsinformatik: Bereitstellung wissenschaftlicher Literatur und Analyse von Interaktionen der Wissensteilung. WIRTSCHAFTSINFORMATIK. **48**, 47–59 (2006)
15. Razzaque, M.A., Milojevic-Jevric, M., Palade, A., Clarke, S.: Middleware for Internet of Things: a survey. IEEE Internet Things J. **3**, 70–95 (2016)
16. Mineraud, J., Mazhelis, O., Su, X., Tarkoma, S.: A gap analysis of internet-of-things platforms. Comput. Commun. **89–90**, 5–16 (2016)
17. Perera, C., Jayaraman, P.P., Zaslavsky, A., Georgakopoulos, D., Christen, P.: Sensor discovery and configuration framework for the Internet of Things paradigm. In: 2014 IEEE World Forum on Internet of Things (WF-IoT) (2014)
18. Zdravković, M., Trajanović, M., Sarraipa, J., Jardim-Gonçalves, R., Lezoche, M., Aubry, A., Panetto, H.: Survey of internet-of-things platforms. In: 6th International Conference on Information Society and Technology (2016)
19. Sommerville, I.: Software Engineering. Pearson Deutschland (2007)

20. Gubbi, J., Buyya, R., Marusic, S., Palaniswami, M.: Internet of Things (IoT): a vision, architectural elements, and future directions. Future Gener. Comput. Syst. **29**, 1645–1660 (2013)
21. Issarny, V., Georgantas, N., Hachem, S., Zarras, A., Vassiliadist, P., Autili, M., Gerosa, M.A., Ben Hamida, A.: Service-oriented middleware for the future internet: state of the art and research directions. J. Internet Serv. Appl. **2**, 23–45 (2011)
22. Botta, A., de Donato, W., Persico, V., Pescapé, A.: Integration of cloud computing and Internet of Things: a survey. Future Gener. Comput. Syst. **56**, 684–700 (2016)
23. Stankovic, J.A.: Research directions for the Internet of Things. IEEE Internet Things J. **1**, 3–9 (2014)
24. Teixeira, T., Hachem, S., Issarny, V., Georgantas, N.: Service Oriented Middleware for the Internet of Things: A Perspective. In: Abramowicz, W., Llorente, I.M., Surridge, M., Zisman, A., Vayssière, J. (eds.) ServiceWave 2011. LNCS, vol. 6994, pp. 220–229. Springer, Heidelberg (2011). doi:10.1007/978-3-642-24755-2_21
25. Jiang, L., Da Xu, L., Cai, H., Jiang, Z., Bu, F., Xu, B.: An IoT-oriented data storage framework in cloud computing platform. IEEE Trans. Ind. Inf. **10**, 1443–1451 (2014)
26. Broring, A., Schmid, S., Schindhelm, C.-K., Khelil, A., Kabisch, S., Kramer, D., Le Phuoc, D., Mitic, J., Anicic, D., Teniente, E.: Enabling IoT ecosystems through platform interoperability. IEEE Softw. **34**, 54–61 (2017)
27. Patti, E., Acquaviva, A.: IoT platform for smart cities: requirements and implementation case studies. In: 2016 IEEE 2nd International Forum on Research and Technologies for Society and Industry Leveraging a Better Tomorrow (RTSI), pp. 1–6. IEEE (2016)
28. Dar, K., Taherkordi, A., Baraki, H., Eliassen, F., Geihs, K.: A resource oriented integration architecture for the Internet of Things: a business process perspective. Pervasive Mob. Comput. **20**, 145–159 (2015)
29. Kreps, J.: Questioning the Lambda Architecture. https://www.oreilly.com/ideas/questioning-the-lambda-architecture
30. Stolpe, M.: The Internet of Things: opportunities and challenges for distributed data analysis. SIGKDD Explor. Newsl. **18**, 15–34 (2016)
31. Pasupuleti, P., Purra, B.S.: Data Lake Development with Big Data. Packt Publishing Ltd., Birmingham (2015)
32. Fang, H.: Managing data lakes in big data era: What's a data lake and why has it became popular in data management ecosystem. In: 2015 IEEE International Conference on Cyber Technology in Automation, Control, and Intelligent Systems (CYBER), pp. 820–824. IEEE (2015)
33. Mashhadi, A., Kawsar, F., Acer, U.G.: Human data interaction in IoT: the ownership aspect. In: 2014 IEEE World Forum on Internet of Things (WF-IoT) (2014)
34. Lewis, J., Fowler, M.: Microservices: a definition of this new architectural term (2014). http://www.martinfowler.com/articles/microservices.html

Towards a Taxonomy of Constraints
in Demand-Side-Management-Methods
for a Residential Context

Dennis Behrens[✉], Thorsten Schoormann, and Ralf Knackstedt

Department of Information Systems, University of Hildesheim, Universitätsplatz 1,
31141 Hildesheim, Germany
dennis.behrens@uni-hildesheim.de

Abstract. To address current challenges in the management of energy grids, Demand-Side-Management (DSM) is one possibility. In this field various approaches exist but consider often different constraints regarding appliances. This paper aims at identifying these constraints through a triangulation: we conducted two literature reviews and several expert interviews simultaneously to derive a taxonomy of (DSM) load constraints. This taxonomy grows during the research process and contains at the end five constraints, a short description, examples of appliances for each constraint and a mathematical representation. This taxonomy can be used for future research, e.g. for designing, evaluating or benchmarking DSM-Methods.

Keywords: Demand-Side-Management · Demand-Side-Management-Methods · Constraints · Taxonomy · Residential context

1 Problem Identification

Due to various current challenges such as climatic changes, growing population and urbanization, improvements in the management of energy grids are required [1]. New technological and political developments contribute to face these challenges by, for example, considering sustainable but volatile energy resources (e.g., photovoltaic, wind or water) or integrating Smart Meters in residential buildings (e.g., [2, 3]). Moreover, the number of Plug in Hybrids (PHEVs) and Electric-Vehicles (EVs) increased which arises new chances for the energy systems. Hence, innovative methods, algorithms and techniques need to be provided. Demand-Side-Management (DSM) (often synonymously but wrongly called Demand Response (DR) [3]) is one possibility which contributes to this by providing various approaches such as managing, shifting and controlling loads, saving energy or reducing peaks in energy grids (e.g., [4, 5]). The main goal of DMS is to achieve a balanced load. Current studies indicate that about 50% of the loads can be shifted [6]. In addition, field tests in the USA indicate that optimization can lead to significant peak reductions [7].

For implementing DSM, different frameworks are available e.g. radius optimization (e.g., number of buildings/flats) (e.g. [8, 9]), centralized controlled optimization (e.g. [10, 11]) or decentral controlled optimization ([12–14]). An overview of methods

© Springer International Publishing AG 2017
W. Abramowicz (Ed.): BIS 2017, LNBIP 288, pp. 283–295, 2017.
DOI: 10.1007/978-3-319-59336-4_20

available can exemplarily be found in [2, 4, 15, 16]. Different methods consider different or no constraints which makes, for example, a benchmark difficult. However, these constrains have to be in consideration because energy loads and peaks cannot be moved to any time, and users have preferences as well as habits of consumption which restrict the movement of energy loads. By conducting an initial literature search (Sect. 3.1), only research could be found which addresses constraints implicitly, e.g. by considering "interruptible tasks" (e.g. [17–20]), "HVAC" (heating ventilation and air conditioning) loads (e.g. [21–23]), load classifications (e.g. [24, 25]) or a utility function of users (e.g. [26–28]). In addition the term "Smart Home" or "Smart Meter" and "Smart Home Management Systems" or similar are mentioned (e.g. [19, 25, 29]). Consequently, this paper aims to answer the following research questions (RQs):

- *RQ 1: Which (load) constraints are relevant to the DSM field?*
- *RQ 2: How these constraints are implemented in common DSM-Methods?*

In order to answer these RQs, our paper is structured as follows: In Sect. 2, we outline our research design for deriving a taxonomy of constraints. In Sects. 3 and 4, we derive and validate relevant constraints from the literature (conceptual) as well as from expert interviews (empirical) in a simultaneously manner. In Sect. 5, we consolidate and present our findings in a taxonomy. In Sect. 6, we discuss our findings, outline limitations and provide implications for future research for DSM.

2   Research Design

This research is allocated in a large design science project which aims at benchmarking different DSM-Approaches. Therefore, it is essential to know which constraints exist. In order to answer the RQs, we triangulate with different research methods: literature reviews (Sect. 3), expert interviews (Sect. 4) and we plan to extract information from websites (Sect. 6). All three methods are suitable to derive a taxonomy [30]. We

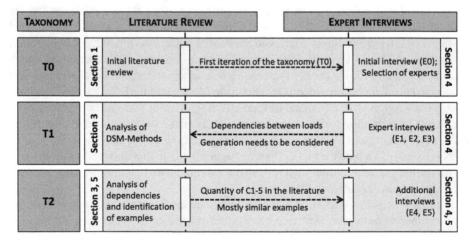

Fig. 1. Research Process of Deriving the Taxonomy

conducted the literature reviews and expert interviews simultaneously. Because of the limited space we cannot describe the process in detail. Hence, we describe the main "lessons learned" and their impacts (Fig. 1).

Initially, we conducted an expert interview (E0) with an energy provider as well as an initial literature search (Sect. 3). Moreover, we selected adequate experts for additional interviews to confirm, validate and extend the initial findings. So, we identified a first taxonomy (**T0**). Based on this, we created a questionnaire for the following interviews. Afterwards, we conducted three interviews and search more literature dealing with DSM-Methods. Two main findings for the taxonomy are, (1) a new restriction dealing with the dependencies between loads and (2) that most algorithms formulate a mathematical model of their loads. Therefore, we added a mathematical model and a new constraint to the taxonomy (**T1**). Furthermore, we found many articles dealing with Smart Home. Due to this, we added "users" to the expert scope. By finishing the literature review on the basis of the DSM-Method-articles, we could identify a diverse quantity of constraints. Thus, we added questions about the significance of constraints. If the interviews were already conducted, we asked the experts later again. Afterwards, we conducted two more interviews, including one Smart Home user. We also evaluated the quantity of literature findings and the rating of our experts. In addition, we asked for more examples as the examples found in our literature were rather limited. These were added to the taxonomy (**T2**).

3   Literature Review

3.1   Search

Literature Search (T0).   Our literature review is based on the methodology from [31]. We selected Google Scholar as a starting source. As possible search items we identified: *Demand Side Management, Demand-Side-Management, Demand Response, Load, Appliance, Device, Task, Restriction* and *Constraint*. Due to the fact that we focus on energy respectively power, we excluded DSM for gas and water. In total, we found 305 articles. We evaluated (a) title, keywords and abstracts, and, afterwards, (b) full texts. These articles were considered for deriving the initial taxonomy (T0).

Literature Search (T1).   As described in Sect. 2, we added more sources to our literature review to validate or extend our initial taxonomy. Therefore, we analyzed the following literature reviews of DSM-Methods as well as the articles used in the reviews: [4, 15, 16] (a total of 247 papers). After evaluating the same way as in the former search, we found a total of 36 relevant articles. We only considered sources which deal with single loads – this means that every load is specified individually in contrary to a summarized amount [32] and a constraint is mentioned somehow.

3.2   Analysis and Synthesis

The results of the literature search (**T0**) are already described in Sect. 1 (summary: no articles address our research aim directly; but some address several constraints indirectly, by considering some constraints but not mentioning them adequately). Based on the literature analysis (**T0** and **T1**), we identified four classes of main constraints. In the following, we describe them and provide some details of the literature findings as well as examples mentioned (Table 1).

Table 1. Literature Review results (**T1**) [C5 is defined in Sect. 4.2]

Ref.	C1	C2	C3	C4	C5	Ref.	C1	C2	C3	C4	C5
[8]	X	X				[50]	X	X			
[33]	X	X		X		[51]	X	X		X	
[34]	X			X		[52]	X				
[35]	X	X		X		[53]		X		X	
[36]	X			X		[54]					
[37]	X	X				[55]	X			X	
[38]	X	X				[56]	X	X		X	
[39]	X	X		X		[57]	X	X		X	
[40]	X			X		[58]					
[41]			X	X		[59]	X	X		X	
[42]				X		[60]	X				
[43]						[61]	X				
[44]	X	X		X		[62]					
[45]	X					[63]	X			X	
[46]						[64]		X			
[47]	X			X		[65]	X			X	
[48]	X					[66]	X				
[49]	X			X		[67]	X			X	
	C1		C2		C3			C4		C5	
Count	27 (75,0%)		12 (33,3%)		1 (2,8%)			18 (50,0%)		0 (0,0%)	

Horizontal Separability of Loads (C1). A load can be separated horizontally. This means according its time of use, for example, pause/interrupt the load. After it has been started, it can be paused and started again as often and as long as necessary. The load profile itself remains unchanged. For many loads this is not suitable but many algorithms do not consider this constraint. Examples are the washing process of a washing machine or a dishwasher. If these types of appliances are turned on, they (mostly) cannot be turned off or paused until they are finished. Of course, some newer products allow a pause during the process, but the load profile might change in this case, for example, if the water needs to be heat up again.

Vertical Separability of Loads (C2). A load can be separated vertically – according its intensity. A certain amount of power can be shifted to another time during the process. The load profile is therefore variable and can be changed. An example is the charging process of an electric vehicle. Most vehicles offer different charging modes, besides the "normal" mode, which means loading with the standard electrical socket, also a fast charging mode can be chosen. Here the time of charging is reduced but the needed power per time is higher and the load profile is therefore changed.

Time Interval of Use (C3). Some loads can only be turned on during specific time intervals; others cannot be turned on during these intervals. Moreover, some loads need to be finished on a specific time. Most of these deadlines or intervals are connected to the user's habits or requirements. For example, an electric vehicle needs to be charged to a certain amount in the morning because the user needs to drive to work. However, the charging process can only be started at the point the user returns at the evening from work. The washing machine instead can perhaps not be started in the night, because the neighbors would be disturbed.

Environmental Effects (C4). As some loads affect the environment (temperature for example) and are also affected by the environment, these loads cannot be handled in the same way as other loads, which do not subject to these effects. The prediction of the load curves, mostly dependent on the users comfort requirements, is flexible. Because many unknown variables do affect the result, the estimation of amount and duration is difficult. Popular examples are so called Heating Ventilation and Air-Conditioning loads. These try to reach a certain temperature but must deal with several variables: How is the temperature at the moment? Is the window open? Because of these effects the amount of energy needed to reach a certain point can only be estimated. In the literature often differential equations are used.

Other subjects mentioned by the authors are utility functions of different users and their comfort levels. Because this also leads to different using times on the one hand and dependences on the other hand, we covered this with the "Time Interval of Use".

In the following Table 2 we summarize the sources and the addressed constraint(s). Because the Reviews consider general DSM-Methods and HVAC papers are very specific, in the second search only one paper is found. Nevertheless the first search identified several papers (e.g. mentioned in Sect. 1) which address this constraint.

Table 2. Overview of the experts ($\mu = mean$; $\sigma = Standard\ Deviation$)

ID	Age	Occupation	Field	Experiences	Length
E0	55	Practitioner	Energy Provider	15 years; diploma	44 min
E1	30	Researcher	PHEV and Batteries	3 years, Master degree, Phd	45 min
E2	30	Practitioner	Electrical Eng.	1 year; Master degree	41 min
E3	32	Practitioner	Electrical Eng.	2 years; Master degree	32 min
E4	40	Smart Home User	IT	8 years in IT; Master degree; 3 years private usage	38 min
E5	26	Practitioner	IT Energy consult	3 years of experiences	35 min
Interview statistics: $\sum$235 min; $\mu = 31,16$; $\sigma = 4,66$ min					

4   Expert Interviews

4.1   Setting

First of all, we were able to speak with an energy supplier. This happened before we had the chance to derive our first taxonomy version. Nevertheless, this interview provided a good impression of how energy providers look at this field and which kind of factors are important. Furthermore, we formulated our expert requirements: (I) connection to the field (occupation, researcher, user) and (II) more than 1 year of experience. We tried to cover different backgrounds and perspectives. In total we interviewed (additionally to E0) five experts which characterized as follows (Table 2):

4.2   Results

All of the experts mentioned that the field is important and needs further research. They rated C2 and C4 as most important and C5 as less important (see Table 3). Moreover they stated some loads might dependent on each other (E1). Therefore we added a fifth constraint to the taxonomy: **Dependencies between Loads (C5)**. This means that appliance b has to start right after finishing appliance a or vice versa. Examples could be washing machine and dryer: the dryer stars right after the washing machine finishes. By searching directly for this topic, we found one paper dealing with dependencies between different loads on a mathematical level [68].

Table 3. Expert rating of the constraints ($\mu = mean$; $\sigma = Standard\ Deviation$)

#	E1	E2	E3	E4	E5	μ	σ	Mean/SD
C1	3	4	2	3	5	3.4	1.02	
C2	4	5	5	5	5	4.8	0.4	
C3	4	3	4	5	4	4	0.63	
C4	4	4	4	5	4	4.2	0.4	
C5	2	3	2	2	1	2	0.63	

E3 and E5 suggested that in C1 the pausing of a load profile might be possible but only a certain time, because the water in the washing machine is too cold afterwards.

In addition, E5 mentioned, that not all loads can be changed freely according their intensity, e.g. a washing machine needs a certain temperature. We, therefore, added a minimum and maximum border in the mathematical model of C2.

E2, E3, E4 and E5 mentioned dependencies between constraints: C1–C3 (E2, E3); C2–C3 (E2); C2–C4 (E3, E4); C3–C4 (E4); C1–C2 (E5). Furthermore the experts mentioned that additional factors need tobe considered: Energy generation (E0, E1, E3, E5, E0), user specific scenarios (E1), energy storages (E1, E2, E3, E4, E5, E0), the cost factor and the amortization of investments (E4, E0) are important. In addition, E0, E4 and E5 mentioned, that a variable pricing needs to be accomplished to enable a higher saving potential. Also additive information as the current location should be considered. Generally, forecasts are essential (E1 and E5). We asked the experts also for additional examples or if the constraints also apply for other loads. They mentioned several, additional appliances which we added to the taxonomy.

5 Taxonomy for Constraints

Consolidating the results from the literature review, the expert interviews and the mathematical representation (see below), we are able to derive an aggregated version of our taxonomy (**T2**) (Fig. 2). It includes a total of five constraints, a short description, examples and the mathematical model.

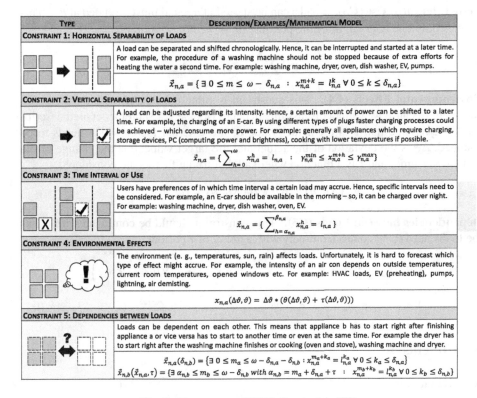

Type	Description/Examples/Mathematical Model
Constraint 1: Horizontal Separability of Loads	
	A load can be separated and shifted chronologically. Hence, it can be interrupted and started at a later time. For example, the procedure of a washing machine should not be stopped because of extra efforts for heating the water a second time. For example: washing machine, dryer, oven, dish washer, EV, pumps.
	$$\vec{x}_{n,a} = \{\exists\, 0 \leq m \leq \omega - \delta_{n,a}\ :\ x_{n,a}^{m+k} = l_{n,a}^k \,\forall\, 0 \leq k \leq \delta_{n,a}\}$$
Constraint 2: Vertical Separability of Loads	
	A load can be adjusted regarding its intensity. Hence, a certain amount of power can be shifted to a later time. For example, the charging of an E-car. By using different types of plugs faster charging processes could be achieved – which consume more power. For example: generally all appliances which require charging, storage devices, PC (computing power and brightness), cooking with lower temperatures if possible.
	$$\vec{x}_{n,a} = \{\sum_{h=0}^{\omega} x_{n,a}^h = l_{n,a}\ :\ \gamma_{n,a}^{min} \leq x_{n,a}^{m+h} \leq \gamma_{n,a}^{max}\}$$
Constraint 3: Time Interval of Use	
	Users have preferences of in which time interval a certain load may accrue. Hence, specific intervals need to be considered. For example, an E-car should be available in the morning – so, it can be charged over night. For example: washing machine, dryer, dish washer, oven, EV.
	$$\vec{x}_{n,a} = \{\sum_{h=\alpha_{n,a}}^{\beta_{n,a}} x_{n,a}^h = l_{n,a}\}$$
Constraint 4: Environmental Effects	
	The environment (e. g., temperatures, sun, rain) affects loads. Unfortunately, it is hard to forecast which type of effect might accrue. For example, the intensity of an air con depends on outside temperatures, current room temperatures, opened windows etc. For example: HVAC loads, EV (preheating), pumps, lightning, air demisting.
	$$x_{n,a}(\Delta\vartheta, \vartheta) = \Delta\vartheta * (\theta(\Delta\vartheta, \vartheta) + \tau(\Delta\vartheta, \vartheta)))$$
Constraint 5: Dependencies between Loads	
	Loads can be dependent on each other. This means that appliance b has to start right after finishing appliance a or vice versa has to start to another time or even at the same time. For example the dryer has to start right after the washing machine finishes or cooking (oven and stove), washing machine and dryer.
	$$\vec{x}_{n,a}(\delta_{n,b}) = \{\exists\, 0 \leq m_a \leq \omega - \delta_{n,a} - \delta_{n,b} : x_{n,a}^{m_a+k_a} = l_{n,a}^{k_a} \,\forall\, 0 \leq k_a \leq \delta_{n,a}\}$$ $$\vec{x}_{n,b}(\vec{x}_{n,a}, \tau) = \{\exists\, \alpha_{n,b} \leq m_b \leq \omega - \delta_{n,b} \text{ with } \alpha_{n,b} = m_a + \delta_{n,a} + \tau\ :\ x_{n,a}^{m_b+k_b} = l_{n,a}^{k_b} \,\forall\, 0 \leq k_b \leq \delta_{n,b}\}$$

Fig. 2. Taxonomy of DSM-Constraints (**T2**)

Mathematical Model. Let N be all considered living units and A_n be all the appliances of unit n and ω the number of time periods over one day (e.g. 15 min slots: $\omega = 24\,\text{h} * 4 = 96$). Moreover let $x^t = \sum_{n \in N} \sum_{a \in A} x_{n,a}^t$ *with* $t \in 0, 1, \ldots, \omega$ be the sum of all appliances of living unit n in the timeslot and $l_{n,a}^k$ be the load profile, $l_{n,a}$ the load sum and $\delta_{n,a}$ the length of load a. Then a possible load representation (based on [69]) could be: $\vec{x}_{n,a} = \{\sum_{h=0}^{\omega} x_{n,a}^h = l_{n,a}\}$. This implies that the load must occur during 0–24 h and can be varied according its intensity between 0 and the load sum.

Horizontal Separability of Loads (C1). The initial formula needs to consider the fixed load profile of load a and hold it together.

$$\vec{x}_{n,a} = \{\exists\, 0 \leq m \leq \omega - \delta_{n,a} : x_{n,a}^{m+k} = l_{n,a}^k \,\forall\, 0 \leq k \leq \delta_{n,a}\}$$

Vertical Separability of Loads (C2). Load a can be shifted according its intensity but only in a predefined intervals, each with min $(\gamma_{n,a}^{min})$ and max $(\gamma_{n,a}^{max})$ borders.

$$\vec{x}_{n,a} = \{\sum\nolimits_{h=0}^{\omega} x_{n,a}^h = l_{n,a} : \gamma_{n,a}^{min} \leq x_{n,a}^{m+h} \leq \gamma_{n,a}^{max}\}$$

Time Interval of Use (C3). Load a can be started at $\alpha_{n,a}$ and must be ended at $\beta_{n,a}$. This implies only one interval, more are possible (is → vector of borders).

$$\vec{x}_{n,a} = \{\sum\nolimits_{h=\alpha_{n,a}}^{\beta_{n,a}} x_{n,a}^h = l_{n,a}\}$$

Environmental Effects (C4). This constraint is often considered with a differential equation. Because we cannot discuss this in detail (look e.g. [31] for a detailed description), we assume that ϑ is the actual value of the external factor which needs to be changed. $\Delta\vartheta$ is the deviation of the actual and the target value. θ is the change of the value (over time) and τ is a function where all disturbing factors are considered.

$$x_{n,a}(\Delta\vartheta, \vartheta) = \Delta\vartheta * (\theta(\Delta\vartheta, \vartheta) + \tau(\Delta\vartheta, \vartheta))$$

Dependencies between Loads (C5). This constraint should be combined with C1 as the loads need to be connected timely with a delay of τ periods.

$$\vec{x}_{n,a}(\delta_{n,b}) = \left\{\exists\, 0 \leq m_a \leq \omega - \delta_{n,a} - \delta_{n,b} : x_{n,a}^{m_a+k_a} = l_{n,a}^{k_a} \forall\, 0 \leq k_a \leq \delta_{n,a}\right\}$$

$$\vec{x}_{n,b}(\vec{x}_{n,a}, \tau) = \{\exists\, \alpha_{n,b} \leq m_b \leq \omega - \delta_{n,b}\, with\, \alpha_{n,b} = m_a + \delta_{n,a} + \tau : x_{n,a}^{m_b+k_b} = l_{n,a}^{k_b} \forall\, 0 \leq k_b \leq \delta_{n,b}\}$$

6   Conclusion

6.1   Discussion

Implications. Based on our taxonomy future research can benefit, for example, (a) having an orientation when designing DSM-Methods, (b) benchmarking DSM-Methods, (c) increasing user acceptance, (d) identifying information needs or (e) making DSM-Methods more realistic.

Limitations. For deriving our taxonomy, we followed [30]. We selected this methodology because it is – in our opinion – adequate for conducting a literature review in the domain of Information Systems and DSM. However, the selection of key words, literature sources, inclusion and exclusion terms, and dimensions regarding the classification is conducted by own decisions and choices which has limitations. An extended search may enable the identification of more relevant literature. Moreover, we initially asked six experts – more experts can enhance and validate the taxonomy.

6.2 Research Agenda

Based on our work various aspects can be investigated in the future, for example:

Evaluate Possible Combinations of Constraints. Some experts mentioned that constraints might be dependent on each other. For example C3 could be combined or perhaps must be combined with C1 and/or C2. As some constraints might depend or are related to others we suggest evaluating all combinations and their impact, for example, how does the combination of C1 and/or C2 to C3 impacts the DSM-Methods.

Measure the Impact of the Constraints. We already identified the quantity of constraints in the literature and the rating of our experts. Furthermore, we plan to identify the impacts of certain constraints to results. Which constraint or which combination of constraints has the biggest impact? We suggest implementing a simulation with typical DSM-Methods and datasets representing different scenarios.

Identify Dependencies of Appliances. Especially constraint five (C5) needs further input, for example, whether two (or more) loads are connected with each other and how strong. We plan a further research project with an "appliance mining" approach. We plan to analyze existing datasets according the occurrence of various loads and identify some kind of correlation between them.

Add a Practical View of the Field. We plan to examine websites, white papers and other publication. Afterwards, these texts can be analyzed with text mining approaches to identify additional constraints, validate the existing ones or find more examples.

Investigate the Transferability to Other Contexts (Commercial, Industrial). This taxonomy only included the residential context. We plan to interview experts from other contexts as well as conduct an additional literature review. Both can benefit from the taxonomy as we have a fixed starting point.

References

1. United Nations: Millenium development goals and beyond (2015). http://www.un.org/millenniumgoals/bkgd.shtml. Accessed 30 Jan 2017
2. Barbato, A., Capone, A.: Optimization models and methods for demand-side management of residential users: a survey. Energies 2014 **7**(9), 5787–5824 (2014)
3. Feuerriegel, S., Bodenbenner, P., Neumann, D.: Value and granularity of ICT and smart meter data in demand response systems. Energ. Econ. **54**, 1–10 (2016)
4. Balijepalli, M.V.S.K., Pradhan, V., Khaparde, S.A., Shereef, R.M.: Review of demand response under smart grid paradigm. In: Proceedings of Innovative Smart Grid Technologies (2011)

5. Gellings, C.W., Chamberlin, J.H.: Demand-Side Management: Concepts and Methods. Prentice Hall, Englewood Cliffs (1987)
6. National Institute of Standards and Technology (NIST), U.S. Dep. of Com.: NIST Framework and Roadmap for Smart Grid Interoperability Standards. Special Publication (2010)
7. Wilkes, J., Moccia, J.: Wind in Power - 2010 European Statistics (2011)
8. Bassamzadeha, N., Ghanemb, R., Luc, S., Kazemitabard, S.J.: Robust scheduling of smart appliances with uncertain electricity prices in a heterogeneous population. Energ. Build. **84**, 537–547 (2014)
9. Stüdli, S., Crisostomi, E., Middleton, R., Braslavsky, J., Shorten, R.: distributed load management using additive increase multiplicative decrease based techniques. In: Rajakaruna, S., Shahnia, F., Ghosh, A. (eds.) Plug In Electric Vehicles in Smart Grids, pp. 173–202. Springer, Singapore (2015)
10. Bashash, S., Fathy, H.K.: Modeling and control of aggregate air conditioning loads for robust renewable power management. Trans. Control Syst. **21**, 1318–1327 (2013)
11. Costanza, E., Fischer, J.E., Colley, J.A., Rodden, T., Ramchurn, S.D., Jennings, N.R.: Doing the laundry with agents: a field trial of a future smart energy system in the home. In: Proceedings of ACM Conference on Human Factors in Computing Systems (2014)
12. Beaude, O., Lasaulce, S., Hennebel, M.: Charging games in networks of electrical vehicles. In: Proceedings of Network Games, Control and Optimization (NetGCooP) (2012)
13. Kwak, J., Varakantham, P., Maheswaran, R., Tambe, M., Jazizadeh, F., Kavulya, G., Klein, L., Becerik-Gerber, B., Hayes, T., Wood, W.: SAVES: a sustainable multiagent application to conserve building energy considering occupants. In: Proceedings of International Conference on Autonomous Agents and Multiagent Systems (2012)
14. Serkan Özdemir, S., Unland, R.: A winner agent in a smart grid simulation platform. In: International Conference on Intelligent Agent Technology (IAT 2015) (2015)
15. Gerwig, C., Behrens, D., Knackstedt, R., Lessing, H.: Demand side management in residential contexts – a literature review. In: Proceedings of the Informatik, Cottbus (2015)
16. Al-Sumaiti, A.S., Ahmed, M.H., Salama, M.M.A.: Smart home activities: a literature review. Electr. Pow. Compo. Syst. **42**, 294–305 (2014)
17. Gatsis, N., Giannakis, G.B.: Residential demand response with interruptible tasks: duality and algorithms. In: Conference on Decision and Control and European Control Conference (CDC-ECC) (2011)
18. Bahrami, S., Parniani, M., Vafaeimehr, A.: A modified approach for residential load scheduling using smart meters. In: International Conference and Exhibition on Innovative Smart Grid Technologies (ISGT Europe) (2012)
19. Kim, S.-J., Giannakis, G.B.: Scalable and robust demand response with mixed-integer constraints. Trans. Smart Grid **4**(4), 2089–2099 (2013)
20. Yang, P., Chavali, P., Gilboa, E., Nehorai, A.: Parallel load schedule optimization with renewable distributed generators in smart grids. In: International Conference on Smart Grid Communications (SmartGridComm) (2012)
21. Karmakar, G., Kabra, A., Ramamritham, K.: Coordinated scheduling of thermostatically controlled real-time systems under peak power constraint. In: Real-Time and Embedded Technology and Applications Symposium (RTAS) (2013)
22. Mangiatordi, F., Pallotti, E., Vecchio, P.D., Leccese, F.: Power consumption scheduling for residential buildings. In: International Conference on Environment and Electrical Engineering (EEEIC) (2012)
23. Thomas, A.G., Jahangiri, P., Wu, D., Cai, C., Zhao, H., Aliprantis, D.C., Tesfatsion, L.: Intelligent residential air-conditioning system with smart-grid functionality. IEEE Trans. Smart Grid **3**(4), 2240–2251 (2012)

24. Chavali, P., Yang, P., Nehorai, A.: A distributed algorithm of appliance scheduling for home energy management system. Trans. Smart Grid **5**, 282–290 (2014)
25. Li, N., Chen, L., Low, S.H.: Optimal demand response based on utility maximization in power networks. In: Proceedings of Power and Energy Society General Meeting (2011)
26. Weckx, S., Driesen, J., D'hulst, R.: Distributed residential load control of binary behaving loads. In: PowerTech (POWERTECH) (2013)
27. Rassaei, F., Soh, W.-S., Chua, K.-C.: Demand response for residential electric vehicles with random usage patterns in smart grids. Trans. Sustain. Energ. **6**(4), 1367–1376 (2015)
28. Rao, M.R., Kuri, J., Prabhakar, T.V.: Towards optimal load management with day ahead pricing. In: Proceedings of the International Conference on Communication Systems and Networks (COMSNETS) (2015)
29. Ozturk, X., Jha, P., Kumar, P., Lee, G.: A gateway sensor network architecture for home energy management. In: International Conference on Collaboration Technologies and Systems (CTS) (2013)
30. Nickerson, R., Muntermann, J., Varshney, U.: Taxonomy development in information systems: a literature survey and problem statement. In: Proceedings of the 16th Americas Conference on Information Systems (AMCIS) (2010)
31. vomBrocke, J., Simons, A., Niehaves, B., Reimer, K., Plattfaut, R., Cleven, A.: Reconstructing the giant: on the importance of rigour in documenting the literature search process. In: Proceedings of European Conference on Information Systems (ECIS) (2009)
32. Behrens, D., Schoormann, T., Knackstedt, R.: Datensets für demand-side-management – literatur-review-Basierte analyse und Forschungsagenda. In: Mayr, H.C., Pinzger, M. (Hrsg.) INFORMATIK 2016, Klagenfurt, Austria (2016)
33. Goudarzi, H., Hatami, S., Pedram, M.: Demand-side load scheduling incentivized by dynamic energy prices. In: Proceedings of the International Conference on Smart Grid Communications (SmartGridComm) (2011)
34. Barbato, A., Capone, A., Carello, G., Delfanti, M., Merlo, M., Zaminga, A.: House energy demand optimization in single and multi-user scenarios. In: International Conference on Smart Grid Communications (SmartGridComm) (2011)
35. Agnetis, A., de Pascale, G., Detti, P., Vicino, A.: Load scheduling for household energy consumption optimization. Trans. Smart Grid **4**(4), 2364–2373 (2013)
36. Deng, R., Yang, Z., Chen, J., Chow, M.Y.: Load scheduling with price uncertainty and temporally-coupled constraints in smart grids. IEEE Trans. Power Syst. **29**(6), 2823–2834 (2014)
37. Hatami, S., Pedram, M.: Minimizing the electricity bill of cooperative users under a quasi-dynamic pricing model. In: IEEE International Conference on Smart Grid Communications (SmartGridComm) (2010)
38. Zhao, Z., Lee, W.C., Shin, Y., Song, K.B.: An optimal power scheduling method for demand response in home energy management system. IEEE Trans. Smart Grid **4**(3), 1391–1400 (2013)
39. Chang, T.H., Alizadeh, M., Scaglione, A.: Real-time power balancing via decentralized coordinated home energy scheduling. IEEE Trans. Smart Grid **4**(3), 1490–1504 (2013)
40. Mohesenian-Rad, A.-H., Wong, V.W.S., Jatskevich, J., Schober, R., Leon-Garcia, A.: Autonomous demand-side management based on game-theoretic energy consumption scheduling for the future smart grid. IEEE Trans. Smart Grid **1**(3), 320–331 (2010)
41. Mohesenian-Rad, A.-H., Wong, V.W.S., Jatskevich, J., Schober, R: Optimal and autonomous incentive-based energy consumption scheduling algorithm for smart grid. In: Proceedings of Innovative Smart Grid Technologies (ISGT) (2010)

42. Zhao, W., Cooper, P., Perez, P., Ding, L.: Cost-driven residential energy management for adaption of smart grid and local power generation. In: Proceedings of the International Symposium of Next Generation Infrastructure (2013)

43. Keerthisinghe, C., Verbic, G., Chapman, A.C.: Evaluation of a multi-stage stochastic optimization framework for energy management of residential PV-storage systems. In: Australasian Universities Power Engineering Conference (AUPEC) (2014)

44. Castillo-Cagigal, M., Gutiérrez, A., Monasterio-Huelin, F., Caamaño-Martín, E., Masa, D., Jiménez-Leube, J.: A semi-distributed electric demand-side management system with PV generation for self-consumption enhancement. Energy Convers. Manage. **52**(7), 2659–2666 (2011)

45. McNamara, P., McLoone, S.: Hierarchical demand response using Dantzig-Wolfe decomposition. In: Proceedings of Innovative Smart Grid Technologies Europe (ISGT EUROPE) (2013)

46. Mohammad, F., Mehrdad, G.: Localized demand-side management in electric power systems. In: Iranian Conference on Smart Grids (ICSG) (2012)

47. Verschae, R., Kato, T., Matsuyama, T.: A distributed coordination framework for on-line scheduling and power demand balancing of households communities. In: European Control Conference (ECC) (2014)

48. Verschae, R., Kato, T., Matsuyama, T.: A distributed coordination framework for on-line scheduling and power demand balancing of households communities. In: European Control Conference (ECC)(2014)

49. Song, L., Schaar, M.V.D.: Non-stationary demand side management method for smart grids. In: International Conference on Acoustics, Speech and Signal Processing (2014)

50. Ashfaq, A., Yingyun, S., Khan, A.Z.: Optimization of economic dispatch problem integrated with stochastic demand side response. In: IEEE International Conference on Intelligent Energy and Power Systems (IEPS) (2014)

51. Lunden, J., Werner, S., Koivunen, V.: Distributed demand-side management with load uncertainty. In: International Conference on Acoustics, Speech and Signal Processing (ICASSP) (2013)

52. De Angelis, F., Boaro, M., Fuselli, D., Squartini, S., Piazza, F., Wei, Q., Wang, D.: Optimal task and energy scheduling in dynamic residential scenarios. In: 9th International Symposium on Neural Networks (2012)

53. Soliman, H.M., Leon-Garcia, A.: Game-theoretic demand-side management with storage devices for the future smart grid. Trans. Smart Grid **5**, 1475–1485 (2014)

54. Liu, P., Fu, Y.: Construction of multiband uncertainty set for building energy management with uncertain loads and solar power. In: PES General Meeting (2014)

55. Miao, H., Huang, X., Chen, G.: A genetic evolutionary task scheduling method for energy efficiency in smart homes. Int. Rev. Electr. Eng. **7**, 5897–5904 (2012)

56. Kim, S-J., Giannakis, G.B.: Efficient and scalable demand response for the smart power grid. In: International Workshop on Computational Advances in Multi-Sensor Adaptive Processing (CAMSAP) (2011)

57. Ali, S.Q., Maqbool, S.D., Ahamed, T.P.I., Malik, N.H.: Pursuit algorithm for optimized load scheduling. In: Power Engineering and Optimization Conference (PEDCO) (2012)

58. Lin, Y.-B.: Optimal threshold policy for in-home smart grid with renewable generation integration. Trans. Parallel Distrib. Syst. **26**, 1096–1105 (2014)

59. Huang, Y., Mao, S., Nelms, R.M.: Smooth electric power scheduling in power distribution networks. In: Globecom Workshops (GC Wkshps) (2012)

60. Lee, J., Park, G.-L., Kim, H.-J.: Multithreaded power consumption scheduler based on a genetic algorithm. In: Kim, T., et al. (eds.) FGCN 2011. CCIS, vol. 265, pp. 47–52. Springer, Heidelberg (2011). doi:10.1007/978-3-642-27192-2_6

61. Hu, T., Chu, X., Zhang, W., Liu, Y.: An optimal day-ahead dispatch strategy for deferrable loads. In: International Conference on Power System Technology (POWERCON) (2014)

62. Zang, C., Qin, X., Li, X., Jin, X., Che, W.: Simulated annealing based fuzzy Markov game energy management for smart grids. In: Control and Decision Conference (CCDC) (2013)

63. Lee, J., Park, G.-L.: Power load distribution for wireless sensor and actuator networks in smart grid buildings. Int. J. Distrib. Sens. Netw. **9**(1) (2013)

64. Bu, H., Nygard, K.E.: Adaptive scheduling of smart home appliances using fuzzy goal programming. In: The Sixth International Conference on Adaptive and Self-Adaptive Systems and Applications (ADAPTIVE) (2014)

65. Maqbool, S.D., Ahamed, T.P.I., Ali, S.Q., Pazheri, F.R., Malik, N.H.: Comparison of pursuit and ε-Greedy algorithm for load scheduling under real time pricing. In: International Conference on Power and Energy (PECon) (2012)

66. Alam, M.R., St-Hilaire, M., Kunz, T.: Cost optimization via rescheduling in smart grids – a linear programming approach. In: International Conference on Smart Energy Grid Engineering (SEGE) (2013)

67. Ali, S.Q., Maqbool, S.D., Ahamed, T.P.I., Malik, N.H.: Load scheduling with maximum demand and time of use pricing for microgrids. In: Global Humanitarian Technology Conference: South Asia Satellite (GHTC-SAS) (2013)

68. Alvarez, C., Malhame, R.P., Gabaldon, A.: A class of models for load management application and evaluation revisited. Trans. Power Syst. **7**, 1435–1443 (1992)

69. Behrens, D., Gerwig, C.: Selbstregulierende Verbraucher im Smart Grid: Design einer Infrastruktur mit Hilfe eines Multi-Agenten-Systems. In: Proceedings of Multikonferenz der Wirtschaftsinformatik (MKWI), Paderborn, Germany (2014)

Decomposition of Tasks in Business Process Outsourcing

Kurt Sandkuhl[1,3(✉)], Alexander Smirnov[2,3], and Nikolay Shilov[2,3]

[1] The University of Rostock, Rostock, Germany
kurt.sandkuhl@uni-rostock.de
[2] SPIIRAS, St. Petersburg, Russia
{smir,nick}@iias.spb.su
[3] ITMO University, St. Petersburg, Russia

Abstract. In industrial areas with a highly competitive environment many enterprises consider outsourcing of IT-services as an option to reduce IT-related costs. In this context, cloud computing architectures and outsourcing of business processes into the cloud are potential candidates to improve resource utilization and to reduce operative IT-costs. In this paper, we focus on a specific aspect of cloud computing and outsourcing: the use of concepts from crowd-sourcing or crowd computing in business process outsourcing (BPO). The approach used in this paper is to bring together techniques from enterprise modeling and from crowd-computing for the purpose of business process decomposition. The contributions of the paper are an analysis of requirements to process decomposition from a business process outsourcing perspective, three different strategies for performing the decomposition and an initial validation of these strategies using an industrial case.

Keywords: Business Process Outsourcing · Crowdsourcing · Enterprise modeling · Task pattern · Process decomposition

1 Introduction

In industrial areas with a highly competitive environment all enterprise departments and functions are expected to contribute to efficient operations and an economic cost structure. In particular the IT-budgets of organizations have been under pressure during the last decade with a clearly expressed expectation towards IT-departments to provide solutions and services of high quality tailored to business demands. Furthermore, many enterprises consider outsourcing of those IT-services as an option to reduce IT-related costs, which can be classified as commodities [1]. In this context, cloud computing architectures and outsourcing of business processes into the cloud are potential candidates to improve resource utilization and to reduce operative IT-costs [2]. Outsourcing of resources, products and competences could also be necessary due to continuous fluctuations and changes in business processes caused by the increased dynamicity of the modern global markets. This often happens for two reasons: the enterprise does not have enough capacities to fulfill the current demand of a certain

© Springer International Publishing AG 2017
W. Abramowicz (Ed.): BIS 2017, LNBIP 288, pp. 296–310, 2017.
DOI: 10.1007/978-3-319-59336-4_21

resource/product/competence to solve all pertinent tasks in time, or the enterprise doesn't have required resource/products/competences corresponding to the current task.

But outsourcing approaches often are criticized for being not sufficiently flexible when automatable and manual tasks have to be combined [3]. In this paper, we focus on a specific aspect of cloud computing and outsourcing: the use of concepts from crowd-sourcing or crowd computing in business process outsourcing (BPO). Crowd computing is informally defined in [4] as "an umbrella term to define a myriad of tools that allow human interaction to exchange ideas, non-hierarchical decision making and full use of mental space of the globe".

More concrete, the aspect addressed in this paper is decomposition of a process into what crowd-sourcing defines as "micro-tasks" (see Sect. 2.2), i.e. smaller portions of the process which could be outsourced to a crowd-member. At first glance, a straightforward answer to this question seems to be that a business process anyhow consists of different activities, which can be considered as micro-tasks. Since business processes often have been modeled with a process modeling language (BPMN [6], EPC [7], flow charts [5] or similar), these process models could also serve as source for the micro-tasks. However, a closer analysis of real-world requirements reveals that constraints with respect to competences of the crowd-member, resources required for the task and the subject of the work have to be taken into account. Techniques for capturing and expressing such constraints are known from enterprise modeling. Thus, the approach used in this paper is to bring together techniques from enterprise modeling and from crowd-computing for the purpose of business process decomposition.

The contributions of the paper are an analysis of requirements to process decomposition from a business process outsourcing perspective, three different strategies for performing the decomposition and an initial validation of these strategies using an industrial case. The remainder of the paper is structured as follows: Sect. 2 briefly introduces the required background from enterprise modeling and crowdsourcing. Section 3 presents the industrial case from utility industries motivating the research and providing examples for business process outsourcing which can be studied to elicit requirements. Section 4 presents our approach for decomposing business processes into micro-tasks which includes three different strategies for different industrial demands. Section 5 evaluates the three strategies. Section 6 summarizes the work and discusses future research.

2 Background

This section briefly summarizes the background for our work from enterprise knowledge modeling and crowdsourcing. In enterprise modeling, our work is based on the task pattern approach (Sect. 2.1) and in crowdsourcing on task decomposition into micro-tasks (Sect. 2.2).

2.1   Enterprise Knowledge Modeling with Task Patterns

In general terms, enterprise modeling is addressing the systematic analysis and modeling of processes, organization structures, products structures, IT-systems or any other perspective relevant for the modeling purpose [8]. Sandkuhl et al. [9] provide a detailed account of enterprise modeling approaches. Enterprise models can be applied for various purposes, such as visualization of current processes and structures in an enterprise, process improvement and optimization, introduction of new IT solutions or analysis purposes. Enterprise knowledge modeling combines and extends approaches and techniques from enterprise modeling. The knowledge needed for performing a certain task in an enterprise or for acting in a certain role has to include the context of the individual, which requires including all relevant perspectives in the same model [10]. A best practice for identifying these perspectives is the so-called "POPS*"-approach proposed by [11]. POPS* is an abbreviation for the perspective of an enterprise to be included in an enterprise model: process (P), organization structure (O), product (P), systems & resources (S) and other aspects required for the modelling purpose (*). The best practice basically recommends to always include the four POPS perspectives in a model because they are mutually reflective: process are performed by the roles captured in the organisation structure, the roles are using systems and resources which at the same time capture information about products; manufacturing and design of products is done in processes by roles using systems, etc. [10].

Patterns are a proven way to capture experts' knowledge in fields where there are no simple "one size fits all" answers [12], such as enterprise modelling. Each pattern poses a specific design problem, discusses the considerations surrounding the problem, and presents an elegant solution that balances the various forces or drivers. The POPS* best practice was applied in the EU-FP6 project MAPPER to capture reusable portions of enterprise knowledge in so called task patterns [13], which proved feasible and economically rewarding [14]. Task patterns always include all four POPS perspectives and are represented in a visual modelling language.

2.2   Crowd Sourcing

Crowdsourcing is an emerging research area and it is usually understood (e.g., [15, 16]) as a form of outsourcing, in which tasks traditionally performed by organizational employees or other companies are sent through the internet to the members of an undefined large of group people (called "crowd"). The research area of massively parallel solution of problems with the help of "crowds" is still actively developing and there are several highly connected research areas: crowdsourcing, crowd computing, human computations, social computing, peer production. Boundaries between them are often blurred. Even more recent is the concept of hybrid crowd, where human solvers are accompanied by hardware and software services. This concept is more general, as it in some sense extends both classical service-oriented approach and crowd computing, but pertains all special effects inherent to crowd computing as a result of the inclusion of an outside human to an organization process in a transient, per-task basis.

In the context of this paper, the decomposition of organizational tasks into smaller portions, also called micro-tasks, is in focus. Micro-tasks are informally described as clearly defined sub-tasks of an organizational activity which are useful for the organization and clearly defined, can be performed independently by a crowd member, are economically affordable and involve no risk. Previous work on decomposition focused on a reference model [17], the use of the crowd for deciding on the decomposition [18] or construction of complex workflows out of micro-tasks [19].

3 Case Study

Research in this paper is motivated by an industrial case from business process outsourcing (BPO). The business service provider (BSP) studied in the case study is a medium-sized enterprise from Germany which offers more than 20 different BPO services to their clients. The target group for these services is medium-sized utility providers and other market roles of the energy sector in Germany, Bulgaria, Macedonia and several other European countries. Many energy distribution companies are outsourcing some business functions and business processes connected to these functions. Examples for typical business functions are meter readings, meter data evaluation, automatic billing, processing and examination of invoices, customer relationship management and order management. The BSP offers the performance of a complete business process for a business function or only of selected tasks of a business process. The IT-basis for these services in our case study is a software product which was developed and is maintained by the BSP. Integrated with a workflow engine and business activity monitoring, this software product provides the business logic for the energy sector, which is implemented using a database-centric approach. In addition to this software product, other cloud-based services for information exchange, document management and security are integrated. Different deployment models are used including a provider-centric model (the software product and the business processes are run at the BSP's computing center), a client-centric model (the software product is installed at the client site and the manual work of the business process is performed at the BSP) and mixed models (e.g. the software product is offered in the cloud, work and process are performed partly at the client and partly at the BSP).

When providing the outsourcing services for their clients, the BSP needs to offer the technical facilities for providing the service (see above) and the workforce for performing activities, which cannot be done in an automated way. Such activities often concern exceptions in the automated part of the processing, which will be illustrated in an example below. This example is depicted in Fig. 1 and includes the POPS perspectives (see Sect. 2.1), which all are required to completely describe what resources, roles, competences and products are needed to perform a given activity. The example was developed using the modelling tool Troux Architect[1] and the GEM[2] modelling language. The relationships in this model are typed, which usually is indicated by text

---

[1] See www.troux.com.

[2] GEM = Generic Enterprise Modeling language.

showing the relationship type. For readability reasons, we had to switch off the visibility of the relationship texts.

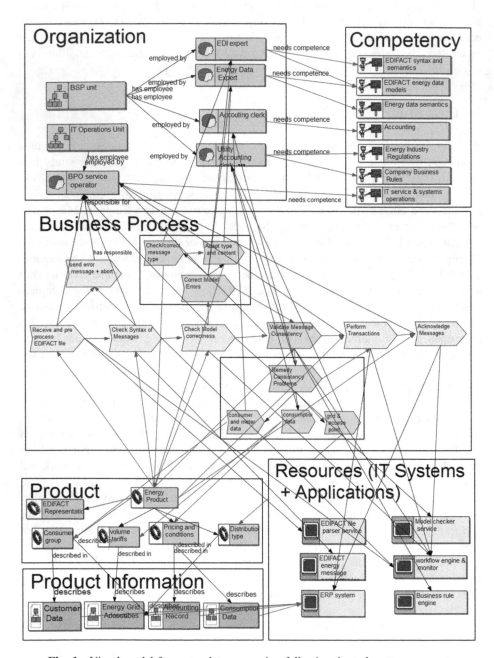

Fig. 1. Visual model for meter data processing following the task pattern concept

The selected example is a process concerned with the communication between market roles in energy sector about meter readings and energy consumption of households. The core process is depicted in the middle of Fig. 1 and includes the activities "receive and pre-process EDIFACT file", "check syntax of messages", "check model correctness", "validate message consistency", "perform transactions" and "acknowledge messages". This process basically shows all steps from receiving an EDIFACT[3] file with often thousands of messages, identifying the different messages by pre-processing the file, checking all messages for syntactical and semantic correctness, validating the soundness of the message content, recording the transaction and acknowledging it. This kind of EDIFACT message exchange is common practice between energy producers, distributors and grid operators, and is subject to regulation.

As long as no exceptions occur, the process can be performed along this "happy path" and is fully automated, i.e. no human actors are involved. But as soon as exceptions occur, the "knowledge workers" need to be involved, which is described by the alternative paths. One possible exception type is syntax errors in the EDIFACT file. In this case the process is aborted and an error message is sent to the sender. Another exception type is caused by a model error. In this case "correct model errors" is performed, which contains several activities: model errors could by caused by wrong/unknown message types or missing data depending on the message type.

Such exceptions usually can only be remedied by human actors. This is done by an "energy data expert" and requires expert knowledge regarding the information model used in EDIFACT messages of the energy sector. In Fig. 1, "correct model error" and its sub-activities are framed by a rectangle indicating that this is a task with refinements.

During validation of message consistency, several exceptions can happen: the address data of the consumer can be incorrect, e.g. if the consumer has several addresses and meters which were mixed up; the consumption data can be implausible; the grid data can be faulty. Some inconsistencies (e.g. of consumer data) can be corrected by an accountant without specific domain knowledge from energy sector. Others require competencies regarding the energy sector and business rules of the company and are performed by a utility accounting clerk. If the errors and inconsistencies can be remedied successfully, the messages can be processed by the ERP system and the transactions can be recorded. Products in the utility sector basically are different tariffs which depend on the customer group, the volume of consumption, the way of distribution (own energy grid or third party grid) and the pricing/payment conditions. The type of product and related product information are relevant for some steps in checking model correctness and crucial for validation of consistency and performing the actual transaction.

Resources required for performing the business process primarily are different information systems and services. A workflow engine controls and monitors the overall process flow. An ERP system for the energy sector manages all product information and performs the transactions. EDIFACT file parser, message parser and model checker

---

[3] EDIFACT = Electronic Data Interchange for Administration, Commerce and Transport (cf. http:// www.unece.org/cefact/edifact/welcome.html).

are implemented as web-services. A rule engine is used to model and execute company business rules. In time periods with high workload or large numbers of exceptions, the BSP is interested in including external workforce in the performance of the BPO service. This could be an application case for using crowdsourcing, although the BSP so far does not have experience in this field. What activity could be outsourced to which outsourcing partner depends on different criteria: for most activities, the competence requirements are decisive, i.e. only a sub-supplier (or crowd-member) with the right competence could possibly be considered as outsourcing partner. Furthermore, the availability of resources and the access to product details or product information can be important. In case of confidential product details, this aspect might be the decisive one when considering outsourcing and in case of use of internal resources provided in a private cloud only, the resource aspect is dominant for decision making.

4 Strategies for Decomposition of Business Process Models

This section proposes and specifies different strategies for the decomposition of business process models into smaller portions, which in crowdsourcing are called micro-tasks (see Sect. 2.2). The purpose of this decomposition is to identify tasks potentially suitable for crowdsourcing including the competences, resources and product information needed for these tasks. All strategies follow the same general approach:

- starting point for each strategy is a business process model which includes the POPS perspectives according to the task pattern concept, e.g. a model like the example illustrated in Fig. 1,
- the aim is to isolate micro-tasks from the process model which also must follow the task pattern concept (see Sect. 2.1), i.e. which have to include the POPS perspectives in order to fully specify activity, competences, resources and product information needed.
- the strategies are based on structural patterns in the process model. In this paper we do not take into account runtime information and we do not analyze the textual parts of the model elements, but only their types.
- the strategies reflect the different priorities from the industrial case, i.e. to consider competence, resource or product information as highest priority
- the structural patterns of each strategy at the same time form the structure of the micro-tasks, i.e. if the model has been fully decomposed into patterns, the required micro-tasks are defined.
- we assume that the resources are not consumed by the activities. This assumption is possible as we focus on business process outsourcing using digital resources (IT systems, services) and not on conventional machinery or consumable resources in manufacturing industry.

The competence-first strategy follows the principle that a portion of a process only can be outsourced, if a supplier or crowd-member can be identified who has all competences required by the role(s) assigned to the process-portion under consideration. Thus, the strategy starts from roles and competences in the process model and attempts to segment the process into portions with as few as possible process steps. These process steps together with the resources, product details, roles and competences assigned to them from the micro-tasks. The resource-first strategy assumes that the availability of all resources required for a process-portion is decisive when outsourcing it. The strategy starts from the resources in the process model and attempts to segment the process into portions with as few as possible process steps.

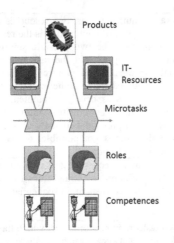

Fig. 2. Symbols used when describing the strategies illustrated in a generic model

The product-first assumes that the access to product information is decisive, starts from here and continues like the resource-first strategy. When discussing the strategies in Sects. 4.1 to 4.3, we use a symbol set which is illustrated by the generic model shown in Fig. 2. In this particular example there is a process with two tasks (which both are potential micro-tasks), each of them is associated with one IT resource and one role. Both of the tasks are associated with one common product. Each role is associated with one competence. For simplicity, only tasks and corresponding resources (product or IT resource or role together with competence) are shown in the figures in Sects. 4.1 to 4.3 illustrating the patterns.

4.1   Resource-First Strategy

The motivation for this strategy is that missing or insufficient capacity of IT resources for business process implementation could require outsourcing of those tasks requiring these resources. Thus, the strategy assumes that the availability of resources is the primary precondition for crowdsourcing, i.e. decomposing a process consisting of tasks into micro-tasks requires that for each micro-task all required resources are available at the outsourcing partner. By examining all possible task - resource multiplicities, the patterns depicted in Table 1 were identified. The patterns below illustrate the possible situations and suggest corresponding outsourcing strategies.

4.2   Product-First Strategy

The product-first strategy is motivated by the fact that different tasks in a business process require different parts of the product or the product information. "Product" in this context can also be a service and "product part" a portion of information required for this service. Outsourcing "product-first" could be motivated by the fact that only certain parts of the product details shall be disclosed to someone outside the enterprise. However, if a task requires certain product details, all these product details have to be

Table 1. Resource-first strategy patterns

#	Diagram	Description
1.		This is the first and the most simple and straightforward pattern when a task of the business process requires one resource. The solution for this pattern is to consider this task as a microtask for outsourcing
2.		In this pattern, a task requires at least two resources and there is a lack for at least one of them which is the reason for outsourcing. There could be two possible solutions here. The first one is to try to split the task into two microtasks so that the one, which requires a missing resource, could be outsourced and the other could remain within the company. This is especially recommended if this microtask deals with some sensitive information, which the company would prefer not to share with third parties. If splitting the task is not possible, the whole tasks is a microtask to be outsourced
3.		In this pattern, two or more sequential tasks require the same resource. The solution is straightforward: the microtask to be outsourced consists of all sequential tasks requiring the resource
4.		In this pattern, two tasks require a resource, which has to be outsourced. However, between these tasks, there is one or more other tasks, which can be performed internally. There could be two possible solutions here: outsource all tasks as one microtask or only those tasks, which require the missing resource. The solution might depend on the sensitivity of the information used in the intermediate tasks: the more sensible the information, the more reasons to keep these tasks inside the company

made available to the corresponding micro-task. Similar to the resource-first strategy, we examined all possible task - product multiplicities. The patterns identified were those already shown in Table 1 for the resource-first strategy, if we substitute the "resource" symbols in these patterns with a "product" symbol. Thus, there are also four patterns (#5 to 8) for the product-first strategy.

4.3 Competence-First Strategy

The competence-first strategy is different from the previous ones because usually in enterprise modelling there is no direct link between competences and tasks, but tasks are related through user roles. The reason for using this strategy is obviously a lack of required competences. Due to the increased complexity caused by the intermediate layer of user roles it has eight patterns (Table 2).

Table 2. Competence-first strategy patterns

#	Diagram	Description
9.		Simple and straightforward pattern when a microtask requires one role and one competence, which has to be outsourced. The solution is the same as for the previous strategies: outsource the appropriate microtask
10.		In this pattern, a microtask requires at least two roles, with only one of the roles requiring a missing competence, which has to be outsourced. There could be two possible solutions here. The first one is to try to split the microtask into two so that the one, which requires a role with missing competence, could be outsourced and the other could remain within the company. If splitting the microtask is not possible, than the whole microtask has to be outsourced
11.		In this pattern, a microtask requires a role that requires several competences, with one of the competences requiring outsourcing. Since the role cannot be split, the only solution here is to outsource the whole microtask
12.		This pattern is also very similar to the previously described strategies: two or more microtasks require a role with a competence, which has to be outsourced. The solution is to outsource the appropriate microtasks.
13.		In this pattern, two sequential microtasks require different roles; however, both of the roles need a missing competence. The solution is to outsource both of the microtasks
14.		This pattern is similar to the previous one; however, each of the two microtasks is also associated with roles, which do not require outsourcing. This condition produces an idea to split the microtasks so that some of the microtasks could be outsourced and others could stay within the company. If splitting of at least one of the microtasks is not possible, than the need to split the other one should also be re-considered
15.		In this pattern, a microtask (in the middle) requires two roles, with one of the roles (e.g. the one on the right) requiring a missing competence, which has to be outsourced. There could be two possible solutions here. The first one is to try to split the microtask into two so that the one which requires the role with missing competence could be outsourced together with the following microtask, and the other could remain within the company with the previous microtask. If splitting the microtask is not possible, than the two microtasks (the one in the middle and the following one) have to be outsourced
16.		In this pattern, two microtasks require roles with a competence that has to be outsourced. However, between these microtasks, there are one or more other microtasks, which can be performed internally. There could be two possible solutions here: outsource all microtasks or only those, which require the missing competence. The reasons to choose one of the solutions could be different and the decision has to be based on the particular circumstances

5   Case Study-Based Evaluation of the Strategies

This section discusses the evaluation of the strategies proposed in Sect. 4. This evaluation focuses on three different questions, which basically address feasibility, usefulness and differences of the strategies:

1. Can the strategies be applied for decomposing real-world business process models into micro-tasks? This includes whether it is possible to identify micro-tasks and whether the business process can be fully decomposed.
2. What differences exist between the different strategies? This question includes (a) differences between individual micro-tasks and (b) usefulness of the sets of micro-tasks for practical use.
3. Are the micro-tasks identified by the strategies applicable for outsourcing tasks in practice? This includes whether the complete business process can be (re-)composed from the micro-tasks and whether isolated performance of micro-tasks would be possible.

In order to answer questions 1 and 2(a), the process model presented in Sect. 3 was decomposed into micro-tasks with all three strategies and the resulting sets of micro-tasks were compared. For questions 2(b) and 3, an expert evaluation of the micro-tasks was performed. Table 3 shows the result of applying the different strategies for decomposing the business process introduced in Sect. 3. The column "Str." indicates, which strategy or strategies produced the micro-task (C = competence-first; P = product-first; R = resource-first). In the column "organization" role and competence are separated with a "/"; in all columns, whenever two or more processes, roles, product details or resources are required, they are separated with a ";".

Table 3 also shows that the three strategies produce different sets of micro-tasks which have some overlap. Resource-first produces the smallest number of micro-tasks with only 7 elements, but with the largest tasks of all strategies (#15). The reason for this is that the resource "workflow engine" is needed for all activities in #15 while each of the individual tasks in #15 also requires at least one additional resource. Production-first and competence-first show only a difference regarding the micro-tasks identified for the tasks included in "remedy consistency problems" (see Fig. 1).

For the expert evaluation, we needed an experienced professional from the BPO domain. One of the BPO product managers of the BSP introduced in Sect. 3 was selected. This expert had more than three years of experience in this position. The expert evaluation included three steps. The first step was to go through all micro-tasks identified by the three strategies, i.e. the set of all sets. The expert had to answer the question for each micro-task whether the micro-task was completely defined, i.e. all resources, competences and product details are included. This check related to the task pattern feature of being "self-contained". As a result, for one micro-task a missing resource was discovered. It showed that this was due to a mistake in the model where the resource was not connected to the activity with the required relationship type. All other micro-tasks were found to be self-contained. The second step was to judge the different sets of micro-tasks produced by the three strategies whether they fully define the business process and which of the sets was considered by the expert as the best one

Table 3. Micro-tasks produced by the different strategies

#	Str.	Process	Organization (role/competence)	Product detail	System (Resource)
1	C P	Receive and pre-process EDIFACT file	BPO service operator/IT service & system operations	EDIFACT representation	EDIFACT file parser service; workflow engine (WE)
2	C R P	Send error message + abort			EDIFACT file parser service; EDIFACT energy message parser
3	C P	Check syntax of message			EDIFACT energy message parser; WE
4	C P	Check model correctness			Model checker service; WE
5	C P	Validate message consistency		Consumer group; distribution types; pricing and conditions	Business Rule engine; WE
6	C P	Perform transactions		Consumer group; volume tariffs; distribution types; pricing and conditions	ERP system; WE
7	C P	Acknowledge message		EDIFACT representation	
8	C R P	Check/correct message type	EDIFACT expert/EDIFACT syntax and semantics	EDIFACT representation	Model checker service; WE
9	C R P	Align data & content	Energy data expert/EDIFACT energy data representation		
10	C R	Validate consumer and meter data	Accounting clerk/accounting	Consumer group; distribution types	ERP system; WE
11	C	Validate consumption data; validate grid and access data	Utility accounting clerk/energy industry regulations; company business rules	Consumer group; distribution types; pricing and conditions	
12	R	Validate consumption data			
13	R	validate grid and access data			
14	P	Validate consumer and meter data; validate consumption data; validate grid and access data	Accounting clerk/accounting; Utility accounting clerk/energy industry regulations; company business rules		
15	R	Receive and pre-process EDIFACT file; Check syntax of message; Check model correctness; Validate message consistency; Perform transactions; Acknowledge message	BPO service operator/IT service & system operations	EDIFACT representation; consumer group; volume tariffs; distribution types; pricing and conditions	EDIFACT file parser service; EDIFACT energy message parser; Model checker service; Business Rule engine; ERP system; (WE)

and why. According to the expect judgment, the set produced by the resource-first strategy is the best one because it keeps the "happy flow" of the business process (i.e. micro-tasks #1 and #3 to #7) together. Further discussion with the expert shows that high throughput is important for this fully automatic flow which is another argument for keeping the happy flow integrated. The only problematic result of the resource-first strategy is according to the expert the separation of "validate consumption data" and "validate grid and access data" (i.1. #12 and #13) since these two steps are tightly interwoven. The comparison of competence-first and product-first shows advantages for competence-first, mainly because competence-first separates the "validation of consumer and meter data" (#10) from the other validation tasks (#11) while product-first keeps them together (#14). "Validation of consumer and meter data" requires less knowledge and could be performed by a less-experienced knowledge worker.

The third step was to ask the expert to select the micro-tasks out of the set of sets which could be outsourced to the crowd, i.e. where the required resources could be provided in a public cloud or via secure access paths to the private cloud, where the product detail of information was not affected by a high level of confidentiality and where it can be expected that a substantial number of people have the required competences. The expert identified only one task suitable for crowdsourcing from his perspective: "validation of consumer and meter data" (#10).

6 Summary and Future Work

Motivated by an industrial case, the paper investigated strategies for decomposing business processes into micro-tasks in the context of crowdsourcing. The three strategies put an emphasis on different priorities of enterprises by starting from resources, competences or product details when deciding on decomposition. The three strategies proved to be feasible and – at least in the industrial case investigated – also produced sound proposals for micro-tasks.

The use of the task pattern approach with the POPS perspectives proved to be both, feasible and useful. In the industrial case it helped to clearly identify the prerequisites for performing micro-tasks in terms of product information, required competences and resources. However, task patterns originally were meant to capture reusable organizational knowledge, while micro-tasks cannot be considered as portions of organizational knowledge since the required context of use, i.e. the overall business process and organization structures, is not fully defined which does not make it an "asset" for the organization.

Although the decomposition of the business process example from Sect. 3 easily could be performed manually, the algorithmic nature of the strategies calls for the implementation of a software tool performing the decomposition. This will be part of the future work and can form the basis for a quantitative evaluation of the strategies, e.g. by using the task pattern collection developed in the MAPPER project.

Future work also needs to investigate potential refinements of the strategies

- runtime information should be taken into account, if available. Many business process models include information about expected execution times of tasks which could be a first hint for refining the strategies,
- the textual parts of the model elements could be analyzed and not only their types. The semantics of the text could indicate how tightly different tasks are related to each other which could recommend to keep them together,
- potentially required changes in strategy have to be investigated for the case that resources are consumed by the activities, like conventional machinery or consumable resources in manufacturing industry.

Our work so far has a number of limitations: we used just one case and one expert for evaluation and motivation purposes. Other cases could show different requirements which may affect the utility of the strategies. The evaluation of the strategies is far complete. We did not include perspectives, like implementability, performance or acceptance, which again might lead to refined strategies or different approaches.

Acknowledgements. This work was partially financially supported by the Project 213 within the research program I.5P of the Russian Academy of Sciences, by Government of Russian Federation, Grant 074-U01. Furthermore, it was partly financed by the German Ministry of Research and Education, research project KOSMOS-2.

References

1. Carr, N.G.: IT doesn't matter. IEEE Eng. Manage. Rev. **32**(1), 24–32 (2004)
2. Krogstie, J.: Model-Based Development and Evolution of Information Systems - A Quality Approach. Springer, London (2012)
3. Thuan, N.H., Antunes, P., Johnstone, D.: Factors influencing the decision to crowdsource. In: Antunes, P., Gerosa, M.A., Sylvester, A., Vassileva, J., Vreede, G.-J. (eds.) CRIWG 2013. LNCS, vol. 8224, pp. 110–125. Springer, Heidelberg (2013). doi:10.1007/978-3-642-41347-6_9
4. Schneider, D., de Souza, J., Moraes, K.: Multidões: a nova onda do CSCW? In: 8th Brazilian Symposium on Collaborative Systems. SBSC (2011)
5. Knuth, D.E.: Computer-drawn flowcharts. Commun. ACM **6**(9), 555–563 (1963)
6. White, S.A.: Introduction to BPMN, vol. 2. IBM Cooperation, Armonk (2004)
7. Scheer, A.-W., Nüttgens, M.: ARIS architecture and reference models for business process management. In: Aalst, W., Desel, J., Oberweis, A. (eds.) Business Process Management. LNCS, vol. 1806, pp. 376–389. Springer, Heidelberg (2000). doi:10.1007/3-540-45594-9_24
8. Vernadat, F.B.: Enterprise Modelling and Integration. Chapman & Hall, London (1996)
9. Sandkuhl, K., Stirna, J., Persson, A., Wißotzki, M.: Enterprise Modeling: Tackling Business Challenges with the 4EM Method (The Enterprise Engineering Series). Springer, Heidelberg (2014). ISBN 978-3662437247
10. Lillehagen, F., Krogstie, J.: Active Knowledge Modelling of Enterprises. Springer, Heidelberg (2009). ISBN 978-3-540-79415-8
11. Lillehagen, F.: The foundations of AKM technology. In: 10th International Conference on Concurrent Engineering (CE) Conference, Madeira, Portugal (2003)

12. Enterprise Integration Patterns. http://www.eaipatterns.com/. Accessed 26 Mar 2012
13. Sandkuhl, K., Smirnov, A., Shilov, N.: Configuration of automotive collaborative engineering and flexible supply networks. In: Cunningham, P., Cunningham, M. (eds.) Expanding the Knowledge Economy – Issues, Applications, Case Studies. IOS Press, Amsterdam (2007). ISBN 978-1-58603-801-4
14. Sandkuhl, K.: Capturing product development knowledge with task patterns: evaluation of economic effects. Q. J. Control Cybern. **39**(1), 259–273 (2010). Systems Research Institute, Polish Academy of Sciences
15. Schenk, E., Guittard, C.: Towards a characterization of crowdsourcing practices. J. Innov. Econ. **7**(1), 93–107 (2011)
16. Howe, J.: The rise of crowdsourcing. Wired Mag. **14**, 1–4 (2006). Dorsey Press
17. Smirnov, A., Ponomarev, A., Shilov, N.: Hybrid crowd-based decision support in business processes. In: CENTERIS 2014, 15–17 October 2014, Lisbon, Portugal, vol. 16, pp. 376–384 (2014)
18. Kulkarni, A., Can, M., Hartmann, B.: Collaboratively crowdsourcing workflows with turkomatic. In: Proceedings of the ACM 2012 Conference on Computer Supported Cooperative Work, pp. 1003–1012. ACM (2012)
19. Little, G., Chilton, L.B., Goldman, M., Miller, R.C.: Turkit: human computation algorithms on mechanical turk. In: Proceedings of the 23nd Annual ACM Symposium on User Interface Software and Technology, pp. 57–66. ACM (2010)

Applications

Analysis Method for Conceptual Context Modeling Applied in Production Environments

Eva Hoos[1,3]($\boxtimes$), Matthias Wieland[2], and Bernhard Mitschang[2,3]

[1] Daimler AG, 71034 Böblingen, Germany
Eva.Hoos@daimler.com
[2] Institute of Parallel and Distributed Systems, University of Stuttgart, 70569
Stuttgart, Germany
matthias.wieland@ipvs.uni-stuttgart.de
[3] Graduate School of Excellence Advanced Manufacturing Engineering,
University of Stuttgart, 70569 Stuttgart, Germany

Abstract. Context-awareness is a well-accepted approach to adapt applications to the needs of a user. Yet, it is hardly used in enterprise information systems, especially in production environments. Production environments are complex due to multi-faced actors, products, manufacturing equipment and processes. Hence, the modeling of context is sophisticated, particularly to determine relevant context. The goal of this paper is to facilitate and support context modeling in production environments. Our contribution is an analysis method for conceptual context modeling suited for Industry 4.0 and an extensible engineering context model as starting point for modeling of different Industry 4.0 use cases. The analysis model consists of a graphical notation for a simplified and abstract context model and a template-based concept to detail the model. Furthermore, we evaluate the approach by modeling a real use case from the car manufacturing industry.

Keywords: Context-awareness · Production environments · Industry 4.0

1 Introduction

Context-aware applications acquire context information and leverage them in order to address the needs of users. There is a wide range of potential for context-aware applications in production environments, such as enhancement of industrial monitoring systems [1] or providing role-based, valuable information to the shop floor worker [2]. Yet, context-awareness is hardly used in existing enterprise information systems, especially in production environments [3,4]. Major reasons are that the modeling of context is too expensive and the benefits are not obvious [4]. With the raise of the new paradigms Industry 4.0 and Internet of Things, more and more context data will be captured [5]. The production environment will be equipped with intelligent things (e.g., cyberphysical systems), which gather, process and provide context information and are highly connected [2]. This increases the complexity of context modeling, since more

© Springer International Publishing AG 2017
W. Abramowicz (Ed.): BIS 2017, LNBIP 288, pp. 313–325, 2017.
DOI: 10.1007/978-3-319-59336-4_22

and more complex entities have to be modeled. Though, only relevant context regarding the domain and use cases have to be considered. Existing context modeling approaches often focus on the technical modeling [6–9]. These models are fine-grained and complex. However, to determine *relevant* context for a use case involving domain experts is necessary. They are no IT-experts and hardly understand these complex models. For that purpose, simplified and abstract context models are necessary. The main goal of this paper is to reduce the complexity in context modeling. The different phases of context modeling process are shown in Fig. 1. Therefore, following contributions are presented in this paper:

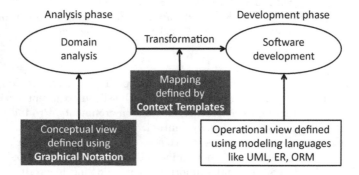

Fig. 1. Phases of context modeling

(I) Analysis method to support domain analysis: Our method serves as starting point to implement a new context-aware application. This is the analysis phase as shown in Fig. 1. To analyze the domain, we introduce a *graphical notation* for conceptual context models. The result of the analysis can be detailed using *context templates* that are transformed thereafter to standard software modeling languages for the development phase, such as UML, ER-diagrams, or to existing context methods on a lower implementation level, such as ORM [6]. Our method aims to bridge the gap between engineers and IT as well as to communicate relevant context and its structure.

(II) Basic, extensible engineering context model: We conceive an *engineering context model* exploiting our analysis method. The engineering context model serves as basis for defining further context-aware applications implementing different use cases in Industry 4.0.

(III) Case-oriented evaluation: Our evaluation is based on a real use case from a German car manufacturer. We applied our method and the engineering context model to define relevant context for a *context-driven documentation application* in the prototype factory.

This paper is structured as follows: Sect. 2 describes the analysis method for conceptual context modeling in production environments. Section 3 introduces the engineering context model which defines an extensible, basic conceptual context model for Industry 4.0 scenarios. In Sect. 4, both the method and the engineering context model are evaluated using a real use cases scenario. Section 5 gives an overview about related work and Sect. 6 summarizes the paper.

2   Analysis Method for Conceptual Context Modeling

In this section, we first give a brief introduction in the area of context and its specialty regarding production environments. Then we introduce our analysis method containing a graphical notation and context templates. The resulting context model is extensible and allows integration with context models designed for different use cases. The context model is simple, abstract and easy to understand because it only focuses on the requirements of the current use case.

2.1   Context in Production Environments

There are lots of definitions in related work what context is. Dey et al. [10] introduce a well-accepted definition: *"Context is any information that can be used to characterize the situation of an entity"* [10]. The central artifact is the *entity*. According to Dey et al., entities can be people, places, and things [10]. Things can be real or virtual. Each entity represents a concrete object of the environment. Context can be characterized by four main *context categories*: identity, time, location, and activity. In each of these categories the entity can have a set of attributes providing data of this category. Furthermore, Zimmermann et al. define [11] *relations* between entities as important context information. Relations constitute a semantical dependency and are often determined by spatial, temporal or individual context of the entities.

Since production environments are complex, lots of actors, parts, manufacturing equipment and processes are involved to manufacture a variety of products. All have to be modeled as context entities. In addition, there is a large amount of data in production environments describing these manufacturing artifacts [12] that have to be associated with context. As a consequence, it is complicated and expensive to design a comprehensive context model from scratch. In general, the context model relies on the use cases that specify the relevant context. The relevant context of manufacturing engineering entities is depending on the context of other entities, since just the interoperation between entities enables to manufacture a product. For example, if an error at a station occurs, the context of all participants in this process is important, such as which shop floor worker worked at the station or on which product the error occurred. Hence, the context of the station is dependent on the context of the shop floor worker and the product. For that purpose, it is important to model *relations* between entities. In order to derive the technical context model for the software development phase, it is important to classify the context category of each relation.

2.2 Graphical Notation for Entity Relation Analysis

To address the issues in conceptual context modeling with respect to production environments, we develop the graphical notation shown in Fig. 2. The graphical notation represents entities, relations between entities, and context categories. The entities are visualized by a rounded square containing an icon illustrating the entity and a text naming the entity. The line between them represents their relation. On the relation, the context category is graphically annotated with an icon and a text naming the relation. The context category itself is visualized as a circle. The purpose of the context category on the relation is to support determination of relations, since Zimmermann et al. point out that relations are dependent on time, location and identity [11]. Furthermore, it enables to specify the necessary context attribute. This is explained in the next subsection.

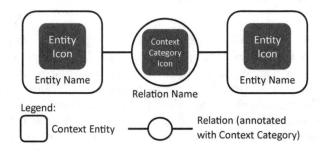

Fig. 2. Graphical notation for conceptual context modeling

Note that this model is used to discuss relevant context with domain experts. In contrast to ER- and UML models, it is simple and reduced to the minimal set of artifacts needed. According to the design principles of Moody [13], we follow the *semiotic clarity* by using two artifacts (relation and entity), *perceptual discriminability* by using only square and cycle and using text only for instance differentiation. Furthermore, for speed up recognition and to improve intelligibility we use *semantic transparency* via icons.

2.3 Context Templates for Transformation

Context templates define the context attributes in each context category and how the context data can be captured and provided. We develop the context template in order to transfer the conceptual view from the analysis phase into an operational view in the software development phase (see Fig. 1) and to decide how to equip the production environment with sensors. The conceptual view specifies the relevant context, whereas the operational view defines needed data acquisition techniques as well as quality and cost factors related to context [5]. The context templates are the basis for the concrete context model. In addition, it is a basis for a cost-benefit analysis. The template can be used to estimate

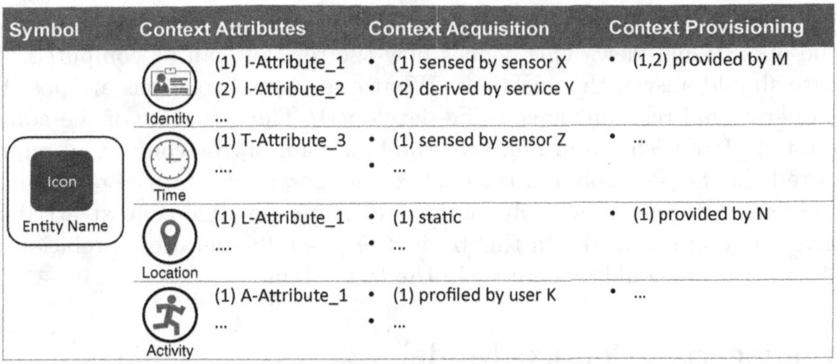

Symbol	Context Attributes	Context Acquisition	Context Provisioning
Icon Entity Name	Identity: (1) I-Attribute_1 (2) I-Attribute_2 ...	• (1) sensed by sensor X • (2) derived by service Y	• (1,2) provided by M
	Time: (1) T-Attribute_3	• (1) sensed by sensor Z • ...	• ...
	Location: (1) L-Attribute_1	• (1) static • ...	• (1) provided by N
	Activity: (1) A-Attribute_1 ...	• (1) profiled by user K • ...	• ...

Fig. 3. Context Entity Template

Conceptual Relation	Category	Context-Attributes Entity 1	Context-Attributes Entity 2	Operational Relation
Relation Name	Context Category X	Entity Name • Context Attribute_1 of Context Category X • Context Attribute_2 of Context Category X • ...	Entity Name • Context Attribute_1 of Context Category X • ...	• Operational _Relation_1 • Operational _Relation_2 • ...

Fig. 4. Context Relation Template

which sensors are useful to be installed. We develop the "Context Entity Template" and "Context Relation Template".

The *context entity template* is based on the schema presented by Perera et al. [5] and is shown in Fig. 3. It consists of four columns: symbol, context attributes, context acquisition and context provisioning. The symbol corresponds to the graphical notation of the entity. The context attributes are classified along the four context categories. This separation is valid for the rest of the template. The categories are visualized by the graphical notation. For each category, the attributes are numbered. The context acquisition column specifies the possible context capturing mechanism for the attributes. These mechanisms are classified by static, sensed, derived, and profiled according to Henricksen [14]. Static context is fixed and does not change over time. For example, ID is a static context attribute. Sensed context is captured by sensors, whereas profiled context is entered by a user. The goal is to reduce manual context acquisition because of efficiency, cost reduction, and quality. Derived context is already processed context, also known as higher-level context [15]. The column context provisioning specifies how the value of the context attributes can be provided to others. This is important according to the Industry 4.0 and Internet of Things paradigms, since intelligent things and machines cannot only gather context information but can also process and provide them to other entities.

The *context relation template* transfers the conceptual view of the relations into the operational view. This defines how the relation can be computed. The template should answer the questions: What operational relations are possible? What operational relations have to be developed? The structure of the context relation template is shown in Fig. 4. In the first column, the conceptual relation is inserted. In the next column, the context category, valid for the relation, has to be given. In the next two columns, the concrete possible context attributes for each entity are defined. On this basis, the possible operational relation can be defined, which should be entered in the last column.

3   Engineering Context Model

In this section, we present the engineering context model using our graphical notation and the context templates. The engineering context model defines the conceptual view which serves as basis to capture and process sensor data. We performed workshops with domain experts and context experts in order to create the engineering context model. In detail, we accomplished the following tasks: (I) identification of entities and design of graphical notation, (II) identification of relations between entities, (III) identification of context categories valid for the relations, (IV) identification of context attributes and their acquisition and providing mechanisms, and (V) refinement of the conceptual relation by defining operational relation.

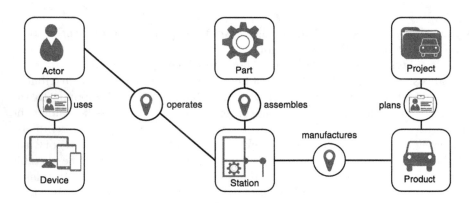

Fig. 5. Engineering context model

The engineering context model is shown in Fig. 5. There are six entities: Actor, Device, Station, Product, Part, and Project. For each entity, we design a symbol as basis for the graphical notation. The *actor* uses a *device*. The *station* assemblies multiple *parts* in order to manufacture a particular *product*. The manufacturing of the product is planned at the *project*. Details about the context attributes and the relations are defined in the context templates. There are many possible context attributes, though, we only consider attributes relevant for the use cases. Due to space reasons, we only present the context template for the entity "Product", since this highlights the novelties raising from IoT and Industry 4.0. The context template is shown in Fig. 6. The product is an intelligent thing, hence, it will be aware of its context and is able to provide its context to other entities. The entity product has the attributes "Product-ID", "Variant" and "Project" belonging to the category identity. These attributes cannot be sensed since they are static attributes, but they can be provided, by e.g., Bluetooth beacons, to others. In contrast "current position" can be sensed by a GPS-receiver and is also provided by Bluetooth beacons. Furthermore, information about the time such as "Creation Time" can be provided by a web service interface.

We present the context relation template in Fig. 7. It consists of five conceptual relations. The conceptual relation "operates" is computed by the "close-to"

Symbol	Context Attributes		Context Acquisition	Context Provisioning
Product	(1) Product-ID (2) Variant (3) Project		• (1,2,3) static	• (1) provided by bluetooth beacon
	(1) Creation Time (2) Creation Duration		• (1,2) sensed by time clock	• (1,2) provided by web service
	(1) Current Position		• (1) sensed by GPS-Sensor	• (1) provided by bluetooth beacon
	(1) State		• (1) Derived by relevant entities	• (1) provided by web service

Fig. 6. Context entity template of product

Conceptual Relation	Category	Context-Attributes Entity 1	Context-Attributes Entity 2	Operational Relation
operates	Location	Actor • Position	Station • Position • Spread	• Close-To (Position, Position) • Close-To (Position, Spread)
uses	Identity	Actor • ID • Name	Device • Category • Capability • ID	• Login • Finger recognition
manufactures	Location	Station • GPS-Position • Spread	Product • GPS-Position	• Close-To (Position, Position) • Close-To (Position, Spread)
plans	Identity	Product • ID	Project • ID	• Manufacturing-Time (Product-ID) • Current-Phase (Time, Product ID)
assembles	Location	Station • Position • Spread	Part • Position	• Close-To (Position, Position) • Close-To (Position, Spread)

Fig. 7. Context relation template of the engineering context model

relation, based on the position attribute (e.g. GPS-Positioning). The "close-to" relation can be calculated by the distance between the two entities.

4 Case-Oriented Evaluation: Context-Driven Problem Documentation

We evaluate our approach including the context modeling method and the engineering context model by applying it to a real use case. First, we present the use case scenario. After that, we show the extended engineering context model. Note that we extend the engineering context model on the conceptual view and not on the technical.

We performed our use case analysis at a German car manufacturer. Our use case scenario is part of the pre-production test in the development phase of a new car. In detail, it takes part in the prototype factory, which produces first prototype versions of cars. Since the manufacturing process in the prototype factory is not as well defined as in series production, problems occur regularly. The documentation of problems is crucial to define an appropriate problem resolving process. If a problem occurs at a station, the shop floor worker has to document it. At each station, a paper-based problem list is pinned to the wall. Whenever the shop floor worker recognizes a problem, she documents it on this list. For example, it may happen that the welding spot cannot be set, because the correct position cannot be reached. In order to provide a solution for this problem, it is necessary to know at which car the problem occurred as well as the involved parts e.g., which are assembled by this welding spot. For further inquiries, the shop floor worker notes her name. The paper-based documentation has following drawbacks: Often, necessary information is missing, since the worker has no time to gather the required information. The documentation is only available once, hence, the distribution of the information is difficult. This prevents the optimization of the process, since the context of the documents cannot be analyzed, e.g. for determining at which station the most problems occur. Hence, the goal is to eliminate manual paper-based documentation by supporting:

- Place and time independent documentation provided by a mobile app on a mobile device. This allows documenting anywhere and anytime using the built-in sensors to capture context [16].
- Automatic gathering of required information for the documentation. This can be reached by using context-awareness. The mobile app acquires context information automatically in order to use them for documentation.

All this is reflected in Fig. 8 and described in the next subsection in detail.

4.1 Document Extension of the Engineering Context Model

We perform the same steps as explained in Sect. 3. First, we specify the problem document as a new context entity and identify the relations between it and

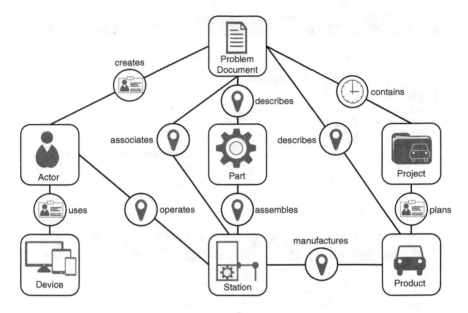

Fig. 8. Extended engineering context model

the already existing context entities. Therefore, we analyze the relevant information needed to describe the problem. Then we describe the relation of this information to the document modeling conceptual relations. As result, we get the extended engineering context model shown in Fig. 8. Here, the actor is the shop floor worker, who "creates" the problem document. The problem occurs at a specific station, hence, the document "associates" the problem with the station. Furthermore, the document "describes" a particular product and particular parts relevant to the problem. For later analysis, it is important to provide information about the project that "contains" the problem and the project phase the problem occurred in.

To detail the problem document entity further and to transfer the conceptual view into the operational view, we create the context template for the entity problem document (see Fig. 9). Note that the "provided context" column is not shown because the document is a virtual entity and all context information can be easily provided to others. In the category *time* the "creation-time" and in the category *location* "creation-position" are specified. The "document context" describes the current context while the problem occurs. This will be determined using the relation defined in the engineering context model. This is refined in the context relation template shown in Fig. 10. One important thing that has to be documented is on which car the problem occurs. Therefore, the relation "describes" is used in the context relation template. To identify the particular car, the location can be used. The "close-to" relation is captured, using the GPS position.

Symbol	Context Attributes		Context Acquisition
		(1) ID	(1) Static
		(2) Title	(2) Static, profiled by the user
Problem Document		(3) Creator	(3) Sensed by actors device
		(1) CreationTime	(1) Sensed by Clock on the actors device
		(1) CreationPosition	(1) Sensed by the GPS-Position of the actors device
		(1) State	(1) Derived by the relation to relevant entities

Fig. 9. Context entity template of document

Conceptual Relation	Category	Context-Attributes Entity 1	Context-Attributes Entity 2	Operational Relation
creates	ID	Actor • ID • Name	Problem Document • Creator	• GetIdentity
describes	Location	Part • Position	Problem Document • CreationPosition	• Close-To • Receive(Bluetooth Signal)
associates	Location	Station • Position • Spread	Problem Document • CreationPosition	• Close-To
contains	Time	Project • ProductID	Problem Document • CreationTime	• CurrentPhase (CreationTime, Product ID)
describes	Location	Product • Position • Spread	Problem Document • CreationPosition	• Close-To

Fig. 10. Context relation template document extension

4.2 Evaluation

We evaluated the presented analysis method by using it in an industrial project. The result was that the graphical notation supports the discussion with the domain experts successfully. Furthermore, it was possible to apply the analysis approach to a real world use-case, the context-driven documentation. The effort for the creation of the new context model for context-driven documentation was lowered by reusing the already available engineering context model. This is enabled by the extensibility of the approach – only one entity and five relations had to be added. Therefore, the overall evaluation was very positive.

5   Related Work

In the following we differentiate three groups with respect to context modeling in production environments. The first group contains related work addressing conceptual modeling of context. Kofod et al. [17] present a generic model for context. It is based on the activity theory and designed to apply case-based reasoning. The focus is on the user context. Relations between different entities are not considered. Reichle et al. [8] introduce a layered approach for context modeling. The three layers are conceptual, exchange and functional layer. The conceptual layer consists of an ontology and a context meta model. The context meta model describes the representation of context. However, they do not consider relations between entities. Costas et al. [18] provide a conceptual foundation for context modeling based on the assumption that context and entities have to be separated. They suggest the modeling of entities using an entity hierarchy and define associations to different context types, which also allows to define relations between context. Nevertheless, there are no mechanism to transfer it to an operational view. All in this group use ontologies to model their conceptual model. However, modeling ontologies is complex and is not suited to discuss this with domain experts, since the graphical representation is complex.

The second group comprises approaches with a more technical perspective. Dey et al. [10] provide a method to support the development of context-aware applications. Since the focus is on software development aspects, there is no support for conceptual context modeling. Henricksen et al. [6] accomplish context modeling using Object-Role-Modeling (ORM). This enables to define the type of context information and relations between entities. However, there is no clear separation between entity and attribute and they are not categorized by context categories. Achilleos et al. [7] design a model-driven approach for context modeling and context-awareness. This approach is based on entities and their attributes are classified according to context categories. Nevertheless, they focus on the context of entities rather than on relations. Therefore, it is not described how to define and operationalize the relations. The last two approaches possess graphical notations, but there are complicated and designed for IT-experts and are not suited for discussion with domain experts.

The last group addresses the integration of context in production environments. Lee et al. [1] aims to improve monitoring systems using context-awareness. They propose a possible solution for the architecture of context-aware monitoring systems and analyze the requirements for context. Alexopoulus et al. [2] present a context-aware information distribution system. They collect sensor data from the shop floor and provide them to the users with respect to the context. Both approaches lack in defining a context model.

6   Summary and Outlook

In this paper, we present an analysis method for conceptual context modeling based on a graphical modeling notation and context templates. This enables the communication with domain experts to define relevant context. We use this method to create an engineering context model that serves as basis for further use cases in this domain. The engineering context model supports context-awareness in production environments in the area of Industry 4.0. As evaluation, we apply the analysis method and the engineering context model in a real use case at a German manufacturer.

In future work, we want to design a technical context model based on this conceptual model. In addition, we plan to develop the applications on top of this engineering context model.

References

1. Lee, A.N., Lastra, J.L.M.: Enhancement of industrial monitoring systems by utilizing context awareness. In: CogSIMA 2013, pp. 277–284 (2013)
2. Alexopoulos, K., et al.: A concept for context-aware computing in manufacturing: the white goods case. Int. J. Comput. Integr. Manufact. **29**, 839–849 (2016)
3. El Kadiri, S., et al.: Current trends on ICT technologies for enterprise information systems. Comput. Ind. **79**, 14–33 (2016)
4. Xie, Y., et al.: Opportunities and challenges for context-aware systems in aerospace industry. J. Enterp. Inf. Manage. **24**(2), 118–125 (2011)
5. Perera, C., et al.: Context aware computing for the internet of things: a survey. IEEE Commun. Surv. Tutorials **16**(1), 414–454 (2014)
6. Henricksen, K., Indulska, J., McFadden, T.: Modelling context information with ORM. In: Meersman, R., Tari, Z., Herrero, P. (eds.) OTM 2005. LNCS, vol. 3762, pp. 626–635. Springer, Heidelberg (2005). doi:10.1007/11575863_82
7. Achilleos, A., et al.: Context modelling and a context-aware framework for pervasive service creation: a model-driven approach. Pervasive Mob. Comput. **6**(2), 281–296 (2010)
8. Reichle, R., Wagner, M., Khan, M.U., Geihs, K., Lorenzo, J., Valla, M., Fra, C., Paspallis, N., Papadopoulos, G.A.: A comprehensive context modeling framework for pervasive computing systems. In: Meier, R., Terzis, S. (eds.) DAIS 2008. LNCS, vol. 5053, pp. 281–295. Springer, Heidelberg (2008). doi:10.1007/978-3-540-68642-2_23
9. Bettini, C., et al.: A survey of context modelling and reasoning techniques. Pervasive Mob. Comput. **6**(2), 161–180 (2010)
10. Dey, A.K., et al.: A conceptual framework and a toolkit for supporting the rapid prototyping of context-aware applications. Hum. Comput. Interact. **16**(2–4), 97–166 (2001)
11. Zimmermann, A., Lorenz, A., Oppermann, R.: An operational definition of context. In: Kokinov, B., Richardson, D.C., Roth-Berghofer, T.R., Vieu, L. (eds.) CONTEXT 2007. LNCS, vol. 4635, pp. 558–571. Springer, Heidelberg (2007). doi:10.1007/978-3-540-74255-5_42
12. Gröger, C., et al.: The data-driven factory - leveraging big industrial data for agile, learning and human-centric manufacturing. In: ICEIS 2016, pp. 40–52 (2016)
13. Moody, D.: The "physics" of notations: toward a scientific basis for constructing visual notations in software engineering. IEEE Trans. Softw. Eng. **35**(6), 756–779 (2009)
14. Henricksen, K., Indulska, J., Rakotonirainy, A.: Modeling context information in pervasive computing systems. In: Mattern, F., Naghshineh, M. (eds.) Pervasive 2002. LNCS, vol. 2414, pp. 167–180. Springer, Heidelberg (2002). doi:10.1007/3-540-45866-2_14
15. Häussermann, K., et al.: Understanding and designing situation-aware mobile and ubiquitous computing systems. World Acad. Sci. Eng. Technol. **14**(1), 329–339 (2010)

16. Hoos, E., Gröger, C., Kramer, S., Mitschang, B.: ValueApping: an analysis method to identify value-adding mobile enterprise apps in business processes. In: Cordeiro, J., Hammoudi, S., Maciaszek, L., Camp, O., Filipe, J. (eds.) ICEIS 2014. LNBIP, vol. 227, pp. 222–243. Springer, Cham (2015). doi:10.1007/978-3-319-22348-3_13
17. Kofod-Petersen, A., Cassens, J.: Using activity theory to model context awareness. In: Roth-Berghofer, T.R., Schulz, S., Leake, D.B. (eds.) MRC 2005. LNCS, vol. 3946, pp. 1–17. Springer, Heidelberg (2006). doi:10.1007/11740674_1
18. Costa, P.D., et al.: Towards conceptual foundations for context-aware applications. In: AAAI Workshop on Modeling and Retrieval of Context, pp. 54–58 (2006)

Evaluating Business Domain-Specific e-Collaboration: The Impact of Integrated e-Collaboration on the Coordination of Component Calculations in Product Costing

Diana Lück[✉] and Christian Leyh

Chair of Information Systems, esp. IS in Manufacturing and Commerce,
Technische Universität Dresden, Dresden, Germany
Diana.Lueck@mailbox.tu-dresden.de,
Christian.Leyh@tu-dresden.de

Abstract. Our concept of Business Domain-Specific e-Collaboration is an approach to integrate e-Collaboration into business domains to enable direct collaboration on business objects in enterprise systems. To validate its practicability and usefulness, we conducted a usability test for a particular use case in the business domain of product costing. In this paper, we present the results of the evaluation for the coordination of component calculations based on Business Domain-Specific e-Collaboration. We reveal how our concept improves the product costing process through higher transparency, an increased speed in processing and improvements regarding consistency.

Keywords: Business Domain-Specific e-Collaboration · Product costing · Enterprise applications · Accounting information systems · Enterprise systems

1 Introduction

In the business domain of product costing, costs of new products whose development cycle has just started are determined. The more complex and uncertain the composition of a product is, the higher the effort for conducting this cost assessment becomes [1]. Collaboration plays an important role in product costing, since the degree of information exchange and communication between the involved parties is very high [2]. In our digitized world, teams are spread all over the world and are becoming increasingly less restricted by limits of time. The evolution of collaboration support has facilitated teams to be spread in different locations without any limits of time. In the 1980s, research started to investigate how technology can be used to support people in their work [3]. Nowadays, terms like Enterprise 2.0 or e-Collaboration stand for business related collaboration techniques [4, 5]. Activities like exchanging information and sharing knowledge in a fast and transparent way have become standard in companies for their employees, as well as for their relationships with customers and partners. Thus, the investigation of e-Collaboration, i.e., its business impacts and benefits, is a common topic in research [6, 7].

© Springer International Publishing AG 2017
W. Abramowicz (Ed.): BIS 2017, LNBIP 288, pp. 326–340, 2017.
DOI: 10.1007/978-3-319-59336-4_23

A prior study of ours, with a focus on the identification of problems and challenges in the product costing process, showed that despite the immense progress in supporting collaboration, severe deficits still exist, especially in the collaborative processes of this field [8]. We discovered that almost 90 percent of the overall workload consists of collaborative work. With the high number of participants and their distribution over numerous divisions and locations, collaboration in product costing turns out to be a very complex task. The study participants expressed that no appropriate IT systems exist in the collaboration process of product costing, and that they are not satisfied with the current solutions and the respective support functionalities. Often, it is unclear who is involved in the costing process, and transparency regarding its progress is missing. Data sources are not integrated, which leads to a high manual effort for data management with media breaks and redundant, inconsistent data. Further results of this study showed that generic e-Collaboration solutions are not specifically designed to integrate collaboration support into the work process. In product costing, a fusion of e-Collaboration and the specific business process with its activities and workflows is needed [8]. A few research projects have been conducted to bring business processes and e-Collaboration together, e.g., working more collaboratively in business process management systems [9] or creating cross-domain collaborative business processes [10], but the focus was different to ours. We have created a concept for e-Collaboration support that is designed specifically for adoption in business domains (see [11]). It has the potential to support all involved participants when and where it is particularly needed in the costing process. With Business Domain-Specific e-Collaboration (BDSpeC) we established a concept to overcome the aforementioned problems and barriers in product costing. Using resources more efficiently, increasing productivity, and saving time and money are just some of the potential benefits of BDSpeC as shown in [11]. In a first evaluation cycle, use cases for collaboration in product costing were assessed regarding integrated e-Collaboration enabled by BDSpeC. We discovered that the coordination of component calculations is a very typical scenario, and that it is a basic use case for collaboration in product costing [12]. In this paper, we present the results of our second evaluation cycle for BDSpeC in which we conducted a usability test with a prototype based on the design concept we developed in a prior research step [13]. Experts tested and assessed a prototype that conveyed the design concept for the coordination of component calculations in product costing enabled by BDSpeC. Therefore, the main research question for this paper is:

How does Business Domain-Specific e-Collaboration influence the coordination of component calculations in product costing?

To answer this question, we present selected results of the prototype evaluation along the following structure: The next section describes the background of the topic. Section 3 presents our research approach followed by a description of the use case, which is the coordination of component calculations, and how this use case is addressed by our concept of BDSpeC (Sect. 4). In Sect. 5, we present how we conducted the usability test and the results of the evaluation. The paper closes with a summary and an outlook.

2 Product Costing in the Discrete Manufacturing Industry

Corporate management covers leading, administering, and directing a company with adequate instruments and methodologies. In managerial accounting, financial and non-financial information are used to establish the foundations for corporate decisions [14]. Product costing is a part of managerial accounting and enables companies to estimate the costs that a product will generate in the future. Since 70% of the costs of goods sold are already set during product development, preliminary costing is specifically crucial, revealing a high potential to influence costs [1].

Providing reliable financial assumptions by calculating realistic costs is the goal in product costing. In the discrete manufacturing industry, products consist of numerous parts, which can be produced in-house, in a different plant of the company, or purchased from suppliers. Each of these products must be assembled, often with complex procedures. Therefore, product costing is a highly relevant task in this industry [2]. New sources of procurement and sales markets evolve daily, and manufacturing processes change constantly due to new innovative technologies. Diverse factors influence the costs of a product. Especially when profit margins are low, like for automotive suppliers, cost calculations need to be exact, because even small deviations from the predicted real costs per piece sum up tremendously, leading to a money-losing business [15]. Thus, accurate cost calculations are essential in the discrete manufacturing industry. Product costing is an interdisciplinary business domain. The sales department communicates with customers regarding new products, and the product costing department is contacted as soon as a cost quote for the product is needed. Product engineering starts to design the product and gives feedback on its possible composition. If parts for the product need to be purchased, the procurement department must negotiate purchase prices for these parts with suppliers, and manufacturing needs to validate the specifications regarding the production before the cost quote can be sent to the customer. Thus, collaboration is essential for the costing process due to the number of participants and the necessary information exchange. In managerial accounting, the amount of data is extensive, and the usage of information technology is substantial [16]. To estimate product costs, mainly spreadsheets are used, created in Microsoft Excel. Hence, problems occur, like the costly manual data administration, inconsistencies, and missing documentation, as well as a low degree of integration [17]. Current IT solutions do not satisfy the needs for collaboration support in product costing, and respective support systems are missing. The result is a lack of transparency as well as high manual efforts. Therefore, the integration of collaboration support into the business activities of product costing is needed, combining e-Collaboration and the work processes of the specific business domain [8].

3 Research Approach

We follow the design science approach according to Hevner et al. in our research project [18]. Our artifact is designed in cycles and the results of each research step are evaluated in an iterative manner.

We started our research project with an exploratory study focusing on problem identification. In this first step, we investigated the collaborative process in product costing, along with its participants and its organizational IT support, to identify relevant problems and challenges. The key findings are summarized in the introduction, and the detailed results are presented in the respective paper [8]. Therefore, with this first study, we addressed the relevance cycle according to Hevner [19]. Since we identified numerous problems, we decided to continue specifically with the investigation of the requirements for collaboration in product costing. In this second research step addressing both the relevance and design cycle according to Hevner [19], we conducted expert interviews to investigate possible solutions for the identified problems. We derived a first approach for BDSpeC in terms of a requirements model for e-Collaboration in product costing by examining the interview protocols in a qualitative data analysis according to Miles et al. [20]. In this paper, we summarize the model shortly in Sect. 4. The detailed requirements analysis can be found in our corresponding research paper [11]. In a further step, we evaluated whether our concept of BDSpeC addresses relevant use cases in product costing, and how our model improves the product costing process [12]. With this iterative step-by-step progress of our research project, we also fulfill the rigor cycle postulated by Hevner [19].

In the research step we present in this paper, we conducted an evaluation of a prototype for the IT support of the use case that showed to be most essential and basic in the use case analysis: the coordination of component calculations (described in more detail in Sect. 4). In a usability test, business experts tested BDSpeC for this scenario. Eight experts from Germany who work in five different international companies from the automotive and machine building industry participated. They all had a professional background in product costing and an extraordinary level of expertise in the field. For the usability test, we implemented a prototype, which was based on a design concept we developed in a prior step [13]. This design concept illustrates how BDSpeC, as a whole, can be integrated into an existing accounting information system used for product costing. Further details on the design concept can be found in the respective paper [13]. For the prototype, we focused on and refined the aspects of the design concept that concern the use case for the coordination of component calculations (see Sect. 4). The usability test was designed for two perspectives: The product manager who determines which person should maintain a certain component, and the costing expert who provides information on a component. Every participant received a test manuscript based on his or her role that explained how to proceed in the prototype. The described steps had to be accomplished within 30 min in teams of two people using one laptop. The usability test concluded with a feedback survey followed by a discussion according to Kriglstein [21]. Details and results of the usability test are presented in Sect. 5.

4   Coordination of Component Calculations in Product Costing Based on Business Domain-Specific e-Collaboration

In a prior research step, we investigated the use cases for collaboration in product costing [12]. We identified three use cases in which the coordination of component calculations turned out to be a basic use case on which all other use cases are built. This use case shows that when the costs for a new product are estimated, several experts usually contribute to the calculation. It is common that the different components of a calculation, e.g., the assembly of a product, are distributed among these experts in order to achieve the most precise information and cost data. They must determine which parts are needed for the respective component, their quantities, and the costs for each part. The distributor, e.g., the project or product manager has the role of a coordinator and ensures that, in the end, the entire product can be calculated by summing up all component calculations considering overhead rates, fixed and variable costs, and direct and indirect portions.

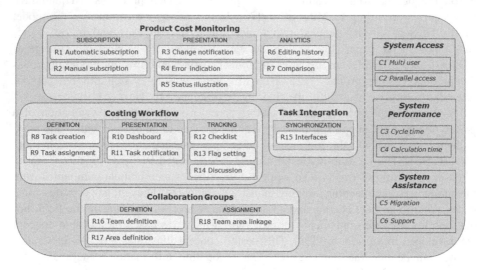

Fig. 1. Requirements model for Business Domain-Specific e-Collaboration in product costing.

To show how BDSpeC addresses this use case, we shortly outline the requirements model for BDSpeC in product costing that we established in a prior research step [11]. BDSpeC is a concept for integrating e-Collaboration into a particular business area and its respective domain-specific application, such as an accounting or enterprise system. Figure 1 illustrates the requirements model for integrated e-Collaboration in product costing. The model consists of four requirement areas that cover 18 requirements (i.e., abbreviated R in the model). Product Cost Monitoring gives an overview of the costing process and helps to keep the participants informed about the status of a process. The Costing Workflow provides self-initiated, ad hoc workflows. Task Integration allows the extension of the task concept of the Costing Workflow to other IT systems used for product costing, and Collaboration Groups authorize the collaboration system. Three system-related prerequisite areas, comprising six constraints (i.e., represented by C in the model), exist to enable integrated e-Collaboration in product costing: System Access, System Performance, and System Assistance. For additional details, see also [11].

In a prior research step, we conducted a use case analysis to investigate whether our concept of BDSpeC addresses relevant aspects of the use case for the coordination of component calculations [12]. This use case is characterized by determining who has the expertise to maintain each of the components of the product calculation, which is executed by elaborating each component and by monitoring when the calculation of all components is complete. From the perspective of the coordinator, BDSpeC offers the initiation of collaboration by creating and assigning new tasks in an ad-hoc manner (R8 Task creation, R9 Task assignment). Since the reference to the component of the calculation that the assignee should work on is directly integrated, manual look-ups of data will no longer be necessary. Furthermore, BDSpeC provides detailed information regarding the status of the process (R5 Status illustration). Displaying the status of a calculation informs team members about all existing tasks and their state of completion. Whenever the status of a task changes, the coordinator receives a notification, enabling a quick reaction (R3 Change notification). From the perspective of the task processor, a notification is sent whenever a new task has been assigned to him or her (R11 Task notification). All this happens automatically, which could enable faster collaboration. By providing a dashboard with details about all tasks, it is easier for team members to keep track of their workload (R10 Dashboard). Being able to monitor the collaboration process professionally while initiating and executing steps with integrated data management has the potential to support the coordination of component calculations to a high degree. Therefore, the process can be automated and becomes more transparent to all participants. For more details, see [12].

5 Evaluation of Coordinating Component Calculations Based on Business Domain-Specific e-Collaboration

To test the design concept for BDSpeC, we developed a prototype for the coordination of component calculations. This prototype was implemented based on the requirements model (see Fig. 1) and the design concept for BDSpeC [11, 13]. Eight experts with a professional background in product costing, who work in the automotive and machine building industry, took part in the usability test of the prototype. Half of the participants assumed the role of a product manager and had to coordinate who needs to maintain which component of a calculation by distributing tasks. The other half assumed the role of a costing expert who had to examine what he or she must accomplish, and execute a task that was assigned to him or her. After the usability test, participants filled out a feedback survey, which was followed by a discussion in order to attain insight about the reasons for the survey ratings. In the following sections, we explain the two scenarios of the usability test and present the results from the feedback session.

5.1 Product Manager Perspective

For the participants with the product manager perspective, the first part of the usability test was about assigning tasks for specific components of a calculation to the responsible colleagues. Therefore, tasks had to be created for a calculation. The test manuscript explained that several aspects of the calculation 'Pump P-100 #Version 1' need to be executed, and that the product manager must ask different colleagues to plan two particular components. In the manuscript, the participant was asked to complete the following steps to create these tasks:

- First, he or she had to open the calculation 'Pump P-100 #Version 1' in the prototype.
- Then, the component 'Casing' had to be selected.
- The button 'Create New Task' had to be clicked and a task description, the assignee, and a due date had to be inserted in the appearing dialog window. The screen to create tasks in the prototype is illustrated in Fig. 2.
- After clicking on the 'Send' button, the task was visible in the side panel of the Calculation View with the status 'Open'.
- To develop routine for the creation of tasks, the test manuscript prompted to request further input for the component 'Drive'. After selecting the corresponding component in the prototype, the button 'Create New Task' had to be clicked, followed by inserting a task description, the assignee, and a due date. The task was shown in the side panel after the 'Send' button was clicked.

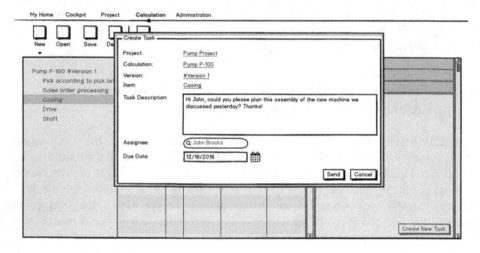

Fig. 2. Create task to request component calculation.

In the next part of the usability test, participants undertaking the product manager perspective should test the monitoring of the progress, while colleagues work on their tasks and edit the calculation.

- As a first step, they had to check all tasks that had been sent to the accountable colleagues. For that purpose, they had to select the root item of the calculation in the Calculation View. At the root level, the tasks for all components of a calculation are displayed as shown in Fig. 3.

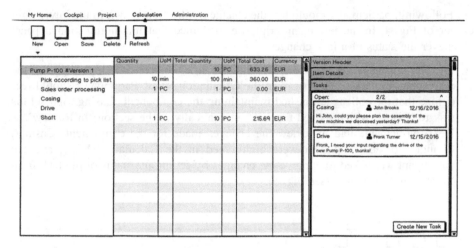

Fig. 3. All tasks created for components of the calculation.

- Next, the monitoring capabilities in the My Home View were presented in the usability test. In the My Home View, the tile of each calculation shows a summary of the existing tasks (see Fig. 4).

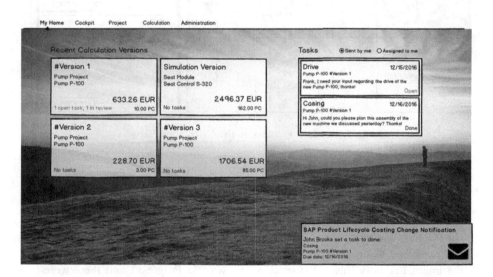

Fig. 4. Monitoring tasks in the My Home View.

- Furthermore, a task list can be found in the My Home View on the right-hand side. The test manuscript explained that one can choose between showing the tasks that you sent to somebody, and the tasks that were assigned to you. Since the product manager perspective should focus on the tasks that he or she delegated, the corresponding option was preselected in the prototype. Figure 4 illustrates the My Home View, showing the task list with two tasks that were created previously.

Following, participants should test the change notification shown in the lower right corner of Fig. 4. In the test manuscript, we explained that this notification appears whenever the status of a task changes.

- The participants were asked to click on the change notification, which triggered the corresponding task to be opened in the Calculation View. Participants should recognize that the colleague finished maintaining the component 'Casing' and set his task to 'Done', which moved the task automatically to the section 'In Review'.
- Next, participants had to review the changes made for the component 'Casing'. Changes to the component were highlighted in the Calculation View, and the participant was asked to approve the changes by setting the status of the task from 'In Review' to 'Done' (see Fig. 5).

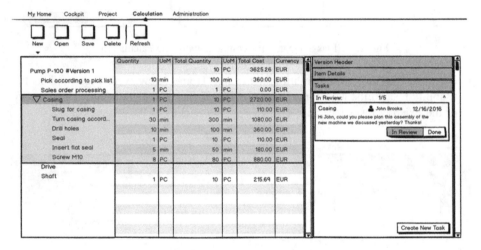

Fig. 5. Reviewing a task in the Calculation View.

The next step was to check the status of all tasks of the calculation again.

- The test manuscript stated to select the root item of the calculation in the Calculation View. Participants were shown that the task for the Casing had the status 'Done'. In this scenario, the task for the drive that was created earlier was shown as 'In Progress'.
- We explained that, in the meantime, other colleagues had created more tasks. For the internal activity 'Insert flat seal', a task was 'In Progress', and there were two additional open tasks. Overall, five tasks existed for the calculation illustrated in the Calculation View.
- Participants were asked to switch to the My Home View to monitor the current status of the calculation again. In the tile of the calculation, only the tasks that were not done yet were summarized. In the task list of the My Home View, participants could only see the two tasks they created.

- The test manuscript finished with the motivation to navigate to the Calculation View, the calculation 'Pump P-100 #Version 1', or the tasks that were created. This should demonstrate the diverse possibilities for monitoring and navigation.

5.2   Costing Expert Perspective

From the perspective of the costing expert, the main purpose of the usability test was to demonstrate how to remain informed about your own workload, and how to complete tasks assigned to you as the costing expert. In the test manuscript, we explained that tasks can be distributed for the components of a calculation. Participants should find out which tasks are assigned to them and what they must work on, since the participants were responsible for specific components of a calculation.

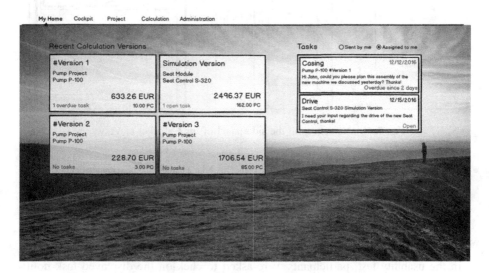

Fig. 6. Insight on tasks in the My Home View.

- The test manuscript stated that the My Home View provides information regarding the workload. It includes a task list in which the user can choose between showing the tasks that you sent to somebody, and the tasks that were assigned to you. In the usability test with the costing expert perspective, the focus was on the tasks that the participant had to work on. Therefore, this option was preselected.
- In the prototype, the My Home View showed two tasks that were assigned to the participant, displayed in Fig. 6.
- In the test manuscript, we asked to click on the overdue task related to the component 'Casing'. The corresponding calculation 'Pump P-100 #Version 1' was opened in the Calculation View and the component was automatically selected.
- Next, the component should be maintained. Therefore, the status of the task should be set to 'In Progress' in the side panel of the Calculation View.

- To edit the component, the participant was asked to click on the component. All relevant data was entered automatically in the prototype. The results of these steps are shown in Fig. 7.
- We prompted to set the status of the task to 'Done' in the side panel.

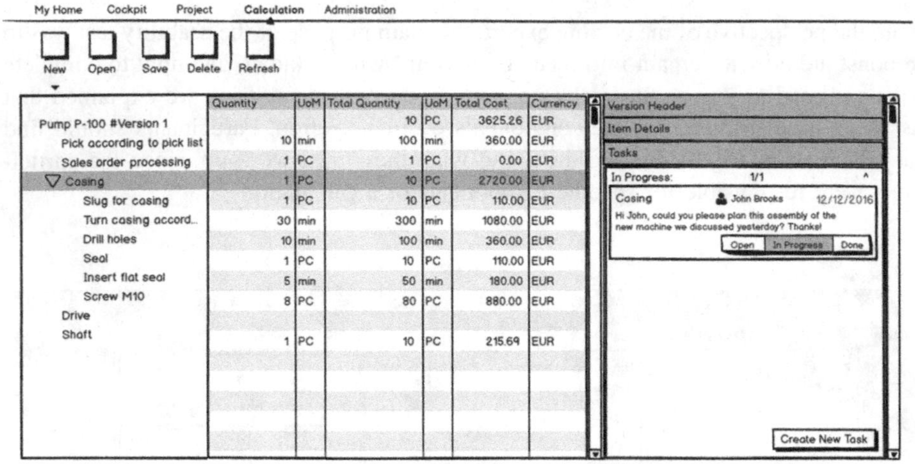

Fig. 7. Status setting in the side panel of the Calculation View.

- Afterwards, participants should go back to the My Home View to point out that the overdue task was not shown in the task list anymore.

Subsequently, the usability test for the costing expert perspective presented the concept of task notifications. Whenever a new task was assigned to a user, he or she was notified, similar to the change notification of the prototype displayed in Fig. 4.

- In the usability test, participants were asked to click on the displayed task notification. The corresponding task was opened in the Calculation View, showing that a task was assigned to the participant to maintain the component 'Impeller' of the calculation '#Version 3' of the calculation 'Pump P-100'.
- As a next step, participants should get an overall impression about the status of the calculation. Therefore, the root item of the calculation had to be selected in the Calculation View, and the participant could see that several tasks were available for this calculation. In addition to the task, there was another open task for the component 'Pick according to pick list'. A task for the component 'Drive' was in progress. Overall, three tasks existed for the calculation (see Fig. 8).
- Next, the test manuscript stated that the task for the component 'Impeller' was not urgent, since its due date was set for next week.
- The participant should return to the My Home View. The script advised that all tasks were also reflected in the tiles showing the recent calculations.

- Finally, we encouraged the participants in the test manuscript to return by clicking on the Calculation View or the calculation 'Pump P-100 #Version 3' to illustrate the integrative navigation that the concept of BDSpeC provided for product costing.

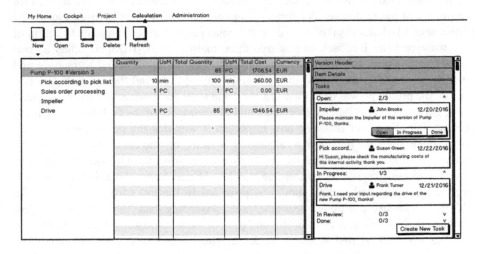

Fig. 8. All tasks for '#Version 3' in Calculation View.

5.3 Validation Results

After the usability test, we conducted a feedback session consisting of a questionnaire that the 8 participants of the usability test had to fill out and a discussion during which everybody could explain their assessments in more detail. In the questionnaire, we asked eleven questions regarding the prototype in general, the visualization, and the navigation. We used a 7-point-likert scale in order to have sufficient room for differentiation, with the options from 1 = Strongly Disagree to 7 = Strongly Agree, according to [22].

Subsequently, the average responses are presented in brackets. Participants agreed that the scenarios of the usability test were realistic (5.5). The prototype was rated to be easy to use (6.5), and that it was possible to complete the scenarios quickly (6.25). Participants stated that the prototype represents a useful support for collaboration (6.375), and that it could improve transparency (6.125) and enhance productivity in product costing (5.625). Also regarding the visualization, our design concept for the coordination of component calculations based on BDSpeC received very good feedback. The approach of connecting tasks with components of a calculation was rated to be useful (6.5), and displaying tasks in the side panel of the Calculation View proved to be effective (6.5). Participants assessed the overview regarding tasks in the My Home View as helpful (5.5), but the discussion demonstrated that they saw a risk of information overload when the number of tasks became very high. A specific aspect of BDSpeC is the integrated navigation between tasks and the components of a calculation. The prototype was rated easy to navigate (6.375), and jumping from tasks to the respective component of the calculation was deemed to be very useful (6.5). Overall, the different aspects of the prototype representing BDSpeC for the coordination of component calculations were assessed as "positive" to "very positive", without exception.

We conducted a follow-up discussion because we were interested in the reasons for the ratings of the prototype. We further deepened the assessment by asking what the participants liked and disliked regarding the concept of BDSpeC presented in form of the prototype. The feedback emphasized three main aspects: BDSpeC offers a great opportunity to give an overview about the collaboration process and to manage and control the whole process. Due to the intuitive navigation and the integrated visualization, the prototype was described as a good and very appropriate tool to quickly send tasks, get insight regarding the workload, and to keep track of the collaborative costing process. Participants emphasized how important it was that the prototype provided a possibility to track the history of a calculation, in terms of who contributed to what task and which input was given by whom. Furthermore, several additional feature ideas were mentioned, especially concerning the My Home View. To avoid information overload, filtering and sorting should be provided in the tasks list, and options like adding comments, accepting, declining and forwarding tasks should be provided. Further notifications in respect of the due date of tasks were suggested, e.g., when a task has not been started within a week, or when a task is due in two days. The discussion further reinforced the strong need for e-Collaboration support in most companies. Nevertheless, there are some individual exceptions. One participant from the automotive industry explained that their processes are so standardized that such an ad-hoc coordination support is only necessary in few special cases that differ from the standard process. However, this could not be confirmed by the other participants.

Concluding the feedback session, the usability test showed that BDSpeC can facilitate the product costing process and improve data consistency due to the direct integration of tasks and components. Data input can be handled directly in the enterprise system used for product costing. Collaboration can be managed and tracked in an integrated manner, and improved transparency as well as easy navigation have the potential to speed up the process.

6 Summary and Outlook

In this paper, we presented an evaluation cycle for our concept of Business Domain-Specific e-Collaboration (BDSpeC). The coordination of component calculations is a basic use case for collaboration in product costing. We conducted a usability test with business experts for BDSpeC applied to this use case. Based on a prototype we implemented, participants tested the integrated coordination of component calculations in a scenario with two typical perspectives: Having to delegate tasks as a product manager, and having to work on tasks as a costing expert.

The evaluation showed that BDSpeC offers substantial support regarding the initiation, the execution, and the monitoring of collaborative aspects of the costing process. Therefore, our concept and our prototype can be seen as a useful and appropriate solution for current problems and challenges in the product costing process. By integrating this concept in an enterprise system, a suitable collaboration support that is directly connected to the process of product costing will be provided for the employees and for the respective business partners involved in the costing process.

As a next step in this research project, we will develop design principles to explore the generalizability of our work for various business domains. Furthermore, we plan an implementation of BDSpeC in a specific enterprise system for product costing. In fact, a fusion of a costing system and a collaboration platform will be established according to the prototype presented in this paper. The goal is to enable employees to use BDSpeC in their everyday work lives, and to evaluate the concept in a real-world setting.

References

1. Saaksvuori, A., Immonen, A.: Product Lifecycle Management. Springer, Berlin (2004)
2. Hansen, D.R., Mowen, M.M., Guan, L.: Cost Management - Accounting and Control. Cengage Learning, Mason (2009)
3. Grudin, J.: Computer-supported cooperative work: history and focus. Comput. **27**(5), 19–26 (1994)
4. McAfee, A.P.: Enterprise 2.0: the dawn of emergent collaboration. MIT Sloan Manage. Rev. **47**(3), 21–28 (2006)
5. Riemer, K., Steinfield, C., Vogel, D.: eCollaboration: on the nature and emergence of communication and collaboration technologies. Electr. Markets **19**(4), 181–188 (2009)
6. Alqahtani, F.H., Watson, J., Partridge, H.: Organizational support and Enterprise Web 2.0 adoption: a qualitative study. In: Proceedings of the 20th Americas Conference on Information Systems (AMCIS 2014), Atlanta (2014)
7. Andriole, S.: Business impact of Web 2.0 technologies. Commun. ACM **53**(12), 67–79 (2010)
8. Lück, D., Leyh, C.: Integrated virtual cooperation in product costing in the discrete manufacturing industry: a problem identification. In: Proceedings of the Multikonferenz Wirtschaftsinformatik 2016 (MKWI 2016), Ilmenau, Germany, pp. 279–290 (2016)
9. Kemsley, S.: Enterprise 2.0 meets business process management. In: Brocke, J.V., Roseman, M. (eds.) Handbook on Business Process Management 1. International Handbooks on Information Systems. Springer, Berlin (2010)
10. Capodieci, A., Del Fiore, G., Mainetti, L.: Adopting collaborative business process patterns for an enterprise 2.0 banking information system. In: Proceedings of the 4th International Conference on Advanced Collaborative Networks, Systems and Applications, pp. 62–71. IARIA XPS Press, Venice (2014)
11. Lück, D., Leyh, C.: Toward Business Domain-Specific eCollaboration: requirements for integrated virtual cooperation in product costing. In: Proceedings of the 22nd Americas Conference on Information Systems (AMCIS 2016), San Diego (2016)
12. Lück, D., Leyh, C.: Enabling Business Domain-Specific e-Collaboration: developing artifacts to integrate e-Collaboration into product costing. In: Proceedings of the 12th International Conference on Design Science Research in Information Systems and Technology (DESRIST 2017), Karlsruhe, Germany (2017)
13. Lück, D.: Enabling Business Domain-Specific eCollaboration - how to integrate virtual cooperation in product costing. In: Proceedings of the 19th International Conference on Enterprise Information Systems (ICEIS 2017), Porto, Portugal (2017)

14. Warren, C.S., Reeve, J.M., Duchac, J.E.: Financial and Managerial Accounting. Cengage Learning, Boston (2014)
15. Drury, C.: Management and Cost. Cengage Learning, London (2008)
16. Fiedler, R., Gräf, J.: Einführung in das Controlling, Methoden, Instrumente und IT-Unterstützung. Oldenbourg Wissenschaftsverlag GmbH, München (2012)
17. Schicker, G., Mader, F., Bodendorf, F.: Product Lifecycle Cost Management (PLCM): status quo, trends und Entwicklungsperspektiven im PLCM – eine empirische Studie. Arbeitspapier Wirtschaftsinformatik II, Universität Erlangen-Nürnberg (2008)
18. Hevner, A.R., March, S.T., Park, J., Ram, S.: Design science in information systems research. MIS Q. **28**(1), 75–105 (2004)
19. Hevner, A.R.: A three cycle view of design science research. Scand. J. Inf. Syst. **19**(2), 1–6 (2007)
20. Miles, M.B., Huberman, A.M., Saldana, J.: Qualitative Data Analysis: A Methods Sourcebook. Sage Publications, London (2013)
21. Kriglstein, S., Leitner, M., Kabicher-Fuchs, S., Rinderle-Ma, S.: Evaluation methods in process-aware information systems research with a perspective on human orientation. Bus. Inf. Syst. Eng. **58**(6), 397–414 (2016)
22. Bethlehem, J., Biffignandi, S.: Handbook on Web Surveys. John Wiley and Sons Inc., Hoboken (2012)

Improving Pavement Anomaly Detection Using Backward Feature Elimination

Jun-Lin Lin[1,2(✉)], Zhi-Qiang Peng[1], and Robert K. Lai[3]

[1] Department of Information Management, Yuan Ze University, 135 Yuan-Tung Road, Chungli, Taoyuan 32003, Taiwan
jun@saturn.yzu.edu.tw, 40141143@gm.nfu.edu.tw
[2] Innovation Center for Big Data and Digital Convergence, Yuan Ze University, 135 Yuan-Tung Road, Chungli, Taoyuan 32003, Taiwan
[3] Department of Computer Science and Engineering, Yuan Ze University, 135 Yuan-Tung Road, Chungli, Taoyuan 32003, Taiwan
krlai@cs.yzu.edu.tw

Abstract. Early approaches for pavement anomaly detection used simple heuristic on a small number of features and their associated thresholds. Such methods may not be suitable for data collected from heterogeneous sources (e.g., different vehicles, pavements, inertial sensors, etc.). Instead of manually selecting a set of features and their thresholds, we propose using backward feature elimination on a large set of features such that the optimal set of features can be determined. Our experimental results show that the features selected by backward feature elimination yield the best performance, compared to using all features from the sampled data of the accelerometer and gyrometer.

Keywords: Accelerometer · Gyrometer · Pothole detection · Backward feature elimination

1 Introduction

In today's society, traveling in automobiles has been and continues to be an important part of our daily lives. When traveling in cars, the presence of pavement anomaly could have an adverse impact on our traveling experience. Pavement anomaly, however, is almost unavoidable. Severe weather conditions (e.g., flooding, sun exposure, and earthquakes), excessive usage intensity (e.g., overloading vehicles) and poor construction quality (e.g., improper filling operations) can all result in pavement anomaly. Thus, monitoring pavement to detect anomaly is crucial to ensure pavement quality.

Traditionally, pavement anomaly is detected by trained inspection personnel via visual inspection. This approach cannot detect pavement anomaly in a timely fashion due to its labor-intensive nature. Participatory sensing offers a feasible solution to this problem [1]. First, with participatory sensing, a larger number of participants can provide data with a better coverage area. Second, pavement anomaly causes abnormal vibration of the vehicles, which can be recorded using inertial sensors. Because inertial sensors

© Springer International Publishing AG 2017
W. Abramowicz (Ed.): BIS 2017, LNBIP 288, pp. 341–349, 2017.
DOI: 10.1007/978-3-319-59336-4_24

are standard in today's smartphones, participatory sensing provides a cost effective solution to this problem.

Therefore, the remaining problem is how to detect pavement anomaly with the data collected from the inertial sensors of a smartphone. In the literature, accelerometers and gyrometers are the two types of inertial sensors commonly used for detecting pavement anomaly. Figure 1 shows the local coordinate system of a vehicle and that of a smartphone. A 3-axis accelerometer measures the acceleration on each axis (denoted as a_x, a_y and a_z), and a gyrometer measures the angular speed along each axis (denoted as r_x, r_y and r_z). For pavement anomaly detection, we collect these six measurements of a vehicle. In this study, we collect the inertial data of a vehicle by placing the smartphone inside the vehicle such that the three axes of the smartphone coincide with the three axes of the vehicle. Thus, no conversion is required to convert the inertial data of the smartphone to that of the vehicle [2–4].

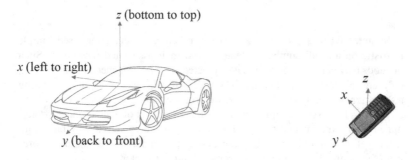

Fig. 1. Local coordinate system of a vehicle (right) and of a smartphone (right).

Pavement anomaly incurs abnormal movement of the vehicle, especially in the vertical direction of the vehicle. Thus, a_z and its related features are often used in pavement anomaly detection. For examples, [5] uses three thresholds, respectively, on three features a_z, a_x/a_z and v_y/a_z to detection potholes, where v_y is the speed of the vehicle; [2] uses two thresholds on a_z, depending on whether v_y is greater than 25 km/h or not. With these approaches, the setting of the thresholds and the selection of features are ad hoc, and may not be suitable when the data is collected from heterogeneous sources, e.g., different cars or various road conditions. Instead of depending on the simple heuristics from a small number of selected features and their respective thresholds, we propose applying backward feature elimination on a large number of features such that the detection model can better adjust to the data under study automatically. Our experimental results show that the proposed method selects 15 out of 24 features, and the selected features yield the best performance, compared to using all 24 features from the accelerometer and gyrometer, or all 15 features from the accelerometer, or five features from z-axis of the accelerometer. Notably, the experiment used three classification algorithms (neural network, support vector machine, and J48 decision tree) from Weka [15] to perform 10-fold cross validation.

The rest of this paper is organized as follows. Section 2 reviews pervious work. Section 3 describes the proposed approach, specifically on data collection, preprocessing, feature calculation, and backward feature elimination. Section 4 shows the performance results, and Sect. 5 concludes this paper.

2   Related Work

In the literature, the data used for pavement anomaly detection can be divided into two groups: image and non-image. Image data is collected using cameras, and image processing techniques are used to detect pavement anomaly. Some research used a specialized camera to take pictures of the pavement from above [6], and others used a low-cost car parking camera [7].

Sensors used for collecting non-image data include GPS, accelerometers, gyrometer, and On-Board Diagnostic System 2 (OBD2). GPS is mainly used to record the location data of pavement anomaly. Clustering algorithms can be applied to the locations of detected pavement anomaly to refine and improve the prediction results [1]. The speed of the vehicle can affect how a vehicle reacts to a pavement anomaly. It can be collected from GPS (less accurate) or OBD2 [8].

As described in Sect. 1, accelerometers are the most common sensors used for pavement anomaly detection [2, 3, 5, 9–11], and some research used both an accelerometer and a gyrometer [12]. Table 1 summarizes the important features derived from the sampled data of an accelerometer.

Table 1. Features derived sampled data of an accelerometer and the speed of a vehicle, where t is an index to time.

Reference	Features
[5]	$a_x, a_z, a_x/a_z, v_y/a_z, v_y$
[2]	a_z, v_y
[13]	$a_x, a_y, a_z, a_z(t) - a_z(t - 1)$, $\text{Stdev}(a_z)$
[3]	$a_z, \max(a_z), \min(a_z), \text{avg}(a_z), v_y$
[14]	$\max(a_z(t) - a_z(t - 1)), \Sigma\left((a_z(t) - a_z(t - 1))^2\right)$

3   Proposed Approach

3.1   Data Collection and Preprocessing

In this study, the data was collected using the accelerometer and gyrometer of a smartphone. The smartphone was placed on the carpet of the front passenger side of a vehicle such that the three axes of the smartphone coincided with the three axes of the vehicle (see Fig. 1). Thus, no transformation was needed to convert the inertial data of the smartphone to that of the vehicle. An Android app was developed to collect the inertial data (including a_x, a_y, a_z, r_x, r_y and r_z) of the smartphone and the time of occurrence of each sample data. Consequently, the collected inertial data includes six time series. Furthermore, the Android app also provides a GUI button to allow the user to record the time of occurrence of an anomaly (i.e., pothole or bump). The vehicle and the smartphone used in this study were

Toyota Yaris and HTC One, respectively. The sample frequencies of the accelerometer and gyrometer were 15 Hz and 30 Hz, respectively.

For each time an anomaly occurred, we form a window of sampled data from each of the six time series of a_x, a_y, a_z, r_x, r_y and r_z. Supposed that an anomaly occurs between times t and $t + 1$, six windows of data are constructed, including the 100 sampled data at or before time t, and 10 sampled data after time t, as shown in Fig. 2. These six windows of sampled data represent a case of an anomaly, and are used to derive features of the anomaly (see Sect. 3.2).

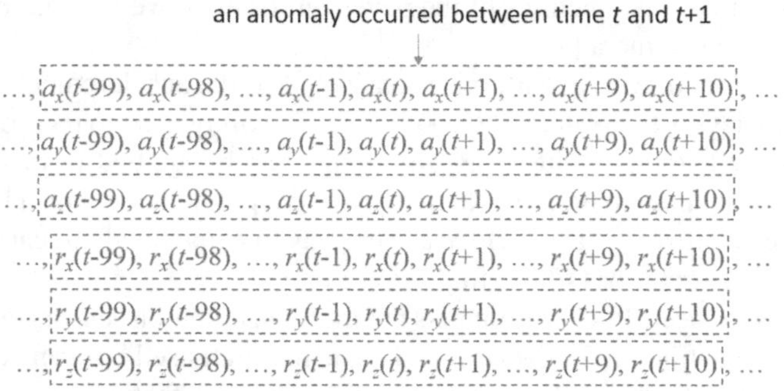

an anomaly occurred between time t and t+1

..., $a_x(t$-99$)$, $a_x(t$-98$)$, ..., $a_x(t$-1$)$, $a_x(t)$, $a_x(t$+1$)$, ..., $a_x(t$+9$)$, $a_x(t$+10$)$, ...

..., $a_y(t$-99$)$, $a_y(t$-98$)$, ..., $a_y(t$-1$)$, $a_y(t)$, $a_y(t$+1$)$, ..., $a_y(t$+9$)$, $a_y(t$+10$)$, ...

..., $a_z(t$-99$)$, $a_z(t$-98$)$, ..., $a_z(t$-1$)$, $a_z(t)$, $a_z(t$+1$)$, ..., $a_z(t$+9$)$, $a_z(t$+10$)$, ...

..., $r_x(t$-99$)$, $r_x(t$-98$)$, ..., $r_x(t$-1$)$, $r_x(t)$, $r_x(t$+1$)$, ..., $r_x(t$+9$)$, $r_x(t$+10$)$, ...

..., $r_y(t$-99$)$, $r_y(t$-98$)$, ..., $r_y(t$-1$)$, $r_y(t)$, $r_y(t$+1$)$, ..., $r_y(t$+9$)$, $r_y(t$+10$)$, ...

..., $r_z(t$-99$)$, $r_z(t$-98$)$, ..., $r_z(t$-1$)$, $r_z(t)$, $r_z(t$+1$)$, ..., $r_z(t$+9$)$, $r_z(t$+10$)$, ...

Fig. 2. From six time series of a_x, a_y, a_z, r_x, r_y and r_z, construct six windows of sampled data for each anomaly occurred between time t and time $t + 1$.

Cases of smooth pavement are constructed as follows. First, a window is constructed for every 120 sampled data in the time series of a_x. That is, the first window contains $a_x(0)$ to $a_x(119)$, the next window contains $a_x(120)$ to $a_x(239)$, and the i-th window contains $a_x(120i - 120)$ to $a_x(120i - 1)$, and so on. Consider a window with a starting time t and an ending time $t + 119$. If no anomaly occurs between time t and time $t + 119$, then another five windows of sampled data between time t and $t + 119$ are constructed from the five time series of a_y, a_z, r_x, r_y and r_z, and these six windows of sampled data represent a case of smooth pavement.

In this study, the preprocessing step yields 151 cases of data (73 for potholes, 29 for bumps, and 49 for smooth pavement), where each contains a set of six windows of sampled data from the time series of a_x, a_y, a_z, r_x, r_y and r_z.

3.2 Derived Features

Let $v_1, v_2, \ldots, v_n$ represent a window of sampled data. Then, seven features derived from $v_1, v_2, \ldots, v_n$ are defined below.

$$f_1(v_1, v_2, \ldots, v_n) = \max\{v_1, v_2, \ldots, v_n\}. \tag{1}$$

$$f_2(v_1, v_2, \ldots, v_n) = \min\{v_1, v_2, \ldots, v_n\}. \tag{2}$$

$$f_3(v_1, v_2, \ldots, v_n) = \max\{v_2 - v_1, v_3 - v_2, \ldots, v_n - v_{n-1}\}. \tag{3}$$

$$f_4(v_1, v_2, \ldots, v_n) = \min\{v_2 - v_1,\ v_3 - v_2, \ldots, v_n - v_{n-1}\}. \tag{4}$$

$$f_5(v_1, v_2, \ldots, v_n) = \max\{|v_2 - v_1|,\ |v_3 - v_2|, \ldots, |v_n - v_{n-1}|\}. \tag{5}$$

$$f_6(v_1, v_2, \ldots, v_n) = \max\{|v_1|, |v_2|, \ldots, |v_n|\}. \tag{6}$$

$$f_7(v_1, v_2, \ldots, v_n) = \max\{|v_2 + v_1|,\ |v_3 + v_2|, \ldots, |v_n + v_{n-1}|\}. \tag{7}$$

As described in Sect. 3.1, each case of data contains a set of six windows of sampled data from the time series of a_x, a_y, a_z, r_x, r_y and r_z. For the three windows from the time series of a_x, a_y and a_z, we apply Eqs. (1)–(5) and yield 15 features, denoted as $f_1(a_x), \ldots, f_5(a_x), f_1(a_y), \ldots, f_5(a_y), f_1(a_z), \ldots, f_5(a_z)$. For the three windows from the time series of r_x, r_y and r_z, we apply Eqs. (5)–(7) and yield 9 features, denoted as $f_5(r_x), f_6(r_x), f_7(r_x), f_5(r_y), f_6(r_y), f_7(r_y), f_5(r_z), f_6(r_z), f_7(r_z)$. Thus, our dataset contains 24 features, plus an attribute indicating pothole, bump, or smooth pavement.

3.3 Backward Feature Elimination

A backward feature elimination process was adopted in this study to select the optimal set of features to build the classification model. At first, all features are used to build a classification model M_0. Then, many classification models are built, each of which uses one less feature than M_0 does, and let M_{best} denote the best of these models. If M_{best} yields better performance than M_0 does, then M_{best} becomes the new M_0, and the same process repeats with the set of features used to build M_{best}. Otherwise, return the set of features used to build M_0, and end the process. The backward feature elimination process is shown below.

```
BackwardFeatureElimination(SetOfAllFeatures)
  RemainFeatures = SetOfAllFeatures;
  M₀ = model built using RemainFeatures;
  R₀ = classification accuracy of M₀;
  repeat
    Rbest = 0;
    for each feature f in RemainFeatures
      M = model built using RemainFeatures\{f};
      R = classification accuracy of M;
      if R ≥ Rbest
        fbest = f;    Rbest = R;    Mbest = M;
      end if
    end for
    if Rbest ≥ R₀
      RemainFeatures = RemainFeatures\{fbest};
      R₀ = Rbest;    M₀ = Mbest;
    else
      return RemainFeatures;
  end.
```

In the backward feature elimination process described above, we use the neural network in Weka [15] to build the classification model. With the data set described in Sects. 3.1 and 3.2, the sequence of features removed by the process is $f_4(a_z), f_4(a_x), f_3(a_z), f_3(a_x), f_5(a_z), f_5(r_y), f_7(r_x), f_6(r_z), f_4(a_y)$. The remaining 15 features are $f_5(a_y), f_5(r_x), f_5(r_z), f_6(r_x), f_6(r_y), f_7(r_y), f_7(r_z), f_1(a_x), f_1(a_y), f_1(a_z), f_2(a_x), f_2(a_y), f_2(a_z), f_3(a_y), f_5(a_x), f_5(a_y), f_5(r_x), f_5(r_z), f_6(r_x), f_6(r_y), f_7(r_y), f_7(r_z)$. The set of these 15 features is denoted as F_{bfe}.

4 Performance Study

In this performance study, we compare the performance of our selected features against that of five sets of features. The first set of features, denoted as F_{all}, includes all 24 features described Sect. 3.2. This set uses features derived from the sampled data of accelerometer and gyrometer. The second set of features, denoted as F_{acc}, includes only the 15 features derived from the sampled data of accelerometer. The third set of features, denoted as F_{accZ}, includes only the five features derived from the sampled data of accelerometer. In the literature of pavement anomaly detection, some used sampled data from both accelerometer and gyrometer, some only used sampled data from accelerometer, and some only used the z-axis sampled data from accelerometer. Thus, this experiment used these three sets of features to represent the three scenarios. Besides, to compare to other feature selection methods, we also rank all 24 features according to information gain or correlation. The fourth set of features, denoted as $F_{infoGain}$, includes the top 15 features with respect to information gain. The fifth set of features, denoted as F_{R2}, includes the top 15 features with respect to correlation.

Three classification algorithms (neural network (NN), support vector machine (SVM), and J48 decision tree) from Weka [15] were used in this study to perform 10-fold cross validation. Default parameter settings of three algorithms in Weka were adopted. The results are shown in Table 2.

Table 2. Classification accuracy.

Feature set	NN	SVM	J48
F_{accZ}	72.8477	72.1854	69.5364
F_{acc}	80.7947	81.457	**78.1457**
F_{all}	83.4437	82.7815	76.8212
F_{bfe}	**87.4172**	**84.106**	75.4967
$F_{infoGain}$	80.1325	81.457	77.4834
F_{R2}	80.7947	81.457	77.4834

J48 decision tree yielded the worst performance among the three algorithms. Because decision tree bases on one feature at a time to build the model, it is not well suited to this problem where the inertial data from all three axes are highly coupled. In contrast, both NN and SVM combine all features into a nonlinear model, making them a better algorithm for this problem.

Consider the performance results of using NN and SVM on F_{accZ}, F_{acc}, F_{all}, or F_{bfe}. F_{bfe} yielded the best performance, F_{all} came in second, F_{acc} third, and F_{acc} was the last. Thus, using only the z-axis sampled data from the accelerometer is not sufficient. Using all three axes of sampled data from the accelerometer improves the performance by about 8% with NN and about 9% with SVM. Adding the sampled data from the gyrometer further improves the performance by about 3% with NN and about 1% with SVM. Finally, backward feature elimination improves the performance further by about 4% with NN and about 2% with SVM.

Finally, the last three rows of Table 2 show that, using NN or SVM, F_{bfe} yielded better results than both $F_{infoGain}$ and F_{R2} did. Thus, the features selected by backward feature elimination are better than that selected according to either information gain or correlation. Table 3 shows the features in F_{bfe}, $F_{infoGain}$ and F_{R2}, where the features in F_{bfe} are marked with asterisks, and the features in $F_{infoGain}$ and F_{R2} are indicated by their ranks in their corresponding sets. By Tables 3(a)–(c), all three sets select features derived from the three axes of the accelerometer. However, Table 3(d) shows that the feature set F_{R2} contains only one feature that is derived from gyrometer's data, but F_{bfe} and $F_{infoGain}$ use five and four features derived from gyrometer's data, respectively.

Table 3. Features in F_{bfe}, $F_{infoGain}$ and F_{R2}.

Set	$f_1(a_z)$	$f_2(a_z)$	$f_3(a_z)$	$f_4(a_z)$	$f_5(a_z)$
F_{bfe}	*	*			
$F_{infoGain}$	3	2		1	7
F_{R2}	10	4	9	12	8

(a) a_z derived features

Set	$f_1(a_y)$	$f_2(a_y)$	$f_3(a_y)$	$f_4(a_y)$	$f_5(a_y)$
F_{bfe}	*	*	*		*
$F_{infoGain}$		14	15	9	10
F_{R2}	13	14	11	6	7

(b) a_y derived features

Set	$f_1(a_x)$	$f_2(a_x)$	$f_3(a_x)$	$f_4(a_x)$	$f_5(a_x)$
F_{bfe}	*	*			*
$F_{infoGain}$		11	12		13
F_{R2}	1	2	5		3

(c) a_x derived features

Features	$f_5(r_x)$	$f_6(r_x)$	$f_7(r_x)$	$f_5(r_y)$	$f_6(r_y)$	$f_7(r_y)$	$f_5(r_z)$	$f_6(r_z)$	$f_7(r_z)$
F_{bfe}	*	*			*	*	*		*
$F_{infoGain}$		4	8					5	6
F_{R2}		15							

(d) r_x, r_y and r_z derived features

5   Conclusions

In this study, we showed that both SVM and NN are effective techniques to construct classification model for pavement anomaly detection. Reference [10] also applied SVM on a large number of features to detect pavement anomaly, but it did not employ feature elimination. We showed that by using backward feature elimination, the performance is further improved.

Although feature elimination has been shown to be an effective step in this study, the algorithm used to perform feature elimination should be chosen carefully. For example, decision tree is not suitable for the problem under study, according to its performance shown in Table 2. Thus, although decision tree is sometimes used as a feature selection technique, it may not be adequate to select features for the problem of pavement anomaly detection. In this study, we applied backward feature elimination, where NN was used to build classification models. Since NN is suitable for the problem under study, according to its performance shown in Table 2, using NN inside backward feature elimination is a good choice.

In this preliminary study, both the amount of collected data and the number of considered features are limited. Experimenting with a small set of data with a large number of features increases the risk of overfitting. Thus, future work is planned along three directions: more data, more features, and more feature selection/elimination methods. First, more data is needed to strengthen the results of this study. Data should be collected from diverse conditions. Second, this study employed only 24 features. More features should be added by adding more data sources (e.g., OBD2, GPS, etc.) and more formulas to derive features. Third, this study only considered backward feature elimination to select features. As the number of features grows, this approach can be very time-consuming. Other options should be explored.

Acknowledgments. This research is supported by the Ministry of Science and Technology, Taiwan, under Grant 105-2632-H-155-022.

References

1. Yi, C.W., Chuang, Y.T., Nian, C.S.: Toward crowdsourcing-based road pavement monitoring by mobile sensing technologies. IEEE Trans. Intell. Transp. Syst. **16**, 1905–1917 (2015)
2. Mohan, P., Padmanabhan, V.N., Ramjee, R.: Nericell: rich monitoring of road and traffic conditions using mobile smartphones. In: Proceedings of the 6th ACM Conference on Embedded Network Sensor Systems, pp. 323–336. ACM, Raleigh (2008)
3. Astarita, V., Caruso, M.V., Danieli, G., Festa, D.C., Giofrè, V.P., Iuele, T., Vaiana, R.: A mobile application for road surface quality control: UNIquALroad. Procedia Soc. Behav. Sci. **54**, 1135–1144 (2012)
4. Wang, H.-W., Chen, C.-H., Cheng, D.-Y., Lin, C.-H., Lo, C.-C.: A real-time pothole detection approach for intelligent transportation system. Math. Probl. Eng. **2015**, 7 (2015)
5. Eriksson, J., Girod, L., Hull, B., Newton, R., Madden, S., Balakrishnan, H.: The pothole patrol: Using a mobile sensor network for road surface monitoring. In: Proceedings of the 6th International Conference on Mobile Systems, Applications, and Services, pp. 29–39. ACM, Breckenridge (2008)
6. Oliveira, H., Correia, P.L.: Automatic road crack detection and characterization. IEEE Trans. Intell. Transp. Syst. **14**, 155–168 (2013)
7. Radopoulou, S.C., Brilakis, I.: Patch detection for pavement assessment. Autom. Constr. **53**, 95–104 (2015)
8. Sapan, B.: Road condition detection using commodity smartphone sensors aided with vehicular data. Department of Computer Science and Engineering, Master thesis. Chalmers University of Technology (2016)

9. Zoysa, K.D., Keppitiyagama, C., Seneviratne, G.P., Shihan, W.W.A.T.: A public transport system based sensor network for road surface condition monitoring. In: Proceedings of the 2007 Workshop on Networked Systems for Developing Regions, pp. 1–6. ACM, Kyoto (2007)
10. Tai, Y., Chan, C., Hsu, J.Y.: Automatic road anomaly detection using smart mobile device. In: 15th Conference on Artificial Intelligence and Applications (TAAI) (2010)
11. Vittorio, A., Rosolino, V., Teresa, I., Vittoria, C.M., Vincenzo, P.G., Francesco, D.M.: Automated sensing system for monitoring of road surface quality by mobile devices. Procedia Soc. Behav. Sci. **111**, 242–251 (2014)
12. Douangphachanh, V., Oneyama, H.: Exploring the use of smartphone accelerometer and gyroscope to study on the estimation of road surface roughness condition. In: 11th International Conference on Informatics in Control, Automation and Robotics (ICINCO), pp. 783–787 (2014)
13. Mednis, A., Strazdins, G., Zviedris, R., Kanonirs, G., Selavo, L.: Real time pothole detection using android smartphones with accelerometers. In: International Conference on Distributed Computing in Sensor Systems and Workshops (DCOSS), pp. 1–6 (2011)
14. Syed, B., Pal, A., Srinivasarengan, K., Balamuralidhar, P.: A smart transport application of cyber-physical systems: road surface monitoring with mobile devices. In: Sixth International Conference on Sensing Technology (ICST), pp. 8–12 (2012)
15. Witten, I.H., Frank, E., Hall, M.A.: Data Mining: Practical Machine Learning Tools and Techniques. Morgan Kaufmann Publishers Inc., San Francisco (2011)

Author Index

Printed in the United States
By Bookmasters